CALCULATIONS FOR A-LEVEL CHEMISTRY

Third Edition

E N Ramsden BSc, PhD, DPhil

Stanley Thornes (Publishers) Ltd

All rights reserved. No part of this publication may be reproduced, stored in a retrieval system or transmitted in any form or by any means, electronic, mechanical, photocopying, recording or otherwise, without the prior written consent of the copyright holders. Applications for such permission should be addressed to the publishers: Stanley Thornes (Publishers) Ltd, Old Station Drive, Leckhampton, CHELTENHAM GL53 0DN, England.

First published in 1982
Second edition published in 1987
Third edition published in 1993 by:
Stanley Thornes (Publishers) Ltd
Ellenborough House, Wellington Street, CHELTENHAM GL50 1YW

98 99 00 / 10 9

A catalogue record of this book is available from the British Library.

ISBN 0-7487-1594-0

ACKNOWLEDGEMENTS

I thank the following examination boards for permission to print questions from recent A-level papers.

The Associated Examining Board; The Joint Matriculation Board; The Northern Ireland Schools Examinations and Assessment Council; The Oxford and Cambridge Schools Examination Board; The Oxford Delegacy of Local Examinations; The University of Cambridge Schools Local Examinations Syndicate; The University of London School Examinations Council; The Welsh Joint Education Committee.

Many numerical values have been taken from the *Chemistry Data Book* by J G Stark and H G Wallace (published by John Murray). For help with definitions and physical constants, reference has been made to *Physico-chemical Quantities and Units* by M L McGlashan (published by The Royal Institute of Chemistry).

I have been fortunate in receiving excellent advice from Professor R R Baldwin, Professor R P Bell, FRS, Dr G H Davies, Professor W C E Higginson, Dr K A Holbrook, Dr R B Moyes and Dr J R Shorter. I thank these chemists for the help they have given me. I am indebted to Mr J P D Taylor for checking the answers to the problems in the First Edition and for many valuable comments and corrections.

I thank Stanley Thornes (Publishers) for their collaboration and my family for their encouragement.

E N Ramsden
Oxford 1993

Typeset by Tech-Set, Gateshead, Tyne & Wear.
Printed in Great Britain by Redwood Books, Trowbridge, Wiltshire

Contents

List of Exercises

Some of the exercises are divided into an easier section (Section 1) and a more advanced section (Section 2). Questions from A-level papers are on the immediately preceding topic(s). Each question is appended with the name of the Examination Board and the year (90 = 1990 etc). p indicates a part question, S an S-level question, AS an Advanced Supplementary question and N a Nuffield syllabus. The most difficult (often S-level) questions are also denoted by an asterisk.

ABBREVIATIONS OF EXAMINATION BOARDS

AEB	Associated Examining Board
C	University of Cambridge Schools Local Examinations Syndicate
JMB	Joint Matriculation Board
L	University of London Schools Examinations Council
NI	Northern Ireland Schools Examinations and Assessment Council
O	Oxford Delegacy of Local Examination
O & C	Oxford and Cambridge Schools Examinations Board
WJEC	Welsh Joint Education Committee

Foreword

It is a common complaint of university and college teachers of physical sciences that many of their incoming students are unable to carry out even simple calculations, although they may appear to have a satisfactory grasp of the underlying subject matter. Moreover, this is by no means a trivial complaint, since inability to solve numerical problems nearly always stems from a failure to understand fundamental principles, rather than from mathematical or computational difficulties. This situation is more likely to arise in chemistry than in physics, since in the latter subject it is much more difficult to avoid quantitative problems and at the same time produce some semblance of understanding.

In attempting to remedy this state of affairs teachers in schools often feel the lack of a single source of well-chosen calculations covering all branches of chemistry. This gap is admirably filled by Dr Ramsden's collection of problems. The brief mathematical introduction serves to remind the student of some general principles, and the remaining sections cover the whole range of chemistry. Each section contains a theoretical introduction, followed by worked examples and a large number of problems, some of them from past examination papers. Since answers are also given, the book will be equally useful in schools and in home study. It should make a real contribution towards improving the facility and understanding of students of chemistry in their last years at school and in the early part of their university or college courses.

R P Bell FRS
Honorary Research Professor, University of Leeds, and
formerly Professor of Chemistry, University of Stirling

Preface

Many topics in Chemistry involve numerical problems. Textbooks are not long enough to include sufficient problems to give students the practice which they need in order to acquire a thorough mastery of calculations. This book aims to fill that need.

Chapter 1 is a quick revision of mathematical techniques, with special reference to the use of the calculator, and some hints on how to tackle chemical calculations. With each topic, a theoretical background is given, leading to worked examples and followed by a large number of problems and a selection of questions from past examination papers. The theoretical section is not intended as a full treatment, to replace a textbook, but is included to make it easier for the student to use the book for individual study as well as for class work. The inclusion of answers is also an aid to private study.

The material will take students up to GCE A- and S-level examinations. It will also serve the needs of students preparing for the Ordinary National Diploma. A few of the topics covered are not in the A-level syllabuses of all the Examination Boards, and it is expected that students will be sufficiently familiar with the syllabus they are following to omit material outside their course if they wish. S-level topics and the more difficult calculations are marked with an asterisk.

Many students are now starting A-level work after GCSE Science: Double Award. The coverage of chemical calculations in the syllabuses for double award science is slighter than it was in the syllabuses for GCE O-level Chemistry and GCSE Chemistry. I have taken this into account in the Third Edition. In previous editions, I assumed that students had mastered some topics before and only an extension was needed for A-level. Since students who have done Double Award Science have spent little time on quantitative work, I have taken a more gradual approach in the Third Edition. I hope that this approach will also suit students with no previous experience of chemical calculations. In the Third Edition I have included more foundation work on formulae, the mole, calculations based on chemical equations and volumetric analysis than in earlier editions. A number of topics which are no longer included in A-level syllabuses have been dropped, and the selection of questions from past papers has been updated.

E N Ramsden
Oxford 1993

1 Basic Mathematics

INTRODUCTION

Calculations are a part of your chemistry course. The time you spend on calculations will be richly rewarded. Your perception of chemistry will become at the same time deeper and more precise. No one can come to an understanding of science without acquiring the sharp, logical approach that is needed for solving numerical problems.

To succeed in solving numerical problems you need two things. The first is an understanding of the chemistry involved. The second is some facility in simple mathematics. Calculations are a perfectly straightforward matter. A numerical problem gives you some data and asks you to obtain some other numerical values. The connection between the data you are given and the information you are asked for is a chemical relationship. You will need to know your chemistry to recognise what that relationship is.

This introduction is a reminder of some of the mathematics which you studied earlier in your school career. It is included for the sake of students who are not studying mathematics concurrently with their chemistry course. A few problems are included to help you to brush up your mathematical skills before you go on to tackle the chemical problems.

USING EQUATIONS

Scientists are often concerned with measuring quantities such as pressure, volume, electric current and electric potential difference. Sometimes they find that when one quantity changes another quantity changes as a result. The related quantities are described as *dependent variables*. The relationship between the variables can be written in the form of a mathematical equation. For example, when a mass of gas expands, its volume increases and its density decreases:

$$\text{Density} \propto 1/\text{Volume}$$

The sign $\propto$ means 'is proportional to', so the expression means 'density is proportional to 1/volume' or 'density is inversely proportional to volume'.

The relationship between the volume and density of a mass of gas is:

$$\text{Density} = \frac{\text{Mass}}{\text{Volume}}$$

This equation can also be written as:

$$\text{Density} = \text{Mass} \div \text{Volume}$$

and as: $\text{Density} = \text{Mass/Volume}.$

If you are given two quantities, mass and volume, you can use the equation to calculate the third, density.

For example, the mass of an aluminium bar is 21.4 g, and its volume is 7.92 cm^3. What is the density of aluminium?

Using the equation

$$\text{Density} = \text{Mass/Volume}$$

$$\text{Density} = 21.4\,\text{g}/7.92\,\text{cm}^3$$

$$= 2.70\,\text{g cm}^{-3}$$

Notice the units. Since mass has the unit gram (g) and volume has the unit cubic centimetre (cm^3), density has the unit gram per cubic centimetre (g cm^{-3}).

REARRANGING EQUATIONS

You might want to use the above equation to find the mass of an object when you know its volume and density. It would help to rearrange the above equation to put mass by itself on one side of the equation, that is in the form

$$\text{Mass} = ?$$

In the equation $\text{Density} = \dfrac{\text{Mass}}{\text{Volume}}$

mass is divided by volume, so to obtain mass by itself you must multiply the right-hand side of the equation by volume. Naturally, you must do the same to the left-hand side. Then

$$\text{Density} \times \text{Volume} = \frac{\text{Mass}}{\cancel{\text{Volume}}} \times \cancel{\text{Volume}}$$

that is $\text{Mass} = \text{Density} \times \text{Volume}$

Perhaps you need to use the equation to find the volume of an object when you know its mass and density. Then you rearrange the equation to put volume alone on one side of the equation. In the equation

$$\text{Mass} = \text{Density} \times \text{Volume}$$

the term volume is multiplied by density. To obtain volume on its own you must divide by density. Doing the same on both sides of the equation gives

$$\frac{\text{Mass}}{\text{Density}} = \frac{\cancel{\text{Density}} \times \text{Volume}}{\cancel{\text{Density}}}$$

or

$$\text{Volume} = \frac{\text{Mass}}{\text{Density}}$$

A TRIANGLE FOR REARRANGING EQUATIONS

A short cut for rearranging equations is to put the quantities into a triangle. For the equation,

$$X = Y \times Z$$

the triangle is

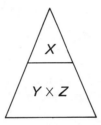

Cover up the letter you want; then what you see is the equation you need to use. If you want Y, cover up Y; then you read X/Z, so you know that $Y = X/Z$. If you cover Z, you read X/Y, so you know that $Z = X/Y$ is the equation you need.

PRACTICE 1

1. Draw a triangle to show the equation

   $$\text{Mass} = \text{Density} \times \text{Volume}$$

   Cover up Density, and write the equation for Density = ?

   Cover up Volume, and write the equation for Volume = ?

2. The equation relating potential difference, V, resistance, R, and current, I, is $V = R \times I$. The equation means just the same written without the multiplication sign as $V = RI$. Rearrange the equation a) in the form $I = ?$, b) in the form $R = ?$.

3. Draw a triangle to show the relationship $V = RI$.
   a) Cover up R. Complete the equation $R = ?$
   b) Cover up I. Complete the equation $I = ?$

4. The concentration of a solution can be expressed

$$\text{Concentration} = \frac{\text{Mass of solute}}{\text{Volume of solution}}$$

Rearrange the equation **a)** into the form Mass of solute = ? and **b)** into the form Volume of solution = ?

5. Rearrange the equation $P = QR$ **a)** into the form $Q = ?$ and **b)** into the form $R = ?$

CROSS-MULTIPLYING

Once you have understood the ideas behind rearranging equations, you can try the method of cross-multiplying. If

$$\frac{a}{b} \diagdown \frac{c}{d}$$

then by cross-multiplying, you obtain

$$ad = bc$$

How can you find out whether this is correct? First multiply both sides of the first equation by d:

$$\frac{ad}{b} = \frac{cd}{d} = c$$

Next multiply both sides by b:

$$\frac{adb}{b} = bc$$

That is $ad = bc$

which is the equation you obtained by cross-multiplying. This shows that cross-multiplying only puts into practice the method of multiplying or dividing both sides of the equation by the same quantity.

Now that you have the equation $ad = bc$

to obtain an equation for a, divide both sides by d; then

$$a = bc/d$$

to obtain an equation for d, divide by a; then

$$d = bc/a$$

and similarly,

$$b = ad/c$$

$$c = ad/b$$

PRACTICE 2

1. The pressure, volume and temperature of a gas are related to the gas constant, R, by the equation

$$\frac{P}{T} = \frac{R}{V}$$

Rearrange the equation by cross-multiplying to obtain equations for **a)** T and **b)** V.

2. The resistance of an electrical conductor is given by

$$R = \rho \times l/A$$

where R = resistance, ρ = resistivity, l = length and A = cross-sectional area. Rearrange the equation to give **a)** an equation of the form $\rho = ?$ and **b)** an equation of the form $A = ?$

3. Rearrange the equation

$$\frac{a \times b}{c} = \frac{p}{q \times r}$$

to give **a)** an equation for p and **b)** an equation for q.

CALCULATIONS ON RATIO

Many of the calculations you meet involve ratios. You have met this type of problem in your maths lessons; do not forget how to solve them when you meet them in chemistry!

EXAMPLE 1 Nancy pays 78p for two toffee bars. How much does Nina have to pay for five of the same bars?

You can tackle this problem by the unitary method:

If 2 toffee bars cost 78p,
1 toffee bar costs 78/2p
and 5 toffee bars cost $5 \times 78/2$p = 195p = £1.95.

EXAMPLE 2 Zinc reacts with dilute acids to give hydrogen. If 0.0400 g of hydrogen is formed when 1.30 g of zinc reacts with an excess of acid, what mass of zinc is needed to produce 6.00 g of hydrogen?

Again, the unitary method will help you.

If 0.0400 g of hydrogen is produced by 1.30 g of zinc,
then 1.00 g of hydrogen is produced by 1.30/0.0400 g of zinc
and 6.00 g of hydrogen are produced by $6.00 \times 1.30/0.0400$ g of zinc

$$= 195 \text{ g of zinc}$$

PRACTICE 3

1. If 0.020 g of a gas has a volume of 150 cm^3 what is the volume of 32 g of the gas (at the same temperature and pressure)?

2. 88 g of iron(II) sulphide is the maximum quantity that can be obtained from the reaction of an excess of sulphur with 56 g of iron.

   What is the maximum quantity of iron(II) sulphide that can be obtained from 7.00 g of iron?

3. A firm obtains 80 tonnes of pure calcium carbonate from 100 tonnes of limestone. What mass of limestone must be quarried to yield 240 tonnes of pure calcium carbonate?

WORKING WITH NUMBERS IN STANDARD FORM

You are accustomed to writing numbers in decimal notation, for example 123 677.54 and 0.001 678. In working with large numbers and small numbers, you will find it convenient to write them in a different way, known as *scientific notation* or *standard form*. This means writing a number as a product of two factors. In the first factor, the decimal point comes after the first digit. The second factor is a multiple of ten. For example, $2123 = 2.123 \times 10^3$ and $0.000\,167 = 1.67 \times 10^{-4}$. 10^3 means $10 \times 10 \times 10$, and 10^{-4} means $1/(10 \times 10 \times 10 \times 10)$. The number 3 or -4 is called the exponent, and the number 10 is the base. 10^3 is referred to as '10 to the power 3' or '10 to the third power'. You will have noticed that, if the exponent is increased by 1, the decimal point must be moved one place to the left.

$$2.5 \times 10^3 = 0.25 \times 10^4 = 25 \times 10^2 = 250 \times 10^1 = 2500 \times 10^0$$

Since $10^0 = 1$, this last factor is normally omitted.

When you multiply numbers in standard form, the exponents are added. The product of 2×10^4 and 6×10^{-2} is given by

$$(2 \times 10^4) \times (6 \times 10^{-2}) = (2 \times 6) \times (10^4 \times 10^{-2})$$
$$= 12 \times 10^2 = 1.2 \times 10^3$$

In division, the exponents are subtracted:

$$\frac{1.44 \times 10^6}{4.50 \times 10^{-2}} = \frac{1.44}{4.50} \times \frac{10^6}{10^{-2}} = 0.320 \times 10^8 = 3.20 \times 10^7$$

In addition and subtraction, it is convenient to express numbers using the same exponents. An example of addition is

$$(6.300 \times 10^2) + (4.00 \times 10^{-1}) = (6.300 \times 10^2) + (0.004\,00 \times 10^2)$$
$$= 6.304 \times 10^2$$

An example of subtraction is

$$(3.60 \times 10^{-3}) - (4.20 \times 10^{-4}) = (3.60 \times 10^{-3}) - (0.420 \times 10^{-3})$$
$$= 3.18 \times 10^{-3}$$

How to enter exponents on a calculator

To enter 1.44×10^6, you enter 1.44; then press the EXP key, then the 6 key.

To enter 4.50×10^{-2}, you enter 4.5; then press the EXP key, then the 2 key, and lastly the $+/-$ key.

To enter 10^{-3}, you enter 1; then press the EXP key, then the 3 key, and lastly the $+/-$ key.

ESTIMATING YOUR ANSWER

One advantage of standard form is that very large and very small numbers can be entered on a calculator. Another advantage is that you can easily estimate the answer to a calculation to the correct order of magnitude (i.e. the correct power of 10).

For example,

$$\frac{2456 \times 0.0123 \times 0.004\,14}{5\,223 \times 60.7 \times 8.51}$$

Putting the numbers into standard form gives

$$\frac{2.456 \times 10^3 \times 1.23 \times 10^{-2} \times 4.14 \times 10^{-3}}{5.223 \times 10^3 \times 6.07 \times 10 \times 8.51}$$

This is approximately

$$\frac{2 \times 1 \times 4}{5 \times 6 \times 8} \times \frac{10^3 \times 10^{-2} \times 10^{-3}}{10^3 \times 10} = \frac{1}{30} \times 10^{-6} = 3 \times 10^{-8}$$

By putting the numbers into standard form, you can estimate the answer very quickly. A complete calculation gives the answer 4.64×10^{-8}. The rough estimate is sufficiently close to this to reassure you that you have not made any slips with exponents of ten.

LOGARITHMS

The logarithm (or 'log') of a number N is the power to which 10 must be raised to give the number.

If $N = 1$,   then since $10^0 = 1$, $\lg N = 0$.

If $N = 100$,   then since $10^2 = 100$, $\lg N = 2$.

If $N = 0.001$, then since $10^{-3} = 0.001$, $\lg N = -3$.

We say that the logarithm of 100 to the base 10 is 2 or $\lg 100 = 2$.

There is another widely used set of logarithms to the base e. They are called natural logarithms as e is a significant quantity in mathematics. It has the value 2.71828 . . . . Natural logarithms are written as $\ln N$. The relationship between the two systems is

$$\ln N = \ln 10 \times \lg N$$

Since $\ln 10 = 2.3026$, for most purposes it is sufficiently accurate to write

$$\ln N = 2.303 \lg N$$

Whenever scientific work gives an equation in which $\ln N$ appears, you can substitute 2.303 times the value of $\lg N$.

To obtain the log of a number, enter the number on your calculator and press the log key. The value of the log will appear in the display. This will happen whether you enter the number in standard form or another form. For example, $\lg 12\,345 = 4.0915$, whether you enter the number as $12\,345$ or as 1.2345×10^4. However, there is a limit to the number of digits your calculator will accept, and you need to enter very large and very small numbers in standard notation.

Operations on logarithms are:

Multiplication. The logs of the numbers are added:

$$\lg (A \times B) = \lg A + \lg B$$

Division. The logs are subtracted:

$$\lg (P/Q) = \lg P - \lg Q$$

Powers. This is a special case of multiplication.

$$\lg A^2 = \lg A + \lg A = 2 \lg A$$

$$\lg A^{-3} = -3 \lg A$$

Roots. It is easy to show that $\lg \sqrt{B} = \frac{1}{2} \lg B$.

Since $B = B^{1/2} \times B^{1/2}$

$$\lg B = \lg B^{1/2} + \lg B^{1/2}$$

$$\lg B^{1/2} = \frac{1}{2} \lg B$$

Similarly, $\lg \sqrt[3]{B} = \frac{1}{3} \lg B$

ANTILOGARITHMS

Your calculator will give you the antilog of a number. You should consult the manual to find out the procedure for your own model of calculator.

Most calculators will give you reciprocals, squares and other powers, square roots and other roots directly. If you have a simpler form of calculator, you can obtain powers and roots by using logarithms.

ROUNDING OFF NUMBERS

Often your calculator will display an answer containing more digits than the numbers you fed into it. Suppose you are given the information that $18.6\,cm^3$ of sodium hydroxide solution exactly neutralise $25.0\,cm^3$ of a solution of hydrochloric acid of concentration $0.100\,mol\,dm^{-3}$. You want to find the concentration of sodium hydroxide solution, and you put the numbers $(25.0\times0.100)/18.6$ into your calculator and obtain a value of $0.134\,408\,6\,mol\,dm^{-3}$. The concentration of the solution is not known as accurately as this, however, because you cannot read the burette as accurately as this.

Since you read the burette to three figures, you quote your answer to three figures. In the number $0.134\,408\,6$, the figures you are sure of are termed the *significant figures*. The significant figures are retained, and the insignificant figures are dropped. This operation is called *rounding off*. If the first number had been $0.134\,708\,6$, it would have been rounded off to 0.135. If the first of the insignificant figures being dropped is 5 or greater, the last of the significant figures is rounded up to the next digit. If the first of the dropped figures is less than 5, the last significant figure is left unaltered.

Some calculations involve several stages. It is sound practice to give one more significant figure in your answer at each stage than the number of significant figures in the data. Then, in the final stage, the answer is rounded off.

If the calculation were $(25.0\times0.100)/26.2 = 0.095\,419\,84\,mol\,dm^{-3}$, would you still round off to 3 significant figures? This would make the answer $0.0954\,mol\,dm^{-3}$. Stated in this way, the answer is claiming an accuracy of 1 part in 954 — about 1 part in 1000. Since the hydrochloric acid concentration is known to about 1 part in 100, the answer cannot be stated to a higher degree of accuracy. You have to use the 3-significant-figure rule sensibly, and say that an error of ±1 in 95 is about as significant as an error of ±1 in 134. The answer should therefore be quoted as $0.095\,mol\,dm^{-3}$.

The number of significant figures is the number of figures which is accurately known. The number 123 has 3 significant figures. The number 1.23×10^4 has 3 significant figures, but 12 300 has 5 significant figures because the final zeros mean that each of these digits is known to be zero and not some other digit. The number 0.001 23 has 3 significant figures. The number 25.1 has 3 significant figures, and the number 25.10 has 4 significant figures as the final 0 states that the value of this number is known to an accuracy of 1 part in 2500.

In addition, the sum is known with the accuracy of the least reliable numbers in the sum. For example, the sum of

$$
\begin{array}{r}
1.4167 \text{ g} \\
+ 100.5 \quad\text{ g} \\
+ \quad 7.12 \quad\text{ g} \\
\hline
\end{array}
$$

is

$$
109.0367 \text{ g}
$$

Since 1 figure is known to only 1 place after the decimal point, the sum also is known to 1 place after decimal point and should be written as 109.0 g. The same guideline is used for subtraction.

In multiplication and division the product or quotient is rounded off to the same number of significant figures as the number with the fewest significant figures. For example, $12\,340 \times 2.7 \times 0.003\,65 = 121.6107$. The product is rounded off to 2 significant figures, 1.2×10^2.

CHOICE OF A CALCULATOR

The functions which you need in a calculator for the problems in this book are:

- Addition, Subtraction, Multiplication and Division
- Squares and other powers (x^2 and x^y keys)
- Square roots and other roots ($\sqrt{x}$ and $x^{1/y}$ keys)
- Reciprocals
- Log_{10} and antilog_{10} (10^x)
- Natural logarithms, $\ln_e$ and antiln_e (e^x)
- Exponent key and $+/-$ key
- Brackets
- Memory

A variety of scientific calculators have these functions and others (such as sin, cos, tan and Σx) which will be useful to you in physics and mathematics problems.

UNITS

There are two sets of units currently employed in scientific work. One is the CGS system, based on the centimetre, gram and second. The other is the Système Internationale (SI) which is based on the metre, kilogram, second and ampere. SI units were introduced in 1960, and in 1979 the Association for Science Education published a booklet called *Chemical Nomenclature, Symbols and Terminology for Use in School Science* that recommended that schools and colleges adopt this system.

Listed below are the SI units for the seven fundamental physical quantities on which the system is based and also a number of derived quantities and their units.

Chemists are still using some of the CGS units. You will find mass in g; volume in cm^3 and dm^3; concentrations in $mol\,dm^{-3}$ or $mol\,litre^{-1}$; conductivity in $\Omega^{-1}cm$ as well as $\Omega^{-1}m$. Pressure is sometimes given in mm mercury and temperatures in $°C$.

Basic SI Units

Physical Quantity	Name of Unit	Symbol
Length	metre	m
Mass	kilogram	kg
Time	second	s
Electric current	ampere	A
Temperature	kelvin	K
Amount of substance	mole	mol
Light intensity	candela	cd

Derived SI Units

Physical Quantity	Name of Unit	Symbol	Definition
Energy	joule	J	$kg\,m^2\,s^{-2}$
Force	newton	N	$J\,m^{-1}$
Electric charge	coulomb	C	$A\,s$
Electric potential difference	volt	V	$J\,A^{-1}s^{-1}$
Electric resistance	ohm	Ω	$V\,A^{-1}$
Area	square metre		m^2
Volume	cubic metre		m^3
Density	kilogram per cubic metre		$kg\,m^{-3}$
Pressure	newton per square metre or pascal		$N\,m^{-2}$ or Pa
Molar mass	kilogram per mole		$kg\,mol^{-1}$

With all these units, the following prefixes (and others) may be used:

Prefix	Symbol	Meaning
deci	d	10^{-1}
centi	c	10^{-2}
milli	m	10^{-3}
micro	μ	10^{-6}
nano	n	10^{-9}
kilo	k	10^3
mega	M	10^6
giga	G	10^9
tera	T	10^{12}

It is very important when putting values for physical quantities into an equation to be consistent in the use of units. If you are, then the units can be treated as factors in the same way as numbers. Suppose you are asked to calculate the volume occupied by 0.0110 kg of carbon dioxide at 27 °C and a pressure of $9.80 \times 10^4\,\mathrm{N\,m^{-2}}$. You know that the gas constant is $8.31\,\mathrm{J\,mol^{-1}\,K^{-1}}$ and that the molar mass of carbon dioxide is $44.0\,\mathrm{g\,mol^{-1}}$. Use the ideal gas equation:

$$PV = nRT$$

The pressure $\quad P = 9.80 \times 10^4\,\mathrm{N\,m^{-2}}$

The constant $\quad R = 8.31\,\mathrm{J\,K^{-1}\,mol^{-1}}$

The temperature $T = 27 + 273 = 300\,\mathrm{K}$

The number of moles

$$n = \text{Mass/Molar mass}$$
$$= 0.0110\,\mathrm{kg}/(44.0 \times 10^{-3}\,\mathrm{kg\,mol^{-1}})$$
$$= 0.250\,\mathrm{mol}$$

Then $\qquad V = \dfrac{0.250\,\mathrm{mol} \times 8.31\,\mathrm{J\,K^{-1}\,mol^{-1}} \times 300\,\mathrm{K}}{9.80 \times 10^4\,\mathrm{N\,m^{-2}}}$

$$= 6.34 \times 10^{-3}\,\mathrm{J\,N^{-1}\,m^2}$$

Since $\qquad \mathrm{J} = \mathrm{N\,m} \qquad$ (1 joule = 1 newton metre)

$$V = 6.34 \times 10^{-3}\,\mathrm{N\,m\,N^{-1}\,m^2}$$
$$= 6.34 \times 10^{-3}\,\mathrm{m^3}$$

Volume has the unit of cubic metre. This calculation illustrates what people mean when they say that SI units form a *coherent system of units*. You can convert from one unit to another by multiplication and division, without introducing any numerical factors.

SOLUTION OF QUADRATIC EQUATIONS

A quadratic equation is the name for an equation of the type

$$ax^2 + bx + c = 0$$

x is the unknown quantity, a and b are the coefficients of x, and c is a constant. The solution of this equation is given by

$$x = \frac{-b \pm \sqrt{b^2 - 4ac}}{2a}$$

There are two solutions to the equation. Often you will be able to decide that one solution is mathematically correct but physically impossible. You may be calculating some physical quantity that cannot possibly be negative so that you will ignore a negative solution and adopt a positive solution.

DRAWING GRAPHS

Here are some hints for drawing graphs.

a) Whenever possible, data should be plotted in a form that gives a straight line graph. It is easier to draw the best straight line through a set of points than to draw a curve.

If the dimensions x and y are related by the expression $y = ax + b$, then a straight line will result when experimental values of y are plotted against the corresponding values of x. The values of x are plotted against the horizontal axis (the x-axis or abscissa), and the corresponding values of y are plotted along the vertical axis (the y-axis or ordinate). The gradient of the straight line obtained $= a$, and the intercept on the y-axis $= b$ (see Fig. 1.1).

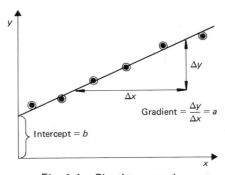

Fig. 1.1 Plotting a graph

b) Choose a scale which will allow the graph to cover as much of the piece of graph paper as possible. There is no need to start at zero. If the points lie between 80 and 100, to start at zero would cramp your graph into a small section at the top of the page (see Fig. 1.2).

Choose a scale which will make plotting the data and reading the graph as simple as possible.

c) Label the axes with the dimensions and the units. Make the scale units as simple as possible. Instead of plotting as scale units $1 \times 10^{-3} \, mol \, dm^{-3}$, $2 \times 10^{-3} \, mol \, dm^{-3}$, $3 \times 10^{-3} \, mol \, dm^{-3}$, etc., plot 1, 2 and 3, etc., and label the axis as $(Concentration/mol \, dm^{-3}) \times 10^3$ (see Fig. 1.3).

The solidus (/) is used because it means 'divided by'. The numbers, 1, 2 and 3, etc. (see Fig. 1.3) are the values of the physical quantity, concentration, divided by the unit, $mol \, dm^{-3}$, and multiplied by the factor 10^3.

d) When you come to draw a straight line through the points, draw the best straight line you can, to pass through, or close to, as many points as possible (see Fig. 1.4). Owing to experimental error, not all the points will fall on the line. A graph of experimental results gives you a better accuracy than calculating a value from just one point. If you are drawing a curve, draw a smooth curve (see Fig. 1.5). Do not join up the points with straight lines. The curve may not pass through every point, but it is more reliable than any one of the points.

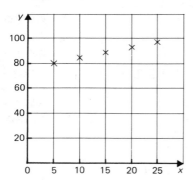

Fig. 1.2   Don't cramp your graph!

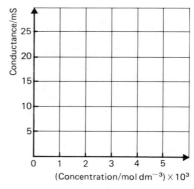

Fig. 1.3   Label the axes

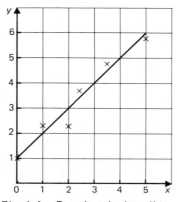

Fig. 1.4   Drawing the best line

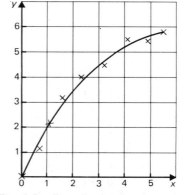

Fig. 1.5   Drawing a smooth curve

A FINAL POINT

Always look critically at your answer. Ask yourself whether it is a reasonable answer. Is it of the right order of magnitude for the data? Is it in the right units? Many errors can be detected by an assessment of this kind.

EXERCISE 1   Practice with Calculations

1. Convert the following numbers into standard form:
   a) 23 678   b) 437.6   c) 0.0169
   d) 0.000 345   e) 672 891

2. Convert each of the following numbers into standard form, enter into your calculator, and multiply by 237. Give the answers in standard form.
   a) 246.8   b) 11 230   c) 267 831   d) 0.051   e) 0.567

3. Find the following quotients:
   a) 2360/0.000 71   b) 28 780/0.106   c) 85.42/460 000
   d) 58/900 670   e) 0.000 88/0.144

4. Find the following sums and differences:
   a) $(2.000 \times 10^4) + (0.10 \times 10^2)$
   b) $48.0 + (5.600 \times 10^3)$
   c) $(1.23 \times 10^5) + (6.00 \times 10^3)$
   d) $(4.80 \times 10^{-4}) - (1.6 \times 10^{-3})$
   e) $(6.300 \times 10^4) - (4.8 \times 10^2)$

5. Make an approximate estimate of the answers to the following:
   a) $\dfrac{4.0 \times 10^3 \times 5.6 \times 10^{-2} \times 7.1 \times 10^6}{8.2 \times 10^{-6} \times 4.9 \times 10^3}$
   b) $567 \times 4183 \times 0.001\,27 \times 0.107$
   c) $\dfrac{496 \times 7124 \times 83\,000 \times 4.7}{7260 \times 41 \times 0.0075}$
   d) $\dfrac{1480 \times 6730 \times 0.173 \times 0.0097}{0.15 \times 0.0088 \times 100\,860 \times 0.10}$
   e) $\dfrac{208 \times 100\,490}{560 \times 0.005\,5 \times 0.000\,49}$

6. Find the logarithms of the following numbers:
   a) 4735   b) 5.072×10^3 c) 0.001 327
   d) 10.076   e) 2.314×10^{-6}

7. Find the antilogarithms of the following:

a) 3.4567 b) 0.0549 c) 7.843 2
d) −6.4712 e) −2.0571

8. Find the reciprocals of the following:

a) 234.5 b) 3 488 123 c) 0.002 477
d) 4.865×10^9 e) 2.645×10^{-5}

9. Solve the quadratic equations:

a) $(x + 3)^2 = 100$ b) $5x^2 - 8 = 18x$
c) $(y + 4)^2 = 18y - 9$ d) $(z - 6)^2 = 4z + 8$
e) $(2x + 4)^2 = 7 - 4x$

2 Formulae and Equations

Calculations are based on formulae and on equations. In order to tackle the calculations in this book you will have to be quite sure you can work out the formulae of compounds correctly, and that you can balance equations. This section is a revision of work on formulae and equations.

FORMULAE

Electrovalent compounds consist of oppositely charged ions. The compound formed is neutral because the charge on the positive ion (or ions) is equal to the charge on the negative ion (or ions). In sodium chloride, NaCl, one sodium ion, Na^+, is balanced in charge by one chloride ion, Cl^-.

This is how the formulae of electrovalent compounds can be worked out:

Compound	*Zinc chloride*
Ions present are	Zn^{2+} and Cl^-
Now balance the charges	One Zn^{2+} ion needs two Cl^- ions
Ions needed are	Zn^{2+} and $2Cl^-$
The formula is	$ZnCl_2$
Compound	*Sodium sulphate*
Ions present are	Na^+ and SO_4^{2-}
Now balance the charges	Two Na^+ balance one SO_4^{2-}
Ions needed are	$2Na^+$ and SO_4^{2-}
The formula is	Na_2SO_4
Compound	*Aluminium sulphate*
Ions present are	Al^{3+} and SO_4^{2-}
Now balance the charges	Two Al^{3+} balance three SO_4^{2-}
Ions needed are	$2Al^{3+}$ and $3SO_4^{2-}$
The formula is	$Al_2(SO_4)_3$
Compound	*Iron(II) sulphate*
Ions present are	Fe^{2+} and SO_4^{2-}
Now balance the charges	One Fe^{2+} balances one SO_4^{2-}
Ions needed are	Fe^{2+} and SO_4^{2-}
The formula is	$FeSO_4$

Compound	*Iron(III) sulphate*
Ions present are	Fe^{3+} and $SO_4{}^{2-}$
Now balance the charges	Two Fe^{3+} balance three $SO_4{}^{2-}$
Ions needed are	$2Fe^{3+}$ and $3SO_4{}^{2-}$
The formula is	$Fe_2(SO_4)_3$

You need to know the charges of the ions in the table below. Then you can work out the formula of any electrovalent compound.

You will notice that the compounds of iron are named iron(II) sulphate and iron(III) sulphate to show which of its valencies iron is using in the compound. This is always done with the compounds of elements of variable valency. For valency and oxidation number, see Chapter 8, p. 72.

Name	Symbol	Charge	Name	Symbol	Charge
Hydrogen	H^+	+1	Hydroxide	OH^-	−1
Ammonium	$NH_4{}^+$	+1	Nitrate	$NO_3{}^-$	−1
Potassium	K^+	+1	Chloride	Cl^-	−1
Sodium	Na^+	+1	Bromide	Br^-	−1
Silver	Ag^+	+1	Iodide	I^-	−1
Copper(I)	Cu^+	+1	Hydrogen-carbonate	$HCO_3{}^-$	−1
Barium	Ba^{2+}	+2	Oxide	O^{2-}	−2
Calcium	Ca^{2+}	+2	Sulphide	S^{2-}	−2
Copper(II)	Cu^{2+}	+2	Sulphite	$SO_3{}^{2-}$	−2
Iron(II)	Fe^{2+}	+2	Sulphate	$SO_4{}^{2-}$	−2
Lead	Pb^{2+}	+2	Carbonate	$CO_3{}^{2-}$	−2
Magnesium	Mg^{2+}	+2			
Zinc	Zn^{2+}	+2			
Aluminium	Al^{3+}	+3	Phosphate	$PO_4{}^{3-}$	−3
Iron(III)	Fe^{3+}	+3			

EQUATIONS

Having symbols for elements and formulae for compounds gives us a way of representing chemical reactions.

EXAMPLE 1   Instead of writing 'Copper(II) carbonate forms copper(II) oxide and carbon dioxide', we can write

$$CuCO_3 \longrightarrow CuO + CO_2$$

The atoms we finish with are the same in number and kind as the atoms we start with. We start with one atom of copper, one atom of carbon and three atoms of oxygen, and we finish with the same. This makes the two sides of the expression equal, and we call it an *equation.* A simple way of conveying a lot more information is to include *state symbols* in the equation. These are (s) = solid, (l) = liquid, (g) = gas, (aq) = in solution in water. The equation

$$CuCO_3(s) \longrightarrow CuO(s) + CO_2(g)$$

tells you that solid copper(II) carbonate dissociates to form solid copper(II) oxide and the gas carbon dioxide.

EXAMPLE 2   The equation

$$Zn(s) + H_2SO_4(aq) \longrightarrow ZnSO_4(aq) + H_2(g)$$

tells you that solid zinc reacts with a solution of sulphuric acid to give a solution of zinc sulphate and hydrogen gas. Hydrogen is written as H_2, since each molecule of hydrogen gas contains two atoms.

EXAMPLE 3   Sodium carbonate reacts with dilute hydrochloric acid to give carbon dioxide and a solution of sodium chloride. The equation could be

$$Na_2CO_3(s) + HCl(aq) \longrightarrow CO_2(g) + NaCl(aq) + H_2O(l)$$

but, when you add up the atoms on the right, you find that they are not equal to the atoms on the left. The equation is not '*balanced*', so the next step is to *balance* it. Multiplying NaCl by two gives

$$Na_2CO_3(s) + HCl(aq) \longrightarrow CO_2(g) + 2NaCl(aq) + H_2O(l)$$

This makes the number of sodium atoms on the right-hand side equal to the number on the left-hand side. But there are two chlorine atoms on the right-hand side, therefore the HCl must be multiplied by two:

$$Na_2CO_3(s) + 2HCl(aq) \longrightarrow CO_2(g) + 2NaCl(aq) + H_2O(l)$$

The equation is now *balanced.*

When you are balancing a chemical equation, the only way you do it is to multiply the number of atoms or molecules. You never try to alter a formula. In the above example, you got two chlorine atoms by multiplying HCl by two, not by altering the formula to HCl_2, which does not exist.

The steps in writing an equation are:

1. Write a word equation.

2. Put in the symbols and formulae (symbols for elements, formulae for compounds and state symbols).

3. Balance the equation.

EXAMPLE 4   When methane burns,

$$Methane + Oxygen \longrightarrow Carbon\ dioxide + Water$$

$$CH_4(g) + O_2(g) \longrightarrow CO_2(g) + H_2O(g)$$

There is one carbon atom on the left-hand side and one carbon atom on the right-hand side. There are four hydrogen atoms on the left-hand side, and therefore we need to put four hydrogen atoms on the right-hand side. Putting $2H_2O$ on the right-hand side will accomplish this:

$$CH_4(g) + O_2(g) \longrightarrow CO_2(g) + 2H_2O(g)$$

There is one molecule of O_2 on the left-hand side and four O atoms on the right-hand side. We can make the two sides equal by putting $2O_2$ on the left-hand side:

$$CH_4(g) + 2O_2(g) \longrightarrow CO_2(g) + 2H_2O(g)$$

This is a balanced equation. The numbers of atoms of carbon, hydrogen and oxygen on the left-hand side are equal to the numbers of atoms of carbon, hydrogen and oxygen on the right-hand side.

EXERCISE 2   Practice with Equations

1. For practice, try writing the equations for the reactions:
   a) Hydrogen + Copper oxide $\longrightarrow$ Copper + Water
   b) Carbon + Carbon dioxide $\longrightarrow$ Carbon monoxide
   c) Carbon + Oxygen $\longrightarrow$ Carbon dioxide
   d) Magnesium + Sulphuric acid $\longrightarrow$ Hydrogen + Magnesium sulphate
   e) Copper + Chlorine $\longrightarrow$ Copper(II) chloride

2. Now try writing balanced equations for the reactions:
   a) Calcium + Water $\longrightarrow$ Hydrogen + Calcium hydroxide solution
   b) Copper + Oxygen $\longrightarrow$ Copper(II) oxide
   c) Sodium + Oxygen $\longrightarrow$ Sodium oxide
   d) Iron + Hydrochloric acid $\longrightarrow$ Iron(II) chloride solution
   e) Iron + Chlorine $\longrightarrow$ Iron(III) chloride

3. Balance these equations:
   a) $Na_2O(s) + H_2O(l) \longrightarrow NaOH(aq)$
   b) $KClO_3(s) \longrightarrow KCl(s) + O_2(g)$
   c) $H_2O_2(aq) \longrightarrow H_2O(l) + O_2(g)$
   d) $Fe(s) + O_2(g) \longrightarrow Fe_3O_4(s)$
   e) $Mg(s) + N_2(g) \longrightarrow Mg_3N_2(s)$
   f) $NH_3(g) + O_2(g) \longrightarrow N_2(g) + H_2O(g)$
   g) $Fe(s) + H_2O(g) \longrightarrow Fe_3O_4(s) + H_2(g)$
   h) $H_2S(g) + O_2(g) \longrightarrow H_2O(g) + SO_2(g)$
   i) $H_2S(g) + SO_2(g) \longrightarrow H_2O(l) + S(s)$

3 Relative Atomic Mass

RELATIVE ATOMIC MASS

Atoms are tiny: one atom of hydrogen has a mass of 1.66×10^{-24} g; one atom of carbon has a mass of 1.99×10^{-23} g. Numbers as small as this are awkward to handle, and, instead of the actual masses, we use relative atomic masses. Since hydrogen atoms are the smallest of all atoms, one atom of hydrogen was originally taken as the mass with which all other atoms would be compared. Then,

$$\text{Original relative atomic mass} = \frac{\text{Mass of one atom of the element}}{\text{Mass of one atom of hydrogen}}$$

Thus, on this scale, the relative atomic mass of hydrogen is 1, and, since one atom of carbon is 12 times as heavy as one atom of hydrogen, the relative atomic mass of carbon is 12.

The modern method of finding relative atomic masses is to use a mass spectrometer. The most accurate measurements are made with volatile compounds of carbon, and it was therefore convenient to change the standard of reference to carbon. There are three isotopes of carbon (with the same number of protons and electrons but different numbers of neutrons, and therefore different masses). It was decided to use the most plentiful carbon isotope, carbon-12. Thus,

$$\text{Relative atomic mass} = \frac{\text{Mass of one atom of an element}}{(1/12) \text{ Mass of one atom of carbon-12}}$$

On this scale, carbon-12 has a relative atomic mass of 12.000 000, carbon has a relative atomic mass of 12.011 11, and hydrogen has a relative atomic mass 1.007 97. Since relative atomic masses are ratios of two masses, they have no units. As this value for hydrogen is very close to one, the value of $H = 1$ is used in most calculations. A table of approximate relative atomic masses is given on page 293. The symbol for relative atomic mass is A_r.

RELATIVE MOLECULAR MASS

You can find the mass of a molecule by adding up the masses of all the atoms in it. You can find the relative molecular mass of a compound by adding the relative atomic masses of all the atoms in a molecule of the compound. For example, you can work out the relative molecular mass of carbon dioxide as follows:

The formula is CO_2.

$$1 \text{ atom of C, relative atomic mass } 12 = 12$$
$$2 \text{ atoms of O, relative atomic mass } 16 = 32$$
$$\text{Total} = 44$$
$$\text{Relative molecular mass of } CO_2 = 44$$

The symbol for relative molecular mass is M_r.

A vast number of compounds consist of ions, not molecules. The compound sodium chloride, for example, consists of sodium ions and chloride ions. You cannot correctly refer to a 'molecule of sodium chloride'. For ionic compounds, the term *formula unit* is used to describe the ions which make up the compound. A formula unit of sodium chloride is NaCl. A formula unit of copper(II) sulphate-5-water is $CuSO_4 \cdot 5H_2O$. It is still correct to use the term relative molecular mass for ionic compounds:

$$\text{Relative molecular mass} = \frac{\text{Mass of one formula unit}}{(1/12) \text{ Mass of one atom of carbon-12}}$$

We work out the relative molecular mass of calcium chloride as follows:

The formula is $CaCl_2$.

$$1 \text{ atom of Ca, relative atomic mass } 40 = 40$$
$$2 \text{ atoms of Cl, relative atomic mass } 35.5 = 71$$
$$\text{Total} = 111$$
$$\text{Relative molecular mass of } CaCl_2 = 111$$

We work out the relative molecular mass of aluminium sulphate as follows:

The formula is $Al_2(SO_4)_3$.

$$2 \text{ atoms of Al, relative atomic mass } 27 = 54$$
$$3 \text{ atoms of S, relative atomic mass } 32 = 96$$
$$12 \text{ atoms of O, relative atomic mass } 16 = 192$$
$$\text{Total} = 342$$
$$\text{Relative molecular mass of } Al_2(SO_4)_3 = 342$$

EXERCISE 3   Problems on Relative Molecular Mass

Work out the relative molecular masses of these compounds:

SO_2	$NaOH$	KNO_3
$MgCO_3$	$PbCl_2$	$MgCl_2$
$Mg(NO_3)_2$	$Zn(OH)_2$	$ZnSO_4$
H_2SO_4	HNO_3	$MgSO_4 \cdot 7H_2O$
$CaSO_4$	Pb_3O_4	P_2O_5
Na_2CO_3	$Ca(OH)_2$	$CuCO_3$
$CuSO_4$	$Ca(HCO_3)_2$	$CuSO_4 \cdot 5H_2O$
$Fe_2(SO_4)_3$	$Na_2CO_3 \cdot 10H_2O$	$FeSO_4 \cdot 7H_2O$

PERCENTAGE COMPOSITION

From the formula of a compound, we can work out the percentage by mass of each element present in the compound.

EXAMPLE 1 Calculate the percentage of silicon and oxygen in silicon(IV) oxide (silica).

METHOD First, work out the relative molecular mass. The formula is SiO_2.

$$1 \text{ atom of silicon, relative atomic mass } 28 = 28$$
$$2 \text{ atoms of oxygen, relative atomic mass } 16 = 32$$
$$\text{Total} = \text{Relative molecular mass} = 60$$

$$\text{Percentage of silicon} = \frac{28}{60} \times 100 = \frac{7}{15} \times 100$$

$$= \frac{7 \times 20}{3} = 47\%$$

$$\text{Percentage of oxygen} = \frac{32}{60} \times 100 = \frac{8}{15} \times 100$$

$$= \frac{8 \times 20}{3} = 53\%$$

ANSWER Silicon(IV) oxide contains 47% silicon and 53% oxygen by mass.

Since every formula unit of silicon(IV) oxide is 47% silicon, and all formula units are identical, bulk samples of pure silicon(IV) oxide all contain 47% silicon. This is true whether you are talking about silicon(IV) oxide found as quartz, or amethyst or crystoballite or sand.

In general,

Percentage of element A =

$$\frac{\text{Relative atomic mass of } A \times \text{No. of atoms of } A \text{ in formula}}{\text{Relative molecular mass of compound}} \times 100$$

EXAMPLE 2   Find the percentage by mass of magnesium, oxygen and sulphur in magnesium sulphate.

METHOD   First calculate the relative molecular mass. The formula is $MgSO_4$.

$$\begin{aligned} \text{1 atom of magnesium, relative atomic mass } 24 &= 24 \\ \text{1 atom of sulphur, relative atomic mass } 32 &= 32 \\ \text{4 atoms of oxygen, relative atomic mass } 16 &= 64 \\ \text{Total } = \text{ Relative molecular mass}, M_r &= 120 \end{aligned}$$

$$\begin{aligned} \text{Percentage of magnesium} &= \frac{A_r(\text{Mg}) \times \text{No. of Mg atoms}}{M_r(\text{MgSO}_4)} \times 100 \\ &= \frac{24}{120} \times 100 \\ &= 20\% \end{aligned}$$

$$\begin{aligned} \text{Percentage of sulphur} &= \frac{A_r(\text{S}) \times \text{No. of S atoms}}{M_r(\text{MgSO}_4)} \times 100 \\ &= \frac{32}{120} \times 100 \\ &= 27\% \end{aligned}$$

$$\begin{aligned} \text{Percentage of oxygen} &= \frac{A_r(\text{O}) \times \text{No. of O atoms}}{M_r(\text{MgSO}_4)} \times 100 \\ &= \frac{16 \times 4}{120} \times 100 \\ &= 53\% \end{aligned}$$

ANSWER   Magnesium 20%; sulphur 27%; oxygen 53%. You can check on the calculation by adding up the percentages to see whether they add up to 100. In this case $20 + 27 + 53 = 100$.

EXAMPLE 3   Calculate the percentage of water in copper sulphate crystals.

METHOD   Find the relative molecular mass. The formula is $CuSO_4 \cdot 5H_2O$.

$$1 \text{ atom of copper, relative atomic mass } 64 = 64 \text{ (approx.)}$$
$$1 \text{ atom of sulphur, relative atomic mass } 32 = 32$$
$$4 \text{ atoms of oxygen, relative atomic mass } 16 = 64$$
$$5 \text{ molecules of water, } 5 \times [(2 \times 1) + 16] = 5 \times 18 = 90$$
$$\text{Total} = \text{Relative molecular mass} = 250$$
$$\text{Mass of water} = 90$$

$$\text{Percentage of water} = \frac{\text{Mass of water in formula}}{\text{Relative molecular mass}} \times 100$$

$$= \frac{90}{250} \times 100$$

$$= 36\%$$

ANSWER   The percentage of water in copper sulphate crystals is 36%.

EXERCISE 4   Problems on Percentage Composition

SECTION 1
Calculators are not needed for these problems.

1. Calculate the percentages by mass of
   a) carbon and hydrogen in ethane, C_2H_6
   b) sodium, oxygen and hydrogen in sodium hydroxide, NaOH
   c) sulphur and oxygen in sulphur trioxide, SO_3
   d) carbon and hydrogen in propyne, C_3H_4.

2. Calculate the percentages by mass of
   a) carbon and hydrogen in heptane, C_7H_{16}
   b) magnesium and nitrogen in magnesium nitride, Mg_3N_2
   c) sodium and iodine in sodium iodide, NaI
   d) calcium and bromine in calcium bromide, $CaBr_2$.

SECTION 2
1. Calculate the percentage by mass of
   a) carbon and hydrogen in pentene, C_5H_{10}
   b) nitrogen, hydrogen and oxygen in ammonium nitrate
   c) iron, oxygen and hydrogen in iron(II) hydroxide
   d) carbon, hydrogen and oxygen in ethanedioic acid, $C_2O_4H_2$.

2. Calculate the percentages of
   a) carbon, hydrogen and oxygen in propanol, C_3H_7OH
   b) carbon, hydrogen and oxygen in ethanoic acid, CH_3CO_2H
   c) carbon, hydrogen and oxygen in methyl methanoate, HCO_2CH_3
   d) aluminium and sulphur in aluminium sulphide, Al_2S_3.

3. Haemoglobin contains 0.33% by mass of iron. There are 2 Fe atoms in 1 molecule of haemoglobin. What is the relative molecular mass of haemoglobin?

4. An adult's bones weigh about 11 kg, and 50% of this mass is calcium phosphate, $Ca_3(PO_4)_2$. What is the mass of phosphorus in the bones of an average adult?

4　The Mole

THE MOLE

Looking at equations tells us a great deal about chemical reactions. For example,

$$Fe(s) + S(s) \longrightarrow FeS(s)$$

tells us that iron and sulphur combine to form iron(II) sulphide, and that one atom of iron combines with one atom of sulphur. Chemists are interested in the exact quantities of substances which react together in chemical reactions. For example, in the reaction between iron and sulphur, if you want to measure out just enough iron to combine with, say, 10 g of sulphur, how do you go about it? What you need to do is to count out equal numbers of atoms of iron and sulphur. This sounds a formidable task, and it puzzled a chemist called Avogadro, working in Italy early in the nineteenth century. He managed to solve this problem with a piece of clear thinking which makes the problem look very simple once you have followed his argument.

Avogadro reasoned in this way:

We know from their relative atomic masses that an atom of carbon is 12 times as heavy as an atom of hydrogen. Therefore, we can say:

If　1 atom of carbon is　　12 times as heavy as 1 atom of hydrogen,
then 1 dozen C atoms are　12 times as heavy as 1 dozen H atoms,
and　1 hundred C atoms are 12 times as heavy as 1 hundred H atoms,
and　1 million C atoms are　12 times as heavy as 1 million H atoms,

and it follows that when we see a mass of carbon which is 12 times as heavy as a mass of hydrogen, the two masses must contain equal numbers of atoms. If we have 12 g of carbon and 1 g of hydrogen, we know that we have the same number of atoms of carbon and hydrogen. The same argument applies to any element. When we take the relative atomic mass of an element in grams:

40 g Calcium	24 g Magnesium	32 g Sulphur	12 g Carbon	1 g Hydrogen

all these masses contain the same number of atoms. This number is 6.022×10^{23}. The amount of an element which contains this number of atoms is called one *mole* of the element. (The symbol for *mole* is *mol*.) The ratio 6.022×10^{23}/mol is called the *Avogadro constant*.

> The mole is defined as the amount of a substance which contains as many elementary entities as there are atoms in 12 grams of carbon-12.

We can count out 6×10^{23} atoms of any element by weighing out its relative atomic mass in grams. If we want to react iron and sulphur so that there is an atom of sulphur for every atom of iron, we can count out 6×10^{23} atoms of sulphur by weighing out 32 g of sulphur and we can count out 6×10^{23} atoms of iron by weighing out 56 g of iron. Since one atom of iron reacts with one atom of sulphur to form one formula unit of iron(II) sulphide, one mole of iron reacts with one mole of sulphur to form one mole of iron(II) sulphide:

$$Fe(s) + S(s) \longrightarrow FeS(s)$$

and 56 g of iron react with 32 g of sulphur to form 88 g of iron(II) sulphide.

Just as one mole of an element is the relative atomic mass in grams, one mole of a compound is the relative molecular mass in grams. If you want to weigh out one mole of sodium hydroxide, you first work out its relative molecular mass.

The formula is NaOH.

$$\text{Relative molecular mass} = 23 + 16 + 1$$
$$= 40$$

If you weigh out 40 g of sodium hydroxide, you have one mole of sodium hydroxide. The quantity $40 \, g \, mol^{-1}$ is the *molar mass* of sodium hydroxide. The molar mass of a compound is the relative molecular mass in grams per mole. The molar mass of an element is the relative atomic mass in grams per mole. The molar mass of sodium hydroxide is $40 \, g \, mol^{-1}$, and the molar mass of sodium is $23 \, g \, mol^{-1}$.

Remember that most gaseous elements consist of molecules, not atoms. Chlorine exists as Cl_2 molecules, oxygen as O_2 molecules, hydrogen as H_2 molecules, and so on. To work out the mass of a mole of chlorine molecules, you must use the relative molecular mass of Cl_2.

$$\text{Relative atomic mass of chlorine} = 35.5$$
$$\text{Relative molecular mass of } Cl_2 = 2 \times 35.5 = 71$$
$$\text{Mass of 1 mole of chlorine, } Cl_2 = 71 \text{ grams.}$$

The noble gases, helium, neon, argon, krypton and xenon, exist as atoms. Since the relative atomic mass of helium is 4, the mass of 1 mole of helium is 4 g.

The *relative molecular mass* of a compound is defined by the expression:

$$\text{Relative molecular mass} = \frac{\text{Mass of one molecule of the compound}}{\frac{1}{12} \text{ Mass of one atom of carbon-12}}$$

The relative molecular mass of a compound expressed in grams is one mole of the compound; thus 44 g of carbon dioxide is one mole of carbon dioxide, and contains 6.022×10^{23} molecules. In the case of ionic compounds, which do not consist of molecules, you refer to a 'formula unit' of the compound. A formula unit of sodium sulphate is $2Na^+SO_4^{2-}$. Then, the relative formula mass in grams, 142 g, is one mole of sodium sulphate.

The term *relative molar mass* embraces relative molecular mass, relative formula mass and relative atomic mass:

$$\text{Relative molar mass} = \frac{\text{Mass of one mole of substance}}{\frac{1}{12} \text{ Mass of one mole of carbon-12}}$$

Since this equation divides mass by mass, relative molar mass is a ratio and does not have units.

The mass of one mole of an element or compound is referred to as its *molar mass*. The molar mass of sodium hydroxide is 40 g mol^{-1}. The molar mass of copper is 63.5 g mol^{-1}. If you have a mass m of a substance which has a molar mass of $M \text{ g mol}^{-1}$, then the amount of substance, n, is given by:

$$n = \frac{m}{M}$$

If the unit of mass is g, and the unit of molar mass is g mol^{-1}, then the amount of substance, n, has the unit mol.

$$\text{Amount of substance (number of moles)} = \frac{\text{Mass}}{\text{Molar mass}}$$

CALCULATION OF MOLAR MASS

The relative molar mass of a compound is the sum of the relative atomic masses of the atoms in a molecule or formula unit of the compound.

EXAMPLE   What is the molar mass of glucose?

Formula $= C_6H_{12}O_6$

Relative molar mass $= (6 \times 12) + (12 \times 1) + (6 \times 16) = 180$

ANSWER   The relative molar mass is 180, and the molar mass is $180\,g\,mol^{-1}$.

EXERCISE 5   Problems on the Mole

SECTION 1

1. State the mass of each element in:
   - a) 0.5 mol chromium
   - b) 1/7 mol iron
   - c) 1/3 mol carbon
   - d) 1/4 mol magnesium
   - e) 1/7 mol nitrogen molecules
   - f) 1/4 mol oxygen molecules.

   Remember that nitrogen and oxygen exist as diatomic molecules, N_2 and O_2.

2. Calculate the amount of each element in:
   - a) 46 g sodium
   - b) 130 g zinc
   - c) 10 g calcium
   - d) 2.4 g magnesium
   - e) 13 g chromium.

3. Find the mass of each element in:
   - a) 10 mol lead
   - b) 1/6 mol copper
   - c) 0.1 mol iodine molecules
   - d) 10 mol hydrogen molecules
   - e) 0.25 mol calcium
   - f) 0.25 mol bromine molecules
   - g) $\frac{3}{4}$ mol iron
   - h) 0.20 mol zinc
   - i) $\frac{1}{2}$ mol chlorine molecules
   - j) 0.1 mol neon.

4. State the amount of substance (mol) in:
   - a) 58.5 g sodium chloride
   - b) 26.5 g anhydrous sodium carbonate
   - c) 50.0 g calcium carbonate
   - d) 15.9 g copper(II) oxide
   - e) 8.00 g sodium hydroxide
   - f) 303 g potassium nitrate
   - g) 9.8 g sulphuric acid
   - h) 499 g copper(II) sulphate-5-water.

5. Given Avogadro's constant is $6 \times 10^{23}\,mol^{-1}$, calculate the number of atoms in:
   - a) 35.5 g chlorine
   - b) 27 g aluminium
   - c) 3.1 g phosphorus
   - d) 336 g iron
   - e) 48 g magnesium
   - f) 1.6 g oxygen
   - g) 0.4 g oxygen
   - h) 216 g silver.

6. How many grams of zinc contain:
   a) 6×10^{23} atoms
   b) 6×10^{20} atoms?

7. How many grams of aluminium contain:
   a) 2×10^{23} atoms
   b) 6×10^{20} atoms?

8. What mass of carbon contains:
   a) 6×10^{23} atoms
   b) 2×10^{21} atoms?

9. Write down:
   a) the mass of calcium which has the same number of atoms as 12 g of magnesium
   b) the mass of silver which has the same number of atoms as 3 g of aluminium
   c) the mass of zinc with the same number of atoms as 1 g of helium
   d) the mass of sodium which has 5 times the number of atoms in 39 g of potassium.

SECTION 2

Use Avogadro constant $= 6 \times 10^{23} \, \text{mol}^{-1}$.

1. Imagine a hardware store is having a sale. The knock-down price of titanium is one billion (10^9) atoms for 1p. How much would you have to pay for 1 milligram (1×10^{-3} g) of titanium?

2. Ethanol, C_2H_6O, is the alcohol in alcoholic drinks. If you have 9.2 g of ethanol, how many moles do you have of
   a) ethanol molecules
   b) carbon atoms
   c) hydrogen atoms
   d) oxygen atoms?

3. A car releases about 5 g of nitrogen oxide, NO, into the air for each mile driven. How many molecules of NO are emitted per mile?

4. How many moles of H_2O are there in 1.00 litre of water?

5. How many moles of Fe_2O_3 are there in 1.00 kg of rust?

6. What is the mass of one molecule of water?

7. What is the amount (mol) of sucrose, $C_{12}H_{22}O_{11}$, in a one kilogram bag of sugar?

5 Equations and the Mole

You will find that the mole concept, which you studied in Chapter 4, helps with all your chemical calculations. In chemistry, calculations are related to the equations for chemical reactions. The quantities of substances that react together are expressed in moles.

CALCULATIONS BASED ON CHEMICAL EQUATIONS

Equations tell us not only what substances react together but also what amounts of substances react together. The equation for the action of heat on sodium hydrogencarbonate

$$2NaHCO_3(s) \longrightarrow Na_2CO_3(s) + CO_2(g) + H_2O(g)$$

tells us that 2 moles of $NaHCO_3$ give 1 mole of Na_2CO_3. Since the molar masses are $NaHCO_3 = 84 \, g \, mol^{-1}$ and $Na_2CO_3 = 106 \, g \, mol^{-1}$, it follows that 168 g of $NaHCO_3$ give 106 g of Na_2CO_3.

The amounts of substances undergoing reaction, as given by the balanced chemical equation, are called the *stoichiometric* amounts. *Stoichiometry* is the relationship between the amounts of reactants and products in a chemical reaction. If one reactant is present in excess of the stoichiometric amount required for reaction with another of the reactants, then the excess of one reactant will be left unused at the end of the reaction.

EXAMPLE 1   How many moles of iodine can be obtained from $\frac{1}{6}$ mole of potassium iodate(V)?

METHOD   The equation

Potassium + Potassium + Hydrogen $\longrightarrow$ Iodine + Potassium
iodate(V)   iodide   ion   ion
                                                       + Water

$$KIO_3(aq) + 5KI(aq) + 6H^+(aq) \longrightarrow 3I_2(aq) + 6K^+(aq) + 3H_2O(l)$$

tells us that 1 mol of KIO_3 gives 3 mol of I_2. Therefore:

ANSWER $\frac{1}{6}$ mol of KIO_3 gives $\frac{1}{6} \times 3$ mol of $I_2 = \frac{1}{2}$ mol of I_2.

EXAMPLE 2   What is the maximum mass of ethyl ethanoate that can be obtained from 0.1 mol of ethanol?

METHOD   Write the equation:

$$\text{Ethanol} + \text{Ethanoic acid} \longrightarrow \text{Ethyl ethanoate} + \text{Water}$$

$$C_2H_5OH(l) + CH_3CO_2H(l) \longrightarrow CH_3CO_2C_2H_5(l) + H_2O(l)$$

1 mol of C_2H_5OH gives 1 mol $CH_3CO_2C_2H_5$
0.1 mol of C_2H_5OH gives 0.1 mol $CH_3CO_2C_2H_5$
The molar mass of $CH_3CO_2C_2H_5$ is $88\,\text{g mol}^{-1}$. Therefore:

ANSWER   0.1 mol of ethanol gives 8.8 g of ethyl ethanoate.

EXAMPLE 3   A mixture of 5.00 g of sodium carbonate and sodium hydrogen-carbonate is heated. The loss in mass is 0.31 g. Calculate the percentage by mass of sodium carbonate in the mixture.

METHOD   On heating the mixture, the reaction

$$2NaHCO_3(s) \longrightarrow Na_2CO_3(s) + CO_2(g) + H_2O(g)$$

takes place. The loss in mass is due to the decomposition of $NaHCO_3$.
Since 2 mol $NaHCO_3$ form 1 mol CO_2 + 1 mol H_2O
2×84 g $NaHCO_3$ form 44 g CO_2 and 18 g H_2O
168 g $NaHCO_3$ lose 62 g in mass.
The observed loss in mass of 0.31 g is due to the decomposition of

$$\frac{0.31}{62} \times 168\,\text{g NaHCO}_3 = 0.84\,\text{g}$$

The mixture contains 0.84 g $NaHCO_3$
The difference, $5.00 - 0.84 = 4.16$ g Na_2CO_3.

ANSWER   Percentage of $Na_2CO_3 = \dfrac{4.16}{5.00} \times 100 = 83.2\%$.

EXERCISE 6   Problems on Reacting Masses of Solids

SECTION 1

1. A sulphuric acid plant uses 2500 tonnes of sulphur dioxide each day. What mass of sulphur must be burned to produce this quantity of sulphur dioxide?

2. An antacid tablet contains 0.1 g of magnesium hydrogencarbonate, $Mg(HCO_3)_2$. What mass of stomach acid, HCl, will it neutralise?

3. Aspirin, $C_9H_8O_4$, is made by the reaction:

$$\text{Salicylic acid} + \text{Ethanoic anhydride} \longrightarrow \text{Aspirin} + \text{Ethanoic acid}$$

$$C_7H_6O_3 + C_4H_6O_3 \longrightarrow C_9H_8O_4 + C_2H_4O_2$$

How many grams of salicylic acid, $C_7H_6O_3$, are needed to make one aspirin tablet, which contains 0.33 g of aspirin?

4. Aluminium sulphate is used to treat sewage. It can be made by the reaction:

$$\text{Aluminium} + \text{Sulphuric} \quad \longrightarrow \quad \text{Aluminium} + \text{Water}$$
$$\text{hydroxide} \quad \text{acid} \qquad\qquad \text{sulphate}$$

a) Balance this *unbalanced* equation for the reaction:

$$Al(OH)_3(s) + H_2SO_4(aq) \longrightarrow Al_2(SO_4)_3(aq) + H_2O(l)$$

b) Say what masses of (i) aluminium hydroxide and (ii) sulphuric acid are needed to make 1.00 kg of aluminium sulphate.

5. Washing soda, $Na_2CO_3 \cdot 10H_2O$, loses some of its water of crystallisation if it is not kept in an air-tight container to form $Na_2CO_3 \cdot H_2O$.

A grocer buys a 10 kg bag of washing soda at 30p/kg. While it is standing in his store room, the bag punctures, and the crystals turn into a powder, $Na_2CO_3 \cdot H_2O$. The grocer sells this powder at 50p/kg. Does he make a profit or a loss?

6. When you take a warm bath, the power station has to burn about 1.2 kg of coal to provide enough electricity to heat the water.

a) If the coal contains 3% sulphur, what mass of sulphur dioxide does the power station emit as a result?

b) Multiply your answer by the number of warm baths you take in a year.

c) This is only a part of your contribution to air pollution. What can be done to reduce this source of pollution — apart from taking cold baths?

7. Nitrogen monoxide, NO, is a pollutant gas which comes out of vehicle exhausts. One technique for reducing the quantity of nitrogen monoxide in vehicle exhausts is to inject a stream of ammonia, NH_3, into the exhaust. Nitrogen monoxide is converted into the harmless products nitrogen and water:

$$4NH_3(g) + 6NO(g) \longrightarrow 5N_2(g) + 6H_2O(l)$$

An average vehicle emits 5 g of nitrogen monoxide per mile. Assuming a mileage of 10 000 miles a year, what mass of ammonia would be needed to clean up the exhaust?

8. Some industrial plants, for example aluminium smelters, emit fluorides. In the past, there have been cases of fluoride pollution affecting the teeth and joints of cattle. The Union Carbide Corporation has invented a process for removing fluorides from waste gases. It involves the reaction:

$$\text{Fluoride ion } (F^-) + \text{Charcoal (C)} \xrightarrow{\substack{\text{high} \\ \text{temperature}}} \text{Carbon tetrafluoride } (CF_4)$$

The product, CF_4, is harmless. The firm claims that 1 kg of charcoal will remove 6.3 kg of fluoride ion. Do you believe this claim? Explain your answer.

9. A factory makes a detergent of formula $C_{12}H_{25}SO_4Na$ from lauryl alcohol, $C_{12}H_{26}O$. To manufacture 11 tonnes of detergent daily, what mass of lauryl alcohol is needed?

10. A mass of 0.65 g of zinc powder was added to a beaker containing silver nitrate solution. When all the zinc had reacted, 2.16 g of silver were obtained.

Calculate

a) the amount of zinc used

b) the amount of silver formed

c) the amount of silver produced by 1 mol of zinc.

d) Write a balanced ionic equation for the reaction.

SECTION 2

1. The element X has a relative atomic mass of 35.5. It reacts with a solution of the sodium salt of Y according to the equation

$$X_2 + 2NaY \longrightarrow Y_2 + 2NaX$$

If 14.2 g of X_2 displace 50.8 g of Y_2, what is the relative atomic mass of Y?

2. TNT is an explosive. The name stands for trinitrotoluene. The compound is made by the reaction

$$\text{Toluene} + \text{Nitric acid} \longrightarrow \text{TNT} + \text{Water}$$
$$C_7H_8(l) + 3HNO_3(l) \longrightarrow C_7H_5N_3O_6(s) + 3H_2O(l)$$

Calculate the masses of a) toluene and b) nitric acid that must be used to make 10.00 tonnes of TNT. (1 tonne = 1000 kg.)

3. A large power plant produces about 500 tonnes of sulphur dioxide in a day. One way of removing this pollutant from the waste gases is to inject limestone. This converts sulphur dioxide into calcium sulphate.

(i)
$$\text{Limestone} + \begin{matrix}\text{Sulphur}\\\text{dioxide}\end{matrix} + \text{Oxygen} \longrightarrow \begin{matrix}\text{Calcium}\\\text{sulphate}\end{matrix} + \begin{matrix}\text{Carbon}\\\text{dioxide}\end{matrix}$$
$$2CaCO_3(s) + 2SO_2(g) + O_2(g) \longrightarrow 2CaSO_4(s) + 2CO_2(g)$$

Another method of removing sulphur dioxide is to 'scrub' the waste gases with ammonia. The product is ammonium sulphate.

(ii)
$$\text{Ammonia} + \begin{matrix}\text{Sulphur}\\\text{dioxide}\end{matrix} + \text{Oxygen} + \text{Water} \longrightarrow \begin{matrix}\text{Ammonium}\\\text{sulphate}\end{matrix}$$
$$4NH_3(g) + 2SO_2(g) + O_2(g) + 2H_2O(l) \longrightarrow 2(NH_4)_2SO_4(aq)$$

a) Calculate the mass of calcium sulphate produced in a day by method (i) and the mass of ammonium sulphate produced in a day by method (ii).

b) Say whether limestone and ammonia are found naturally occurring or are manufactured.

c) Which do you think is the more valuable by-product, calcium sulphate or ammonium sulphate? Explain your answer.

4. What mass of glucose must be fermented to give 5.00 kg of ethanol?
$$C_6H_{12}O_6(aq) \longrightarrow 2C_2H_5OH(aq) + 2CO_2(g)$$

5. What mass of sulphuric acid can be obtained from 1000 tonnes of an ore which contains 32.0% of FeS_2?

6. What mass of silver chloride can be precipitated from a solution which contains 1.000×10^{-3} mol of silver ions?

7. When potassium iodate(V) is allowed to react with acidified potassium iodide, iodine is formed.
$$KIO_3(aq) + KI(aq) + 6H^+(aq) \longrightarrow 3I_2(aq) + 6K^+(aq) + 3H_2O(l)$$
What mass of KIO_3 is required to give 10.00 g of iodine?

8. How many tonnes of iron can be obtained from 10.00 tonnes of
a) Fe_2O_3 and b) Fe_3O_4?

9. What mass of quicklime can be obtained by heating 75.0 g of limestone, which is 86.8% calcium carbonate?

10. What mass of barium sulphate can be precipitated from a solution which contains 4.000 g of barium chloride?

11. A mixture of calcium and magnesium carbonates weighing 10.0000 g was heated until it reached a constant mass of 5.0960 g. Calculate the percentage composition of the mixture, by mass.

12. A mixture of anhydrous sodium carbonate and sodium hydrogencarbonate weighing 10.0000 g was heated until it reached a constant mass of 8.7080 g. Calculate the composition of the mixture in grams of each component.

USING THE MASSES OF THE REACTANTS TO WORK OUT THE EQUATION FOR A REACTION

The equation for a reaction can be used to enable you to calculate the masses of chemicals taking part in the reaction. The converse is also true. If you know the mass of each substance taking part in a reaction, you can calculate the number of moles of each substance taking part in the reaction, and this will tell you the equation.

EXAMPLE 1 Iron burns in chlorine to form iron chloride. An experiment showed that 5.60 g of iron combined with 10.65 g of chlorine. Deduce the equation for the reaction.

METHOD 5.60 g of iron combine with 10.65 g of chlorine

Relative atomic masses are: $Fe = 56$, $Cl = 35.5$
Amount (mol) of iron $= 5.60/56 = 0.10$
Amount (mol) of chlorine molecules $= 10.65/71.0 = 0.15$

The equation must be: $Fe + 1.5Cl_2 \longrightarrow$

or $2Fe + 3Cl_2 \longrightarrow$

To balance the equation, the right-hand side must read $2FeCl_3$. Therefore,

ANSWER $2Fe(s) + 3Cl_2(g) \longrightarrow 2FeCl_3(s)$

EXAMPLE 2 17.0 g of sodium nitrate react with 19.6 g of sulphuric acid to give 12.6 g of nitric acid. Deduce the equation for the reaction.

METHOD Relative molecular masses are: $NaNO_3 = 85$, $H_2SO_4 = 98$, $HNO_3 = 63$

Number of moles of $NaNO_3 = 17.0/85 = 0.2$
Number of moles of $H_2SO_4 = 19.6/98 = 0.2$
Number of moles of $HNO_3 = 12.6/63 = 0.2$

0.2 mol $NaNO_3$ reacts with 0.2 mol H_2SO_4 to form 0.2 mol of HNO_3
1 mol $NaNO_3$ reacts with 1 mol H_2SO_4 to form 1 mol of HNO_3

$$NaNO_3(s) + H_2SO_4(l) \longrightarrow HNO_3(l)$$

The equation must be balanced by inserting $NaHSO_4$ on the right-hand side:

ANSWER $NaNO_3(s) + H_2SO_4(l) \longrightarrow HNO_3(l) + NaHSO_4(s)$

EXERCISE 7   Problems on Deriving Equations

1. To a solution containing 2.975 g of sodium persulphate, $Na_2S_2O_8$, is added an excess of potassium iodide solution. A reaction occurs, in which sulphate ions are formed and 3.175 g of iodine are formed. Deduce the equation for the reaction.

2. Aminosulphuric acid, H_2NSO_3H, reacts with warm sodium hydroxide solution to give ammonia and a solution which contains sulphate ions. 0.540 g of the acid, when treated with an excess of alkali, gave 153 cm^3 of ammonia at 60 °C and 1 atm. Deduce the equation for the reaction.

3. A solution containing 5.00×10^{-3} mol of sodium thiosulphate was shaken with 1 g of silver chloride. 0.717 5 g of silver chloride dissolved, and analysis showed that 5.00×10^{-3} mol of chloride ions were present in the resulting solution. Derive an equation for the reaction.

4. An unsaturated hydrocarbon of molar mass 80 g mol^{-1} reacts with bromine. If 0.250 g of hydrocarbon reacts with 1.00 g of bromine, what is the equation for the reaction?

5. Given that 1.00 g of phenylamine, $C_6H_5NH_2$, reacts with 5.16 g of bromine, derive an equation for the reaction.

PERCENTAGE YIELD

There are many reactions which do not go to completion. Reactions between organic compounds do not often give a 100% yield of product. The actual yield is compared with the yield calculated from the molar masses of the reactants. The equation

$$\text{Percentage yield} = \frac{\text{Actual mass of product}}{\text{Calculated mass of product}} \times 100$$

is used to give the percentage yield.

EXAMPLE   From 23 g of ethanol are obtained 36 g of ethyl ethanoate by esterification with ethanoic acid in the presence of concentrated sulphuric acid. What is the percentage yield of the reaction?

METHOD   Write the equation:

$$CH_3CO_2H(l) + C_2H_5OH(l) \longrightarrow CH_3CO_2C_2H_5(l) + H_2O(l)$$

46 g of C_2H_5OH forms 108 g of $CH_3CO_2C_2H_5$

23 g of C_2H_5OH should give $\dfrac{23}{46} \times 88$ g $= 44$ g of $CH_3CO_2C_2H_5$

Actual mass obtained $= 36$ g

$$\text{Percentage yield} = \frac{\text{Actual mass of product}}{\text{Calculated mass of product}} \times 100$$

ANSWER $\text{Percentage yield} = \dfrac{36}{44} \times 100 = 82\%$

EXERCISE 8   Problems on Percentage Yield

1. Phenol, C_6H_5OH, is converted into trichlorophenol, $C_6H_2Cl_3OH$. If 488 g of product are obtained from 250 g of phenol, calculate the percentage yield.

2. 29.5 g of ethanoic acid, CH_3CO_2H, are obtained from the oxidation of 25.0 g of ethanol, C_2H_5OH. What percentage yield does this represent?

3. 0.8500 g of hexanone, $C_6H_{12}O$, is converted into its 2,4-dinitrophenyl-hydrazone. After isolation and purification, 2.1180 g of product, $C_{12}H_{18}N_4O_4$, are obtained. What percentage yield does this represent?

4. Benzaldehyde, C_7H_6O, forms a hydrogensulphite compound of formula $C_7H_7SO_4Na$. From 1.210 g of benzaldehyde, a yield of 2.181 g of the product was obtained. Calculate the percentage yield.

5. 100 cm^3 of barium chloride solution of concentration 0.0500 mol dm^{-3} were treated with an excess of sulphate ions in solution. The precipitate of barium sulphate formed was dried and weighed. A mass of 1.1558 g was recorded. What percentage yield does this represent?

LIMITING REACTANT

In a chemical reaction, the reactants are often added in amounts which are not stoichiometric. One or more of the reactants is in excess and is not completely used up in the reaction. The amount of product is determined by the amount of the reactant that is not in excess and is used up completely in the reaction. This is called the *limiting reactant*. You first have to decide which is the limiting reactant before you can calculate the amount of product formed.

EXAMPLE   5.00 g of iron and 5.00 g of sulphur are heated together to form iron(II) sulphide. Which reactant is present in excess? What mass of product is formed?

METHOD   Write the equation:

$$Fe(s) + S(s) \longrightarrow FeS(s)$$

1 mole of Fe + 1 mole of S form 1 mole of FeS

∴   56 g Fe and 32 g S form 88 g FeS

  5.00 g Fe is 5/56 mol = 0.0893 mol Fe

  5.00 g S is 5/32 mol = 0.156 mol S

There is insufficient Fe to react with 0.156 mol S; iron is the limiting reactant.

0.0893 mol Fe forms 0.0893 mol FeS = 0.0893 × 88 g = 7.86 g

ANSWER   Mass formed = 7.86 g.

EXERCISE 9   Problems on Limiting Reactant

1. In the blast furnace, the overall reaction is

$$2Fe_2O_3(s) + 3C(s) \longrightarrow 3CO_2(g) + 4Fe(s)$$

   What is the maximum mass of iron that can be obtained from 700 tonnes of iron(III) oxide and 70 tonnes of coke? (1 tonne = 1 000 kg.)

2. In the manufacture of calcium carbide

$$CaO(s) + 3C(s) \longrightarrow CaC_2(s) + CO(g)$$

   What is the maximum mass of calcium carbide that can be obtained from 40 kg of quicklime and 40 kg of coke?

3. In the manufacture of the fertiliser ammonium sulphate

$$H_2SO_4(aq) + 2NH_3(g) \longrightarrow (NH_4)_2SO_4(aq)$$

   What is the maximum mass of ammonium sulphate that can be obtained from 2.0 kg of sulphuric acid and 1.0 kg of ammonia?

4. In the Solvay process, ammonia is recovered by the reaction

$$2NH_4Cl(s) + CaO(s) \longrightarrow CaCl_2(s) + H_2O(g) + 2NH_3(g)$$

   What is the maximum mass of ammonia that can be recovered from 2.00×10^3 kg of ammonium chloride and 500 kg of quicklime?

5. In the Thermit reaction

$$2Al(s) + Cr_2O_3(s) \longrightarrow 2Cr(s) + Al_2O_3(s)$$

   Calculate the percentage yield when 180 g of chromium are obtained from a reaction between 100 g of aluminium and 400 g of chromium(III) oxide.

6 Finding Formulae

EMPIRICAL FORMULAE

The formula of a compound is determined by finding the mass of each element present in a certain mass of the compound.

> Remember,
>
> $$\frac{\text{Amount (in moles) of substance}}{} = \frac{\text{Mass of substance}}{\text{Molar mass of the substance}}$$

EXAMPLE 1 Given that 127 g of copper combine with 32 g of oxygen, what is the formula of copper oxide?

Elements	Copper		Oxygen
Symbols	Cu		O
Masses	127 g		32 g
Relative atomic masses	63.5		16
Amounts	$\dfrac{127}{63.5}$		$\dfrac{32}{16}$
	= 2 mol		= 2 mol
Divide through by 2	= 1 mol	to	1 mol
Ratio of atoms	= 1 atom	to	1 atom
Formula		CuO	

We divide through by two to obtain the simplest formula for copper oxide which will fit the data. *The simplest formula which represents the composition of a compound is called the empirical formula.*

EXAMPLE When 127 g of copper combine with oxygen, 143 g of an oxide are formed. What is the empirical formula of the oxide?

METHOD You will notice here that the mass of oxygen is not given to you. You obtain it by subtraction.

$$\text{Mass of copper} = 127\,\text{g}$$

$$\text{Mass of oxide} = 143\,\text{g}$$

$$\text{Mass of oxygen} = 143 - 128 = 16\,\text{g}$$

Now you can carry on as before:

Elements	Copper	Oxygen
Symbols	Cu	O
Masses	127 g	16 g
Relative atomic masses	63.5	16
Amounts	$\dfrac{127}{63.5}$	$\dfrac{16}{16}$
	= 2	= 1
Ratio of atoms	= 2 to	1

ANSWER   Empirical formula Cu_2O.

EXAMPLE 3   Find n in the formula $MgSO_4 \cdot nH_2O$. A sample of 7.38 g of magnesium sulphate crystals lost 3.78 g of water on heating.

METHOD

Compounds present	Magnesium sulphate	Water
Mass	3.60 g	3.78 g
Molar mass	$120\,g\,mol^{-1}$	$18\,g\,mol^{-1}$
Amount	3.60/120	3.78/18
	= 0.030 mol	= 0.21 mol
Ratio of amounts	$\dfrac{0.030}{0.030}$	$\dfrac{0.21}{0.030}$
	= 1	= 7

ANSWER   The empirical formula is $MgSO_4 \cdot 7H_2O$.

MOLECULAR FORMULAE

The molecular formula is a simple multiple of the empirical formula. If the empirical formula is CH_2O, the molecular formula may be CH_2O or $C_2H_4O_2$ or $C_3H_6O_3$ and so on. You can tell which molecular formula is correct by finding out which gives the correct molar mass.

For methods of finding molar masses, see Chapters 9, 10 and 11. The molar mass is a multiple of the empirical formula mass.

EXAMPLE   A compound has the empirical formula CH_2O and molar mass $180\,g\,mol^{-1}$. What is its molecular formula?

METHOD   Empirical formula mass = $30\,g\,mol^{-1}$
Molar mass = $180\,g\,mol^{-1}$

The molar mass is 6 times the empirical formula mass. Therefore the molecular formula is 6 times the empirical formula. Therefore:

ANSWER   The empirical formula is $C_6H_{12}O_6$.

EXERCISE 10   Problems on Formulae

SECTION 1

1. 0.72 g of magnesium combine with 0.28 g of nitrogen.
   How many moles of magnesium does this represent?
   How many moles of nitrogen atoms combine?
   How many moles of magnesium combine with one mole of nitrogen atoms?
   What is the formula of magnesium nitride?

2. 1.68 g of iron combine with 0.64 g of oxygen.
   How many moles of iron does this mass represent?
   How many moles of oxygen atoms combine?
   How many moles of iron combine with one mole of oxygen atoms?
   What is the formula of this oxide of iron?

3. Calculate the empirical formula of the compound formed when 2.70 g of aluminium form 5.10 g of its oxide.
   What is the mass of aluminium?
   What is the mass of oxygen (not oxide)?
   How many moles of aluminium combine?
   How many moles of oxygen atoms combine?
   What is the ratio of moles of aluminium to moles of oxygen atoms?
   What is the formula of aluminium oxide?

4. Barium chloride forms a hydrate which contains 85.25% barium chloride and 14.75% water of crystallisation. What is the formula of this hydrate?
   What is the mass of barium chloride in 100 g of the hydrate?
   What is the mass of water in 100 g of the hydrate?
   What is the relative molecular mass of barium chloride?
   What is the relative molecular mass of water?
   How many moles of barium chloride are present in 100 g of the hydrate?
   How many moles of water are present in 100 g of the hydrate?
   What is the ratio of moles of barium chloride to moles of water?
   What is the formula of barium chloride hydrate?

5. Calculate the empirical formula of the compound formed when 414 g of lead form 478 g of a lead oxide.
   What mass of lead is present?
   How many moles of lead are present?
   What mass of oxygen (not oxide) is present?
   How many moles of oxygen atoms are present?
   What is the formula of this oxide of lead?

6. Calculate the empirical formulae of the following compounds:
   a) 0.62 g of phosphorus combined with 0.48 g of oxygen
   b) 1.4 g of nitrogen combined with 0.30 g of hydrogen
   c) 0.62 g of lead combined with 0.064 g of oxygen

d) 3.5 g of silicon combined with 4.0 g of oxygen

e) 1.10 g of manganese combined with 0.64 g of oxygen

f) 4.2 g of nitrogen combined with 12.0 g of oxygen

g) 2.6 g of chromium combined with 5.3 g of chlorine

7. Find the molecular formula for each of the following compounds from the empirical formula and the relative molecular mass:

	Empirical formula	M_r		Empirical formula	M_r
A	CF_2	100	E	CH_2	42
B	C_2H_4O	88	F	CH_3O	62
C	CH_3	30	G	CH_2Cl	99
D	CH	78	H	C_2HNO_2	213

8. A metal M forms a chloride of formula MCl_2 and relative molecular mass 127. The chloride reacts with sodium hydroxide solution to form a precipitate of the metal hydroxide. What is the relative molecular mass of the hydroxide?

a 56   b 71   c 73   d 90   e 146

9. A porcelain boat was weighed. After a sample of the oxide of a metal M, of $A_r = 119$, was placed in the boat, the boat was reweighed. Then the boat was placed in a reduction tube, and heated while a stream of hydrogen was passed over it. The oxide was reduced to the metal M. The boat was allowed to cool with hydrogen still passing over it, and then reweighed. Then it was reheated in hydrogen, cooled again and reweighed. The following results were obtained.

$$\text{Mass of boat} = 6.10\,\text{g}$$

$$\text{Mass of boat + metal oxide} = 10.63\,\text{g}$$

$$\text{Mass of boat + metal (1)} = 9.67\,\text{g}$$

$$\text{Mass of boat + metal (2)} = 9.67\,\text{g}$$

a) Explain why hydrogen was passed over the boat while it was cooling.

b) Explain why the boat + metal was reheated.

c) Find the empirical formula of the metal oxide.

d) Write an equation for the reaction of the oxide with hydrogen.

SECTION 2

1. Find the empirical formulae of the compounds formed in the reactions described below:

a) 10.800 g magnesium form 18.000 g of an oxide

b) 3.400 g calcium form 9.435 g of a chloride
c) 3.528 g iron form 10.237 g of a chloride
d) 2.667 g copper form 4.011 g of a sulphide
e) 4.662 g lithium form 5.328 g of a hydride.

2. Calculate the empirical formulae of the compounds with the following percentage composition:

a) 77.7% Fe   22.3% O   b) 70.0% Fe   30.0% O
c) 72.4% Fe   27.6% O   d) 40.2% K   26.9% Cr   32.9% O
e) 26.6% K   35.4% Cr   38.0% O   f) 92.3% C   7.6% H
g) 81.8% C   18.2% H

3. Samples of the following hydrates are weighed, heated to drive off the water of crystallisation, cooled and reweighed. From the results obtained, calculate the values of a–f in the formulae of the hydrates:

a) 0.869 g of $CuSO_4 \cdot aH_2O$ gave a residue of 0.556 g
b) 1.173 g of $CoCl_2 \cdot bH_2O$ gave a residue of 0.641 g
c) 1.886 g of $CaSO_4 \cdot cH_2O$ gave a residue of 1.492 g
d) 0.904 g of $Pb(C_2H_3O_2)_2 \cdot dH_2O$ gave a residue of 0.774 g
e) 1.144 g of $NiSO_4 \cdot eH_2O$ gave a residue of 0.673 g
f) 1.175 g of $KAl(SO_4)_2 \cdot fH_2O$ gave a residue of 0.639 g.

4. An organic compound, X, which contains only carbon, hydrogen and oxygen, has a molar mass of about 85 g mol^{-1}. When 0.43 g of X is burnt in excess oxygen, 1.10 g of carbon dioxide and 0.45 g of water are formed.

a) What is the empirical formula of X?
b) What is the molecular formula of X?

5. A liquid, Y, of molar mass 44 g mol^{-1} contains 54.5% carbon, 36.4% oxygen and 9.1% hydrogen.

a) Calculate the empirical formula of Y, and
b) deduce its molecular formula.

6. An organic compound contains 58.8% carbon, 9.8% hydrogen and 31.4% oxygen. The molar mass is 102 g mol^{-1}.

a) Calculate the empirical formula, and
b) deduce the molecular formula of the compound.

7. An organic compound has molar mass 150 g mol^{-1} and contains 72.0% carbon, 6.67% hydrogen and 21.33% oxygen. What is its molecular formula?

7 Reacting Volumes of Gases

GAS MOLAR VOLUME

A surprising feature of reactions between gases was noticed by a French chemist called *Gay-Lussac* in 1808. *Gay-Lussac's law states that when gases combine they do so in volumes which bear a simple ratio to one another and to the volume of the product if it is gaseous, provided all the volumes are measured at the same temperature and pressure.* For example, when hydrogen and chlorine combine, $1\,dm^3$ (or litre) of hydrogen will combine exactly with $1\,dm^3$ of chlorine to form $2\,dm^3$ of hydrogen chloride. When nitrogen and hydrogen combine, a certain volume of nitrogen will combine with three times that volume to form twice its volume of ammonia.

The Italian chemist Avogadro gave an explanation in 1811. His suggestion, known as *Avogadro's hypothesis*, is that: *Equal volumes of all gases (at the same temperature and pressure) contain the same number of molecules.* It follows from Avogadro's hypothesis that, whenever we see an equation representing a reaction between gases, we can substitute volumes of gases in the same ratio as numbers of molecules. Thus

$$N_2(g) + 3H_2(g) \longrightarrow 2NH_3(g)$$

means that since

a molecule of nitrogen + 3 molecules of hydrogen form 2 molecules of ammonia

then 1 volume of nitrogen + 3 volumes of hydrogen form 2 volumes of ammonia.

For example, $1\,dm^3$ of nitrogen + $3\,dm^3$ of hydrogen form $2\,dm^3$ of ammonia.

Since equal volumes of gases (at the same temperature and pressure) contain the same number of molecules, if you consider the Avogadro constant (L), L molecules of carbon dioxide, L molecules of hydrogen, L molecules of oxygen and so on, then all the gases will occupy the same volume. The volume occupied by L molecules of gas, which is one mole of each gas, is called the *gas molar volume*.

In stating the volume of a gas, one needs to state the temperature and pressure at which the volume was measured. It is usual to give the volume at $0°C$ and 1 atmosphere pressure ($273\,K$ and $1.01 \times 10^5\,N\,m^{-2}$). These conditions are called standard temperature and pressure (s.t.p.). Chapter 9 deals with calculating the volume of a gas at s.t.p. from the volume measured under experimental conditions.

> One mole of gas occupies 22.4 dm^3 at 0°C and 1 atmosphere.

Since one mole of gas occupies 22.4 dm^3 at s.t.p., *the gas molar volume is 22.4 dm^3 at s.t.p.* This makes calculations on reacting volumes of gases very simple. An equation which shows how many moles of different gases react together also shows the ratio of the volumes of the different gases that react together. For example, the equation

$$2NO(g) + O_2(g) \longrightarrow 2NO_2(g)$$

tells us that 2 moles of NO + 1 mole of O_2 form 2 moles of NO_2

$\therefore$ 44.8 dm^3 of NO + 22.4 dm^3 of O_2 form 44.8 dm^3 of NO_2

In general, 2 volumes of NO + 1 volume of O_2 form 2 volumes of NO_2.

EXAMPLE 1 What is the volume of oxygen needed for the complete combustion of 2 dm^3 of propane?

METHOD Write the equation:

$$C_3H_8(g) + 5O_2(g) \longrightarrow 3CO_2(g) + 4H_2O(g)$$

1 mole of C_3H_8 needs 5 moles of O_2

1 volume of C_3H_8 needs 5 volumes of O_2. Therefore:

ANSWER 2 dm^3 of propane need 10 dm^3 of oxygen.

EXAMPLE 2 What volume of hydrogen is obtained when 3.00 g of zinc react with an excess of dilute sulphuric acid at s.t.p.?

METHOD Write the equation:

$$Zn(s) + H_2SO_4(aq) \longrightarrow H_2(g) + ZnSO_4(aq)$$

1 mole of Zn forms 1 mole of H_2

65 g of Zn form 22.4 dm^3 of H_2 (at s.t.p.)

3.00 g of Zn form $\dfrac{3.00}{65} \times 22.4$ dm^3 $= 1.03$ dm^3 H_2

ANSWER 3.00 g of zinc give 1.03 dm^3 of hydrogen at s.t.p.

EXERCISE 11   Problems on Reacting Volumes of Gases

SECTION 1

1. The problem is to find the percentage by volume composition of a mixture of hydrogen and ethane. When 75 cm^3 of the mixture was burned in an excess of oxygen, the volume of carbon dioxide produced was 60 cm^3 (all volumes at s.t.p.).

    a) Write an equation for the combustion of ethane.

    b) Say what volume of ethane would give 60 cm^3 of carbon dioxide.

    c) Calculate the percentage of ethane in the mixture.

2. 25 cm^3 of carbon monoxide were ignited with 25 cm^3 of oxygen. All gas volumes were measured at the same temperature and pressure. The reduction in the total volume was

    a 2.5 cm^3 b 10.0 cm^3 c 12.5 cm^3 d 15.0 cm^3 e 25.0 cm^3

3. Ethene reacts with oxygen according to the equation

$$C_2H_4(g) + 3O_2(g) \longrightarrow 2CO_2(g) + 2H_2O(l)$$

15.0 cm^3 of ethene were mixed with 60.0 cm^3 of oxygen and the mixture was sparked to complete the reaction. If all the volumes were measured at s.t.p., the volume of the products would be:

    a 15 cm^3 b 30 cm^3 c 45 cm^3 d 60 cm^3 e 75 cm^3

4. The table gives the formulae and relative molecular masses of some gases.

Formula	Ne	C_2H_2	O_2	Ar	NO_2	SO_2	SO_3
M_r	20	26	32	40	46	64	80
Volume (cm^3) occupied by 1 g of gas at s.t.p.	1120	861	700	560	485	350	280

    a) Plot a graph of volume (on the vertical axis) against M_r (on the horizontal axis).

    b) Use the graph to predict the volumes occupied at s.t.p. by
      i) 1 g of fluorine, F_2
      ii) 1 g of Cl_2.

    c) What is the relative molecular mass of a gas which occupies 508 cm^3 per gram at s.t.p.? If the gas contains only carbon and oxygen, what is its formula?

SECTION 2

1. Ethene, $H_2C{=}CH_2$, and hydrogen react in the presence of a nickel catalyst to form ethane.

    a) Write a balanced equation for the reaction.

    b) If a mixture of 30 cm^3 of ethene and 20 cm^3 of hydrogen is passed over a nickel catalyst, what is the composition of the final mixture? (Assume that the reaction is complete and that all gas volumes are at s.t.p.)

2. What volume of oxygen (at s.t.p.) is required to burn exactly:

    a) 1 dm^3 of methane, according to the reaction

$$CH_4(g) + 2O_2(g) \longrightarrow CO_2(g) + 2H_2O(g)$$

b) 500 cm^3 of hydrogen sulphide, according to the reaction
$$2H_2S(g) \ + \ 3O_2(g) \ \longrightarrow \ 2SO_2(g) \ + \ 2H_2O(g)$$

c) 250 cm^3 of ethyne, according to the equation
$$2C_2H_2(g) \ + \ 5O_2(g) \ \longrightarrow \ 4CO_2(g) \ + \ 2H_2O(g)$$

d) 750 cm^3 of ammonia, according to the reaction
$$4NH_3(g) \ + \ 5O_2(g) \ \longrightarrow \ 4NO(g) \ + \ 6H_2O(g)$$

e) 1 dm^3 of phosphine, according to the reaction
$$PH_3(g) \ + \ 2O_2(g) \ \longrightarrow \ H_3PO_4(s)?$$

3. 1 dm^3 of H_2S and 1 dm^3 of SO_2 were allowed to react, according to the equation
$$2H_2S(g) \ + \ SO_2(g) \ \longrightarrow \ 2H_2O(l) \ + \ 3S(s)$$
What volume of gas will remain after the reaction?

4. 100 cm^3 of a mixture of ethane and ethene at s.t.p. were treated with bromine. 0.357 g of bromine was used up. Calculate the percentage by volume of ethene in the mixture.

5. Hydrogen sulphide burns in oxygen in accordance with the following equation:
$$2H_2S(g) \ + \ 3O_2(g) \ \longrightarrow \ 2H_2O(g) \ + \ 2SO_2(g)$$
If 4 dm^3 of H_2S are burned in 10 dm^3 of oxygen at 1 atmosphere pressure and $120 \,^\circ C$, what is the final volume of the mixture?

a 6 dm^3 b 8 dm^3 c 10 dm^3 d 12 dm^3 e) 14 dm^3

FINDING FORMULAE BY COMBUSTION

The formula of a hydrocarbon can be found from the results of a combustion experiment. A hydrocarbon in the vapour phase is burned in an excess of oxygen to form carbon dioxide and water vapour. When the mixture of gases is cooled to room temperature, water vapour condenses to occupy a very small volume. The gaseous mixture consists of carbon dioxide and unused oxygen. The volume of carbon dioxide can be found by absorbing it in an alkali. From the volumes of gases, the equation for the reaction and the formula of the hydrocarbon can be found.

The combustion method can be used for other compounds also, e.g. ammonia (see Example 3).

EXAMPLE 1   When 100 cm^3 of a hydrocarbon X burn in 500 cm^3 of oxygen, 50 cm^3 of oxygen are unused, 300 cm^3 of carbon dioxide are formed, and 300 cm^3 of steam are formed. Deduce the equation for the reaction and the formula of the hydrocarbon.

METHOD

$$X + O_2(g) \longrightarrow CO_2(g) + H_2O(g)$$

$$100 \text{ cm}^3 \quad 450 \text{ cm}^3 \qquad 300 \text{ cm}^3 \quad 300 \text{ cm}^3$$

The volumes of gases reacting tell us that

$$X + 4\tfrac{1}{2}O_2(g) \longrightarrow 3CO_2(g) + 3H_2O(g)$$

To balance the equation, X must be C_3H_6. Then,

$$C_3H_6(g) + 4\tfrac{1}{2}O_2(g) \longrightarrow 3CO_2(g) + 3H_2O(g)$$

ANSWER

$$2C_3H_6(g) + 9O_2(g) \longrightarrow 6CO_2(g) + 6H_2O(g)$$

EXAMPLE 2 10 cm³ of a hydrocarbon, C_aH_b, are exploded with an excess of oxygen. A contraction of 35 cm³ occurs, all volumes being measured at room temperature and pressure. On treatment of the products with sodium hydroxide solution, a contraction of 40 cm³ occurs. Deduce the formula of the hydrocarbon.

METHOD Write the equation:

$$C_aH_b(g) + (a + b/4)O_2(g) \longrightarrow aCO_2(g) + b/2 \; H_2O(l)$$

Volume of hydrocarbon $= 10 \text{ cm}^3$

Volume of $CO_2 = a \times$ Volume of C_aH_b

From reaction with NaOH, volume of $CO_2 = 40 \text{ cm}^3$

Therefore $a = 4$

Let the volume of unused oxygen be c cm³.

Final volume $=$ Initial volume $- 35 \text{ cm}^3$

Note that $H_2O(l)$ is a liquid at room temperature and pressure, and does not contribute to the final volume of gas.

$$40 + c = 10 + 40 + 5b/2 + c - 35$$

$$25 = 5b/2$$

$$b = 10$$

ANSWER The formula is C_4H_{10}.

EXAMPLE 3 20 cm³ of ammonia are burned in an excess of oxygen at 110 °C. 10 cm³ of nitrogen and 30 cm³ of steam are formed. Deduce the formula for ammonia, given that the formula of nitrogen is N_2, and the formula of steam is H_2O.

METHOD Let the formula of ammonia be N_aH_b.

The equation for combustion is

$$N_aH_b + b/4 \; O_2 \longrightarrow a/2 \; N_2 + b/2 \; H_2O$$

Volume of N_aH_b = 20 cm³

Volume of N_2 = $a/2 \times 20$ = $10a$ cm³ = 10 cm³ ∴ $a = 1$

Volume of $H_2O(g)$ = $b/2 \times 20$ = $10b$ cm³ = 30 cm³ ∴ $b = 3$

ANSWER   The formula is NH_3.

EXERCISE 12   Problems on Finding Formulae by Combustion

1. 10 cm³ of a hydrocarbon C_xH_y were exploded with an excess of oxygen. There was a contraction of 30 cm³. When the product was treated with a solution of sodium hydroxide, there was a further contraction of 30 cm³. Deduce the formula of the hydrocarbon. All gas volumes are at s.t.p.

2. 10 cm³ of a hydrocarbon C_aH_b were exploded with excess oxygen. A contraction of 25 cm³ occurred. On treating the product with sodium hydroxide, a further contraction of 40 cm³ occurred. Deduce the values of a and b in the formula of the hydrocarbon. All measurements of gas volumes are at s.t.p.

3. 10 cm³ of a hydrocarbon C_4H_8 were exploded with an excess of oxygen. A contraction of a cm³ occurred. On adding sodium hydroxide solution, a further contraction of b cm³ occurred. What are the volumes, a and b? All gas volumes are at s.t.p.

4. When North Sea gas burns completely, it forms carbon dioxide and water and no other products. When 250 cm³ of North Sea gas burn, they need 500 cm³ of oxygen, and they form 250 cm³ of carbon dioxide and 500 cm³ of steam (the volumes being measured under the same conditions). Deduce the equation for the reaction and the formula of North Sea gas.

EXERCISE 13   Problems on Reactions Involving Solids and Gases

SECTION 1

1. In the reaction between marble and hydrochloric acid, the equation is

$$CaCO_3(s) + 2HCl(aq) \longrightarrow CaCl_2(aq) + CO_2(g) + H_2O(l)$$

What mass of marble would be needed to give 11.00 g of carbon dioxide?

What volume would this gas occupy at s.t.p.?

2. Zinc reacts with aqueous hydrochloric acid to give hydrogen.

$$Zn(s) + 2HCl(aq) \longrightarrow H_2(g) + ZnCl_2(aq)$$

What mass of zinc would be needed to give 100 g of hydrogen? What volume would this gas occupy at s.t.p.?

3. Sodium hydrogencarbonate decomposes on heating, with evolution of carbon dioxide:

$$2NaHCO_3(s) \longrightarrow Na_2CO_3(s) + CO_2(g) + H_2O(g)$$

What volume of carbon dioxide (at s.t.p.) can be obtained by heating 4.20 g of sodium hydrogencarbonate? If 4.2 g of sodium hydrogen-carbonate react with an excess of dilute hydrochloric acid, what volume of carbon dioxide (at s.t.p.) is evolved?

4. Many years ago, bicycle lamps used to burn the gas ethyne, C_2H_2. The gas was produced by allowing water to drip on to calcium carbide. The *unbalanced* equation for the reaction is:

Calcium carbide + Water $\longrightarrow$ Ethyne + Calcium hydroxide

$$CaC_2(s) + H_2O(l) \longrightarrow C_2H_2(g) + Ca(OH)_2(s)$$

a) Balance the equation.

b) Calculate the mass of calcium carbide which would be needed to produce 467 cm^3 of ethyne (at s.t.p.).

5. Dinitrogen oxide, N_2O, is commonly called *laughing gas.* It can be made by heating ammonium nitrate, NH_4NO_3. The *unbalanced* equation for this reaction is:

$$NH_4NO_3(s) \longrightarrow N_2O(g) + H_2O(l)$$

a) Balance the equation.

b) Calculate the mass of ammonium nitrate that must be heated to give
   i) 8.8 g of laughing gas
   ii) 11.2 dm^3 of laughing gas (at s.t.p.).

SECTION 2

1. a) Analysis of an oxide of potassium shows that 1.42 g of this oxide contains 0.64 g of oxygen. What is its empirical formula?

   b) This oxide reacts with carbon dioxide to form oxygen and potassium carbonate, K_2CO_3. Write an equation for the reaction.

   c) This reaction is sometimes used as a means of regenerating oxygen in submarines. What volume of oxygen (at s.t.p.) could be obtained from 1.00 kg of this oxide of potassium?

2. A cook is making a small cake. It needs 500 cm^3 (at s.t.p.) of carbon dioxide to make the cake rise. The cook decides to add baking powder, which contains sodium hydrogencarbonate. This generates carbon dioxide by thermal decomposition:

$$2NaHCO_3(s) \longrightarrow CO_2(g) + Na_2CO_3(s) + H_2O(l)$$

What mass of sodium hydrogencarbonate must the cook add to the cake mixture?

3. A certain industrial plant emits 90 tonnes of nitrogen monoxide, NO, daily through its chimneys. The firm decides to remove nitrogen monoxide from its waste gases by means of the reaction

$$\text{Methane} + \begin{array}{c}\text{Nitrogen}\\\text{monoxide}\end{array} \longrightarrow \text{Nitrogen} + \begin{array}{c}\text{Carbon}\\\text{dioxide}\end{array} + \text{Water}$$

$$CH_4(g) + 4NO(g) \longrightarrow 2N_2(g) + CO_2(g) + 2H_2O(l)$$

If methane (North Sea gas) costs 0.50p per cubic metre, what will this clean-up reaction cost the firm to run? (Ignore the cost of installing the process, which will in reality be high.)

(*Hint*   Tonnes of NO ... moles of NO ... moles of CH_4 ... volume of CH_4 ... cost of CH_4 (1 m^3 = 1000 dm^3.)

4. In the Solvay process

$$NaCl(aq) + NH_3(g) + H_2O(l) + CO_2(g) \longrightarrow NaHCO_3(s) + NH_4Cl(aq)$$

what volume of carbon dioxide (at s.t.p.) is required to produce 1.00 kg of sodium hydrogencarbonate?

5. Find the volume of ethyne (at s.t.p.) that can be prepared from 10.0 g of calcium carbide by the reaction

$$CaC_2(s) + 2H_2O(l) \longrightarrow Ca(OH)_2(aq) + C_2H_2(g)$$

6. Find the mass of phosphorus required for the preparation of 200 cm^3 of phosphine (at s.t.p.) by the reaction

$$P_4(s) + 3NaOH(aq) + 3H_2O(l) \longrightarrow 3NaH_2PO_4(aq) + PH_3(g)$$

7. Calculate the mass of ammonium chloride required to produce 1.00 dm^3 of ammonia (at s.t.p.) in the reaction

$$2NH_4Cl(s) + Ca(OH)_2(s) \longrightarrow 2NH_3(g) + CaCl_2(s) + 2H_2O(g)$$

8. What mass of potassium chlorate(V) must be heated to give 1.00 dm^3 of oxygen at s.t.p.? The reaction is

$$2KClO_3(s) \longrightarrow 2KCl(s) + 3O_2(g)$$

9. What volume of chlorine (at s.t.p.) can be obtained from the electrolysis of a solution containing 60.0 g of sodium chloride?

10. What volume of oxygen (at s.t.p.) is needed for the complete combustion of 1.00 kg of octane? The reaction is

$$2C_8H_{18}(l) + 25O_2(g) \longrightarrow 16CO_2(g) + 18H_2O(g)$$

EXERCISE 14   Questions from A-level Papers

1. a) Describe the chemistry involved in the rusting of iron, and explain *two ways* (different in principle) by which it may be prevented.

   b) Aerials in portable radios are made of a mixed oxide of calcium and iron known as 'Ferrite'. It contains 18.5% calcium and 51.9% iron by mass. Calculate the empirical formula of 'Ferrite' and hence deduce the oxidation number of the iron it contains.   (C92)

*2. When chlorine is passed over heated sulphur in the absence of air, an orange liquid A that contains 47.5% of sulphur, by mass, is formed. Its relative molecular mass is about 135. When A is treated with ammonia dissolved in benzene, an explosive orange-yellow solid B is obtained. Compound B contains 30.4% of nitrogen by mass, the remainder being sulphur. Its relative molecular mass is about 185. By X-ray diffraction, it has been shown that all the bonds in this compound are the same length.

When B is reduced by tin(II) chloride in a methanol/benzene solution, a compound C that contains 29.8% of nitrogen and 68.1% of sulphur, by mass, is formed.

   a) What is the likely formula of substance A? Draw a diagram to show the likely shape of its molecule.

   b) Suggest structures for substances B and C, explaining clearly how you arrived at your conclusions.

   c) Write an equation for the conversion of substance B into substance C.

   d) Suggest a reason why substance B is explosive.   (C91S)

*3. A bright red solid A dissolves in water to give a highly acidic solution. When a solution of A is made alkaline with aqueous sodium hydroxide, the solution turns yellow. On heating 1.0 g of A, 0.76 g of a green powder B is formed and 168 cm³ of oxygen (measured at s.t.p.) are given off.

An orange solid C may be obtained from the solution made by dissolving 1.0 g of A in 5.0 cm³ of 2.0 mol dm^{-3} ammonia.

When 1.26 g of C are warmed, a violent reaction takes place, giving off 112 cm³ (measured at s.t.p.) of an inert gas, as well as steam, and leaving behind the same green powder B that was made by heating A.

If A is dissolved in cold, concentrated hydrochloric acid, and concentrated sulphuric acid is then gradually added, a dark, red-brown oil D separates out. D has a boiling point of 117 °C and rapidly reacts with water. D contains 45.8% of chlorine by mass.

Identify A, B, C and D, give equations for all the reactions involved and show that these equations are consistent with the quantitative data.   (C92S)

4. a) A number of compounds resulting from Man's chemical activities are known to cause atmospheric pollution.

In a copy of the table below, give the names and formulae of any *four* atmospheric pollutants, stating the effect of the pollution from each.

Name and formula of pollutant	Effect of pollutant
i)	
ii)	
iii)	
iv)	

b) For *one* of the pollutants in your table suggest a method of reducing its release into the atmosphere.

c) A farmer uses 2000 kg of ammonium nitrate fertiliser per annum. Of this, 5% is leached into neighbouring streams.
   i) Calculate the mass of ammonia required to produce 2000 kg of ammonium nitrate, assuming ALL THE NITROGEN is derived from ammonia.
   ii) If the Local Water Authority does not allow agricultural discharge of nitrogen as nitrate or ammonium ions to exceed the equivalent of 500 mg of NH_3 per dm^3, calculate the minimum volume of water required to take up the fertiliser leached into the streams. (AEB91)

5. a) Both silver nitrate solution and iron(III) chloride solution give brown precipitates when sodium hydroxide solution is added. In an investigation, the brown precipitates were filtered off but then were confused.
   i) Write the formula for each of the brown precipitates.
   ii) Devise a procedure for distinguishing between the precipitates, using the same test on each precipitate. Give the reagent(s) used and state the observation(s) with each precipitate.

b) The following method was used to determine the percentage by mass of iodine in an iodoalkane. A 2.37 g sample of the iodoalkane was boiled with sodium hydroxide solution, the resulting mixture cooled, acidified with dilute nitric acid and treated with excess silver nitrate solution. The precipitate obtained was filtered off, washed and dried. It was found to weigh 3.28 g.
   i) Why was dilute nitric acid added before adding silver nitrate solution?
   ii) Calculate the percentage by mass of iodine in the iodoalkane.
   iii) A different iodoalkane contains, by mass, 90.07% iodine, 8.51% carbon and 1.42% hydrogen. It has a relative molecular mass of 282. Calculate its molecular formula. (JMB91,p)

6. a) When a hydrocarbon fuel is burned with the correct amount of air required for combustion, carbon monoxide is generally present in the exhaust gases.

   i) Write the equation for the complete combustion of a hydrocarbon of formula C_xH_y.

   ii) Suggest *two* reasons why the formation of carbon monoxide is undesirable.

   iii) Adding an excess of air successfully reduces the formation of carbon monoxide. Suggest an important disadvantage of doing this.

   b) Processes involving the roasting of a metal sulphide ore produce high concentrations of sulphur dioxide in the exhaust gas. Chemists have devised several ways of solving this pollution problem, and an extension to existing plant can produce saleable by-products from the sulphur dioxide.

   How might political or economic factors interfere with the implementation of these pollution remedies in certain countries?

   c) One solution to the pollution problems referred to in (b) is to oxidise the sulphur dioxide catalytically and to use the resulting sulphur trioxide to make sulphuric acid. What mass of sulphuric acid ($M_r = 98$) would be produced from one tonne of pyrites (FeS_2, $M_r = 120$) if all the sulphur were converted into sulphuric acid?

   d) There are other solutions to these pollution problems. In the Resox process, sulphur dioxide ($M_r = 64$) is reduced to sulphur by pulverised coal in the presence of steam as a catalyst. The sulphur produced can be sold. The consumable needs of the process are stated to be 'about 0.2 kg coal and 0.05 kW h electricity per kilogram of inlet sulphur dioxide'.

   i) Write an equation for this chemical process.

   ii) From this equation, estimate the coal consumption of the process per kilogram of sulphur dioxide, thus verifying (or otherwise) the stated claim. Assume for simplicity that the coal consists only of carbon.

   iii) Suggest a reason why the electricity is needed.

   iv) The exhaust gases from roasting a metal sulphide ore contain 10% by volume of sulphur dioxide. What volume of exhaust gas before treatment will contain '1 kg of inlet sulphur dioxide'? The molar volume under the conditions used is 60 litres mol^{-1}.

   (JMB91)

7. Fertilisers often contain phosphorus compounds, the proportion of phosphorus being expressed in terms of % P_2O_5 by mass.

   a) Why would this NOT be the actual phosphorus compound in the fertiliser?

b) If a fertiliser was listed as 10% P_2O_5 by mass but actually contained calcium phosphate, $Ca_3(PO_4)_2$, what would the composition be if expressed as % calcium phosphate by mass? (Relative atomic masses: $O = 16$; $P = 31$; $Ca = 40$) (L91,p)

*8. a) Compare and contrast the chemistry of aluminium(III) and chromium(III).

b) Draw the shapes of the ions, NO_2^- and SO_4^{2-} and predict ways in which each of these ions could co-ordinate to a transition metal ion.

c) During an investigation of the reaction of cobalt(II) sulphate with aqueous ammonia and sodium nitrite, a research chemist isolated two isomers, A and B, with the following composition: Co, 20.6%; H, 5.2%; N, 29.4%; O, 33.6%; S, 11.2%. Furthermore, the reaction of 0.1 g of either A or B with a slight excess of an aqueous solution of barium chloride gave a precipitate of 0.0816 g of barium sulphate.
Propose structures for the isomers A and B and suggest a structure for another isomer. (JMB90,S)

*9. Caesium iodide reacts with chlorine at room temperature to give a compound, S, containing only caesium, iodine and chlorine. 3.31 g of S was dissolved in water, and sulphur dioxide was passed through the boiling solution until no further change took place. The resulting mixture was acidified with nitric acid, boiled to expel excess sulphur dioxide and cooled to room temperature. The addition of aqueous silver nitrate produced a precipitate of mass 5.22 g which was partly soluble in concentrated aqueous ammonia, leaving a pale yellow residue of mass 2.35 g. The yellow residue gave off a purple vapour when heated with concentrated sulphuric acid.

a) Deduce the formula of S.

b) Draw a diagram to show the electronic structure of the anion in S.

c) Suggest, giving reasons, a shape for the anion in S.

d) What is the oxidation number of iodine in S?

e) Assuming sulphate ion to be one product, write a balanced redox equation for the reaction between the anion in S and sulphur dioxide.

f) Explain the fact that compounds similar to S with other iodides of Group 1 are less stable.

g) Write balanced equations to explain:
   i) the partial solubility in aqueous ammonia of the precipitate formed with silver nitrate solution;
   ii) the reaction which gave rise to the purple vapour. (L92,S)

*10. a) Propose shapes for the molecules $B(CH_3)_3$, $C(CH_3)_4$, and $O(CH_3)_2$ and discuss their potential Lewis acid–base properties.

b) Identify the compounds A, B, C and D in the following reaction sequence. Deduce probable shapes for the compound A and for the two phosphorus species present in B.

The reaction of PF_3 with chlorine gave a gas A with the percentage composition Cl, 44.7; F, 35.8; P, 19.5; $M_r = 159$. On standing for several hours, samples of A deposited a crystalline white solid B with the same empirical formula as A but with $M_r = 318$. B contains one cation and one anion per formula unit. Reaction of B with sodium fluoride gave a gas C and a solid D. Elemental analysis of C gave Cl, 74.0%; F, 9.9%; P, 16.1%; $M_r = 192$. Elemental analysis of D gave F, 67.9%, Na, 13.7%; P, 18.4%; D contains one cation and one anion per formula unit.

c) Discuss how you would expect the solid B to interact with $N(CH_3)_3$.

(JMB91,S)

8 Volumetric Analysis

CONCENTRATION

One way of stating the concentration of a solution is to state the *mass* of solute present in 1 cubic decimetre of solution. The mass of solute is usually expressed in grams. A solution made by dissolving 5 grams of solute and making up to 1 cubic decimetre of solution has a concentration of $5\,\mathrm{g\,dm^{-3}}$ (5 grams per cubic decimetre).

Other units of volume are the cubic centimetre, cm^3, the cubic metre, m^3, and the litre, l. The litre has the same volume as the cubic decimetre.

$$10^3\,cm^3 \;=\; 1\,dm^3 \;=\; 1l \;=\; 10^{-3}\,m^3 \quad (10^3 = 1000; \; 10^{-3} = 0.001)$$

A more common way of stating concentration in chemistry is to state the *molar concentration* of a solution. This is the *amount in moles* of a substance present per cubic decimetre (litre) of solution.

What is the *molar concentration* of a solution of $80\,g$ of sodium hydroxide in $1\,dm^3$ of solution? The amount in moles of NaOH in $80\,g$ of sodium hydroxide can be calculated from its molar mass.

$$\text{Molar mass of NaOH} \;=\; 23 + 16 + 1 \;=\; 40\,\mathrm{g\,mol^{-1}}$$

$$\text{Amount of NaOH} \;=\; \frac{\text{Mass}}{\text{Molar mass}}$$

$$=\; \frac{80\,g}{40\,\mathrm{g\,mol^{-1}}}$$

$$=\; 2\,\text{mol}$$

The concentration of the solution is given by:

$$\text{Concentration of solution (mol\,dm}^{-3}) \;=\; \frac{\text{Amount of solute (mol)}}{\text{Volume of solution (dm}^3)}$$

For this solution,

$$\text{Concentration} \;=\; \frac{2\,\text{mol}}{1\,dm^3} \;=\; 2\,\mathrm{mol\,dm^{-3}}$$

If **3** moles of sodium hydroxide are dissolved in $500\,cm^3$ of solution,

$$\text{Concentration} \;=\; \frac{3\,\text{mol}}{0.5\,dm^3} \;=\; 6\,\mathrm{mol\,dm^{-3}}$$

The symbol M is often used for $mol\,dm^{-3}$. This solution can be described as a 6 M sodium hydroxide solution.

The concentration in $mol\,dm^{-3}$ used to be referred to as the *molarity* of a solution. (In strict SI units, concentration is expressed in $mol\,m^{-3}$.)

Figure 8.1 gives more examples.

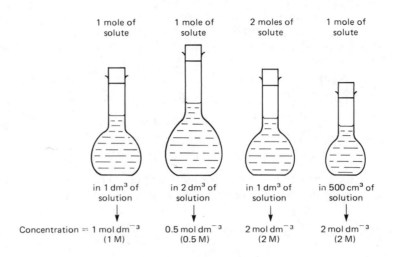

Fig. 8.1   How to calculate concentration

A useful rearrangement of the expression in the box on the previous page is:

$$\boxed{\text{Amount of solute (mol)} = \text{Volume (dm}^3) \times \text{Concentration (mol\,dm}^{-3})}$$

For example, the amount in moles of solute in $2.5\,dm^3$ of a $2.0\,mol\,dm^{-3}$ solution is given by:

Amount of solute $= 2.5\,dm^3 \times 2.0\,mol\,dm^{-3} = 5.0\,mol$

EXAMPLE 1   Calculate the concentration in $mol\,dm^{-3}$ of a solution containing 36.5 g of hydrogen chloride in $4.00\,dm^3$ of solution.

METHOD   Molar mass of HCl $= (35.5 + 1.0) = 36.5\,g\,mol^{-1}$

Amount (mol) present in $36.5\,g = 1.00\,mol$

Volume $= 4.00\,dm^3$

$$\text{Concentration of solution} = \frac{\text{Amount of solute (mol)}}{\text{Volume of solution (dm}^3)}$$

$$= \frac{1.00\,mol}{4.00\,dm^3}$$

$$= 0.25\,mol\,dm^{-3}$$

ANSWER The concentration is 0.25 mol dm^{-3}. (It is a 0.25 M solution.)

EXAMPLE 2 Calculate the amount of solute (mol) in 250 cm^3 of a solution of sodium hydroxide which has a concentration of 2.00 mol dm^{-3}.

METHOD

$$\text{Concentration of solution} = 2.00 \text{ mol dm}^{-3}$$

$$\text{Volume of solution} = 250 \text{ cm}^3 = 0.250 \text{ dm}^3$$

$$\text{Amount of solute} = \underset{(\text{dm}^3)}{\text{Volume}} \times \underset{(\text{mol dm}^{-3})}{\text{Concentration}}$$

$$= 2.00 \times 0.250$$

$$= 0.500 \text{ ml}$$

ANSWER The solution contains 0.500 mol of solute.

EXERCISE 15   Problems on Concentration

1. Calculate the concentration in mol dm^{-3} of
   a) 3.65 g of hydrogen chloride in 2.00 dm^3 of solution
   b) 73.0 g of hydrogen chloride in 2.00 dm^3 of solution
   c) 49.0 g of sulphuric acid in 2.00 dm^3 of solution
   d) 49.0 g of sulphuric acid in 250 cm^3 of solution
   e) 2.80 g of potassium hydroxide in 500 cm^3 of solution
   f) 28.0 g of potassium hydroxide in 4.00 dm^3 of solution
   g) 5.30 g of anhydrous sodium carbonate in 200 cm^3 of solution
   h) 53.0 g of anhydrous sodium carbonate in 2.50 dm^3 of solution.

2. Calculate the amount in moles of solute in
   a) 250 cm^3 of sodium hydroxide solution containing 1.00 mol dm^{-3}
   b) 500 cm^3 of sodium hydroxide solution containing $0.250 \text{ mol dm}^{-3}$
   c) 250 cm^3 of 0.0200 M calcium hydroxide solution
   d) 2.00 dm^3 of 1.25 M sulphuric acid (1.25 mol dm^{-3})
   e) 125 cm^3 of aqueous nitric acid, having a concentration of $0.400 \text{ mol dm}^{-3}$
   f) 200 cm^3 of ammonia solution, having a concentration of $0.125 \text{ mol dm}^{-3}$
   g) 123 cm^3 of aqueous hydrochloric acid of concentration 3.00 mol dm^{-3}
   h) 1500 cm^3 of potassium hydroxide solution of concentration $0.750 \text{ mol dm}^{-3}$.

3. What mass of the solute must be used in order to prepare the required solutions listed below?
   a) 500 cm^3 of 0.100 mol dm^{-3} $H_6C_4O_4$(aq) from $H_6C_4O_4$(s)
   b) 250 cm^3 of 0.200 mol dm^{-3} Na_2CO_3(aq) from Na_2CO_3(s)
   c) 750 cm^3 of 0.100 mol dm^{-3} $H_2C_2O_4$(aq) from $H_2C_2O_4 \cdot 2H_2O$(s)
   d) 2.50 dm^3 of 0.200 mol dm^{-3} $NaHCO_3$(aq) from $NaHCO_3$(s)
   e) 500 cm^3 of 0.100 mol dm^{-3} $Na_2B_4O_7$(aq) from $Na_2B_4O_7 \cdot 10H_2O$(s)

4. What volumes of the following concentrated solutions are required to give the stated volumes of the more dilute solutions?
   a) 2.00 dm^3 of 0.500 mol dm^{-3} H_2SO_4(aq) from 2.00 mol dm^{-3} H_2SO_4(aq)
   b) 1.00 dm^3 of 0.750 mol dm^{-3} HCl(aq) from 10.0 mol dm^{-3} HCl(aq)
   c) 250 cm^3 of 0.250 mol dm^{-3} NaOH(aq) from 5.50 mol dm^{-3} NaOH(aq)
   d) 500 cm^3 of 1.25 mol dm^{-3} HNO_3(aq) from 3.25 mol dm^{-3} HNO_3(aq)
   e) 250 cm^3 of 2.00 mol dm^{-3} KOH(aq) from 2.60 mol dm^{-3} KOH(aq)

TITRATION

A solution of known concentration is called a *standard solution*. Such a solution can be used to find the concentrations of solutions of other reagents.

In *volumetric analysis*, the concentration of a solution is found by measuring the volume of solution that will react with a known volume of a standard solution. The procedure of adding one solution to another in a measured way until the reaction is complete is called *titration*. Volumetric analysis is often referred to as *titrimetric analysis* or *titrimetry*.

ACID-BASE TITRATIONS

A standard solution of acid can be used to find the concentration of a solution of alkali. A known volume of alkali is taken by pipette, a suitable indicator is added, and the alkali is titrated against the standard acid until the equivalence point is reached. The number of moles of acid used can be calculated and the equation used to give the number of moles of alkali neutralised.

EXAMPLE 1 *Standardising sodium hydroxide solution*
What is the concentration of a solution of sodium hydroxide, 25.0 cm^3 of which requires 20.0 cm^3 of hydrochloric acid of concentration 0.100 mol dm^{-3} for neutralisation?

METHOD   In tackling this calculation,

a) Write the equation. Find the number of moles of acid needed to neutralise one mole of alkali.

$$NaOH(aq) \ + \ HCl(aq) \longrightarrow NaCl(aq) \ + \ H_2O(l)$$

1 mole of NaOH needs 1 mole of HCl for neutralisation.

b) Use the expression

Amount of solute (mol) $=$ Volume $(dm^3) \times$ Concentration $(mol \, dm^{-3})$

to find the number of moles of the reagent of known concentration, in this case HCl.

$$
\begin{aligned}
\text{Amount (mol) of HCl} \ &= \ \text{Volume } (dm^3) \times \text{Concn } (mol \, dm^{-3}) \\
&= \ 20.0 \times 10^{-3} \times 0.100 \ = \ 2.00 \times 10^{-3} \, mol
\end{aligned}
$$

From equation: No. of moles of NaOH $=$ No. of moles of HCl

$$= \ 2.00 \times 10^{-3} \, mol$$

But: Amount (mol) of NaOH $=$ Volume $(dm^3) \times$ Concn $(mol \, dm^{-3})$

$$= \ 25.0 \times 10^{-3} \times c$$

(where $c =$ concn)

Equate these two values: $2.00 \times 10^{-3} \ = \ 25.0 \times 10^{-3} \times c$

$$
\begin{aligned}
c \ &= \ (2.00 \times 10^{-3})/(25.0 \times 10^{-3}) \\
&= \ 0.080 \, mol \, dm^{-3}
\end{aligned}
$$

ANSWER   The concentration of sodium hydroxide is $0.080 \, mol \, dm^{-3}$.

EXAMPLE 2   *Standardising hydrochloric acid*
Sodium carbonate (anhydrous) is used as a primary standard in volumetric analysis. A solution of sodium carbonate of concentration $0.100 \, mol \, dm^{-3}$ is used to standardise a solution of hydrochloric acid. $25.0 \, cm^3$ of the standard solution of sodium carbonate require $35.0 \, cm^3$ of the acid for neutralisation. Calculate the concentration of the acid.

METHOD   a) Write the equation:

$$Na_2CO_3(aq) \ + \ 2HCl(aq) \longrightarrow 2NaCl(aq) \ + \ CO_2(g) \ + \ H_2O(l)$$

1 mole of Na_2CO_3 neutralises 2 moles of HCl.

b) Find the amount (mol) of the standard reagent used.

$$
\begin{aligned}
\text{Amount (mol) of } Na_2CO_3(aq) \ &= \ \text{Volume } (dm^3) \times \text{Concn } (mol \, dm^{-3}) \\
&= \ 25.0 \times 10^{-3} \times 0.100 \\
&= \ 2.50 \times 10^{-3} \, mol
\end{aligned}
$$

From equation: No. of moles of HCl $= 2 \times$ No. of moles of Na_2CO_3

$$= 5.00 \times 10^{-3}\, mol$$

But: Amount (mol) of HCl(aq) $=$ Volume $(dm^3) \times$ Concn $(mol\, dm^{-3})$

$$= 35.0 \times 10^{-3} \times c$$

(where $c =$ concn)

Equate these two values: $5.00 \times 10^{-3} = 35.0 \times 10^{-3} \times c$

$$c = (5.00 \times 10^{-3})/(35.0 \times 10^{-3})$$

$$= 0.143\, mol\, dm^{-3}$$

ANSWER   The concentration of hydrochloric acid is $0.143\, mol\, dm^{-3}$.

EXAMPLE 3   *Calculating the percentage of sodium carbonate in washing soda crystals*
5.125 g of washing soda crystals are dissolved and made up to $250\, cm^3$ of solution. A $25.0\, cm^3$ portion of the solution requires $35.8\, cm^3$ of $0.0500\, mol\, dm^{-3}$ sulphuric acid for neutralisation. Calculate the percentage of sodium carbonate in the crystals.

METHOD   a) Write the equation:

$Na_2CO_3(aq) + H_2SO_4(aq) \longrightarrow Na_2SO_4(aq) + CO_2(g) + H_2O(l)$

1 mole of Na_2CO_3 neutralises 1 mole of H_2SO_4.

b) Calculate the amount, in moles, of the standard reagent.

Amount (mol) of $H_2SO_4 = 35.8 \times 10^{-3} \times 0.0500 = 1.79 \times 10^{-3}\, mol$

Amount (mol) of $Na_2CO_3 = 1.79 \times 10^{-3}\, mol$

But: Amount of $Na_2CO_3 = 25.0 \times 10^{-3} \times c\, mol$

(where $c =$ concn)

Equate these two values: $1.79 \times 10^{-3} = 25.0 \times 10^{-3} \times c$

$$c = (1.79 \times 10^{-3})/(25.0 \times 10^{-3})$$

$$= 0.0716\, mol\, dm^{-3}$$

Amount (mol) of Na_2CO_3 in whole solution $=$ Volume $\times$ Concn

$$= 250 \times 10^{-3} \times 0.0716$$

$$= 0.0179\, mol$$

Mass of $Na_2CO_3 =$ Amount (mol) $\times$ Molar mass $= 0.0179 \times 106\, g$

$$= 1.90\, g$$

$$\% \text{ of } Na_2CO_3 = \frac{\text{Mass of sodium carbonate}}{\text{Mass of crystals}} \times 100$$

$$= \frac{1.90}{5.125} \times 100 = 37.1\%$$

ANSWER   Washing soda crystals are 37.1% sodium carbonate.

EXAMPLE 4   *Estimating ammonium salts*
A sample containing ammonium sulphate was warmed with $250 \, cm^3$ of $0.800 \, mol \, dm^{-3}$ sodium hydroxide solution. After the evolution of ammonia had ceased, the excess of sodium hydroxide solution was neutralised by $85.0 \, cm^3$ of hydrochloric acid of concentration $0.500 \, mol \, dm^{-3}$. What mass of ammonium sulphate did the sample contain?

METHOD   a) There are two reactions taking place:

i) the reaction between the ammonium salt and the alkali:

$$(NH_4)_2SO_4(s) + 2NaOH(aq) \longrightarrow 2NH_3(g) + Na_2SO_4(aq) + 2H_2O(l)$$

ii) the reaction between the excess alkali and the hydrochloric acid:

$$NaOH(aq) + HCl(aq) \longrightarrow NaCl(aq) + H_2O(l)$$

b) Pick out the substance for which you have the information you need to calculate the number of moles. As you know its volume and concentration, you can calculate the number of moles of HCl. This will tell you the number of moles of NaOH left over after reaction i). Subtract this from the number of moles of NaOH added to the ammonium salt to obtain the number of moles of NaOH used in reaction i). This will give you the number of moles of $(NH_4)_2SO_4$ with which it reacted.

Amount (mol) of HCl $= 85.0 \times 10^{-3} \times 0.500 = 0.0425 \, mol$

Amount (mol) of NaOH left over from reaction i) $= 0.0425 \, mol$

Amount (mol) of NaOH added $= 250 \times 10^{-3} \times 0.800 = 0.200 \, mol$

Amount (mol) of NaOH used in reaction i) $= 0.200 - 0.0425$
$$= 0.1575 \, mol$$

No. of moles of $(NH_4)_2SO_4 = 0.5 \times$ No. of moles of NaOH
$$= 0.0788 \, mol$$

Molar mass of $(NH_4)_2SO_4 = 132 \, g \, mol^{-1}$

Mass of ammonium sulphate $= 0.0788 \times 132 = 10.4 \, g$

ANSWER   The sample contained 10.4 g of ammonium sulphate.

EXERCISE 16   Problems on Neutralisation

SECTION 1

Calculators are not needed for these problems.

The following are problems on neutralisation. Show, giving your working, whether each of these statements is true or false.

1. 1 mol of HCl will neutralise
   a) 5 dm^3 of KOH(aq) of concentration 0.2 mol dm^{-3}. True or False?
   b) 2 dm^3 of NaOH(aq) of concentration 0.2 mol dm^{-3}
   c) 2 dm^3 of KOH(aq) of concentration 0.5 mol dm^{-3}
   d) 0.5 dm^3 of NaOH(aq) of concentration 1 mol dm^{-3}
   e) 250 cm^3 of Na$_2$CO$_3$(aq) of concentration 2 mol dm^{-3}
   f) 200 cm^3 of Na$_2$CO$_3$(aq) of concentration 4 mol dm^{-3}

2. 1 mol of H$_2$SO$_4$ will neutralise
   a) 500 cm^3 of NaOH(aq) of concentration 4 mol dm^{-3}. True or False?
   b) 1 dm^3 of KOH(aq) of concentration 1 mol dm^{-3}
   c) 400 cm^3 of NaOH(aq) of concentration 5 mol dm^{-3}
   d) 500 cm^3 of Na$_2$CO$_3$(aq) of concentration 1 mol dm^{-3}
   e) 2 dm^3 of Na$_2$CO$_3$(aq) of concentration 0.5 mol dm^{-3}
   f) 4 dm^3 of KOH(aq) of concentration 0.25 mol dm^{-3}

3. 5 mol of NaOH will neutralise
   a) 2 dm^3 of HCl(aq) of concentration 2 mol dm^{-3}. True or False?
   b) 250 cm^3 of HCl(aq) of concentration 10 mol dm^{-3}
   c) 250 cm^3 of H$_2$SO$_4$(aq) of concentration 10 mol dm^{-3}
   d) 500 cm^3 of H$_2$SO$_4$(aq) of concentration 5 mol dm^{-3}
   e) 2500 cm^3 of HNO$_3$(aq) of concentration 2 mol dm^{-3}
   f) 2 dm^3 of HNO$_3$(aq) of concentration 2 mol dm^{-3}

4. 0.5 mol of Na$_2$CO$_3$ will neutralise
   a) 1 dm^3 of HCl(aq) of concentration 1 mol dm^{-3}. True or False?
   b) 1 dm^3 of H$_2$SO$_4$(aq) of concentration 1 mol dm^{-3}
   c) 500 cm^3 of HCl(aq) of concentration 1 mol dm^{-3}
   d) 250 cm^3 of HNO$_3$(aq) of concentration 2 mol dm^{-3}
   e) 200 cm^3 of H$_2$SO$_4$(aq) of concentration 2.5 mol dm^{-3}
   f) 500 cm^3 of HNO$_3$(aq) of concentration 2 mol dm^{-3}

5. Sodium hydroxide is sold commercially as solid *lye*. A 1.20 g sample of lye required 45.0 cm^3 of 0.500 mol dm^{-3} hydrochloric acid to neutralise it. Calculate the percentage by mass of NaOH in lye.

6. Vinegar is a solution of ethanoic acid. A 10.0 cm^3 portion of a certain brand of vinegar needed 55.0 cm^3 of 0.200 mol dm^{-3} sodium hydroxide solution to neutralise the ethanoic acid in it.

| Ethanoic + Sodium | $\longrightarrow$ | Sodium + Water |
| acid | | ethanoate |

$$CH_3CO_2H(aq) + NaOH(aq) \longrightarrow CH_3CO_2Na(aq) + H_2O(l)$$

   a) Calculate the concentration of ethanoic acid in the vinegar in mol dm^{-3}.

   b) Given that the density of this vinegar is 1.06 g cm^{-3}, calculate the concentration of ethanoic acid in percentage by mass.

7. Salt is a necessary ingredient of our diet. In certain illnesses, the salt balance can be lost, and a doctor or nurse must give salt intravenously. They inject *normal saline*, which is a 0.85% solution of sodium chloride in water (0.85 g of solute per 100 g of water). What is the molar concentration of normal saline?

8. A chip of marble weighing 2.50 g required 28.0 g of 1.50 mol dm^{-3} hydrochloric acid to react with all the calcium carbonate it contained. What is the percentage of calcium carbonate in this sample of marble?

   a) Write the balanced equation for the reaction.

   b) Find how many moles of HCl were used ... then how many moles of CaCO$_3$ reacted ... what mass of CaCO$_3$ ... and finally the percentage of CaCO$_3$.

9. A mixture of gases coming from a coke-producing plant contains ammonia. The mixture is bubbled through dilute sulphuric acid to remove the ammonia.

   a) Write a balanced equation for the reaction which occurs.

   b) What volume of ammonia (at s.t.p.) could be removed by 50 dm^3 of 1.50 mol dm^{-3} sulphuric acid?

   c) What use could be made of the product?

10. Nitrosoamines can cause cancer at sufficiently high concentrations. In 1979, a brand of American beer was found to contain 6 p.p.b. (parts per billion) of dimethylnitrosoamine. By 1981, the firm had reduced the level to 0.2 p.p.b.

   a) What was the mass of dimethylnitrosoamine in one 250 cm^3 can of beer in 1979? (1 billion $= 10^9$.)

   b) What fraction of the 1979 level was still present in 1981?

11. If a person's blood sugar level falls below 60 mg per 100 cm^3, insulin shock can occur. The density of blood is 1.2 g cm^{-3}.

   a) What is the percentage by mass of sugar in the blood at this level?

   b) What is the molar concentration of sugar, C$_6$H$_{12}$O$_6$, in the blood?

12. A blood alcohol level of 150–200 mg alcohol per 100 cm^3 of blood produces intoxication. A blood alcohol level of 300–400 mg per 100 cm^3 produces unconsciousness. At a blood alcohol level above 500 mg per 100 cm^3, a person may die. What is the molar concentration of alcohol (ethanol, C_2H_5OH) at the lethal level?

13. An experiment was done to find the percentage composition of an alloy of sodium and lead. The alloy reacts with water:

Alloy + Water $\longrightarrow$ Sodium + Hydrogen + Lead
                                     hydroxide

$2Na \cdot Pb(s) + 2H_2O(l) \longrightarrow 2NaOH(aq) + H_2(g) + 2Pb(s)$

3.00 g of the alloy were added to about 100 cm^3 of water. When the reaction was complete, the sodium hydroxide formed was titrated against 1.00 mol dm^{-3} hydrochloric acid. The volume of acid required to neutralise the sodium hydroxide was 12.0 cm^3. Calculate

a) the amount in moles of HCl used

b) the amount in moles of NaOH neutralised

c) the amount in moles of Na in 3.00 g of the alloy

d) the mass in grams of Na in 3.00 g of alloy

e) the percentage composition by mass.

SECTION 2

1. 0.500 g of impure ammonium chloride is warmed with an excess of sodium hydroxide solution. The ammonia liberated is absorbed in 25.0 cm^3 of 0.200 mol dm^{-3} sulphuric acid. The excess of sulphuric acid requires 5.64 cm^3 of 0.200 mol dm^{-3} sodium hydroxide solution for titration. Calculate the percentage of ammonium chloride in the original sample.

2. A 1.00 g sample of limestone is allowed to react with 100 cm^3 of 0.200 mol dm^{-3} hydrochloric acid. The excess acid required 24.8 cm^3 of 0.100 mol dm^{-3} sodium hydroxide solution. Calculate the percentage of calcium carbonate in the limestone.

3. An impure sample of barium hydroxide of mass 1.6524 g was allowed to react with 100 cm^3 of hydrochloric acid of concentration 0.200 mol dm^{-3}. When the excess of acid was titrated against sodium hydroxide, 10.9 cm^3 of sodium hydroxide solution were required. 25.0 cm^3 of the sodium hydroxide required 28.5 cm^3 of the hydrochloric acid in a separate titration. Calculate the percentage purity of the sample of barium hydroxide.

4. A household cleaner contains ammonia. A 25.37 g sample of the cleaner is dissolved in water and made up to 250 cm^3. A 25.0 cm^3 portion of this solution requires 37.3 cm^3 of 0.360 mol dm^{-3} sulphuric acid for neutralisation. What is the percentage by mass of ammonia in the cleaner?

5. A fertiliser contains ammonium sulphate and potassium sulphate. A sample of 0.225 g of fertiliser was warmed with sodium hydroxide solution. The ammonia evolved required 15.7 cm³ of 0.100 mol dm⁻³ hydrochloric acid for neutralisation. Calculate the percentage of ammonium sulphate in the sample.

6. Calculate the number of carboxyl groups in the compound $C_6H_8O_6$, given that 0.440 g of it neutralised 37.5 cm³ of sodium hydroxide of concentration 0.200 mol dm⁻³.

7. Sodium carbonate crystals (27.823 0 g) were dissolved in water and made up to 1.00 dm³. 25.0 cm³ of the solution were neutralised by 48.8 cm³ of hydrochloric acid of concentration 0.100 mol dm⁻³. Find n in the formula $Na_2CO_3 \cdot nH_2O$.

OXIDATION-REDUCTION REACTIONS

Oxidation–reduction (or 'redox') reactions involve a transfer of electrons. The oxidising agent accepts electrons, and the reducing agent gives electrons. In working out the equation for a redox reaction, a good method is to work out the 'half-reaction equation' for the oxidising agent and the 'half-reaction equation' for the reducing agent, and then add them together.

Examples of half-reaction equations

a) Iron(III) salts are reduced to iron(II) salts. The equation is

$$Fe^{3+} \longrightarrow Fe^{2+}$$

For the equation to balance, the charge on the right-hand side (RHS) must equal the charge on the left-hand side (LHS). This can be accomplished by inserting an electron on the LHS:

$$Fe^{3+} + e^- \longrightarrow Fe^{2+}$$

b) When chlorine acts as an oxidising agent, it is reduced to chloride ions:

$$Cl_2 \longrightarrow 2Cl^-$$

To obtain a balanced half-reaction equation, $2e^-$ must be inserted on the LHS:

$$Cl_2 + 2e^- \longrightarrow 2Cl^-$$

c) Sulphites can be oxidised to sulphates:

$$SO_3^{2-} \longrightarrow SO_4^{2-}$$

To balance the equation with respect to mass, an extra oxygen atom is needed on the LHS. If H_2O is introduced on the LHS to supply this oxygen, the equation becomes

$$SO_3^{2-} + H_2O \longrightarrow SO_4^{2-} + 2H^+$$

To balance the equation with respect to charge, $2e^-$ are needed on the RHS:

$$SO_3^{2-} + H_2O \longrightarrow SO_4^{2-} + 2H^+ + 2e^-$$

d) Potassium manganate(VII) is an oxidising agent. In acid solution, it is reduced to a manganese(II) salt:

$$MnO_4^- + H^+ \longrightarrow Mn^{2+}$$

To balance the equation with respect to mass, $8H^+$ are needed to combine with 4 oxygen atoms:

$$MnO_4^- + 8H^+ \longrightarrow Mn^{2+} + 4H_2O$$

To balance the equation with respect to charge, $5e^-$ are needed on the LHS:

$$MnO_4^- + 8H^+ + 5e^- \longrightarrow Mn^{2+} + 4H_2O$$

It is a good idea to make a final check. Charge on LHS $= -1 + 8 - 5 = +2$. Charge on RHS $= +2$. The equation is balanced.

e) Potassium dichromate(VI) is an oxidising agent in acid solution, being reduced to a chromium(III) salt:

$$Cr_2O_7^{2-} + H^+ \longrightarrow Cr^{3+}$$

To balance the equation for mass, $14H^+$ are needed:

$$Cr_2O_7^{2-} + 14H^+ \longrightarrow 2Cr^{3+} + 7H_2O$$

To balance the equation for charge, $6e^-$ are needed on the LHS:

$$Cr_2O_7^{2-} + 14H^+ + 6e^- \longrightarrow 2Cr^{3+} + 7H_2O$$

A final check shows that the charge on the LHS $= -2 + 14 - 6 = +6$.

Charge on RHS $= 2(+3) = +6$. The equation is balanced.

You may like to practise with the half-reaction equations on p. 85.

Using half-reaction equations to obtain the equation for a redox reaction

a) In the reaction between iodine and thiosulphate ions, the two half-reaction equations are

$$I_2 + 2e^- \longrightarrow 2I^- \qquad [1]$$

$$2S_2O_3^{2-} \longrightarrow S_4O_6^{2-} + 2e^- \qquad [2]$$

Adding [1] and [2] gives

$$I_2 + 2e^- + 2S_2O_3^{2-} \longrightarrow 2I^- + S_4O_6^{2-} + 2e^-$$

Deleting the $2e^-$ term from both sides of the equation gives

$$I_2 + 2S_2O_3^{2-} \longrightarrow 2I^- + S_4O_6^{2-}$$

A check shows that the charges on the LHS and the RHS are both -4.

b) When potassium manganate(VII) oxidises an iron(II) salt to an iron (III) salt, the equations for the half-reactions are

$$MnO_4^- + 8H^+ + 5e^- \longrightarrow Mn^{2+} + 4H_2O \qquad [3]$$

$$Fe^{2+} \longrightarrow Fe^{3+} + e^- \qquad [4]$$

One manganate(VII) ion needs 5 electrons, and one iron(II) ion gives only one. Equation [4] must therefore be multiplied by 5:

$$5Fe^{2+} \longrightarrow 5Fe^{3+} + 5e^- \qquad [5]$$

Equations [3] and [5] can now be added to give

$$MnO_4^- + 8H^+ + 5Fe^{2+} \longrightarrow Mn^{2+} + 4H_2O + 5Fe^{3+}$$

c) When potassium manganate(VII) oxidises sodium ethanedioate, the equation for the manganate(VII) half-reaction is [3] as in Example 2, and the equation for the reduction of ethanedioate is

$$C_2O_4^{2-} \longrightarrow 2CO_2 + 2e^- \qquad [6]$$

d) One manganate(VII) ion needs $5e^-$, and one ethanedioate ion gives $2e^-$. Multiplying equation [3] by 2 and equation [6] by 5 and adding gives

$$2MnO_4^- + 16H^+ + 5C_2O_4^{2-} \longrightarrow 2Mn^{2+} + 10CO_2 + 8H_2O$$

e) Potassium dichromate(VI) oxidises iron(II) salt to iron(III) salts. The equations for the two half-reactions are

$$Cr_2O_7^{2-} + 14H^+ + 6e^- \longrightarrow 2Cr^{3+} + 7H_2O \qquad [7]$$

$$Fe^{2+} \longrightarrow Fe^{3+} + e^- \qquad [8]$$

One dichromate ion will oxidise six iron(II) ions:

$$Cr_2O_7^{2-} + 14H^+ + 6Fe^{2+} \longrightarrow 2Cr^{3+} + 6Fe^{3+} + 7H_2O$$

You may like to try the problem on balancing equations on p. 85 before going on to tackle the numerical problems.

There is another method of balancing redox equations. It is explained in the following section on oxidation numbers.

Oxidation numbers

It is helpful to discuss oxidation–reduction reactions in terms of the change in the *oxidation number* of each reactant. In the reaction

$$Cu(s) \ + \ \tfrac{1}{2}O_2(g) \longrightarrow Cu^{2+}O^{2-}(s)$$

copper is oxidised and oxygen is reduced. It is said that the oxidation number of copper increases from zero to $+2$, and the oxidation number of oxygen decreases from zero to -2. The following rules are followed in assigning oxidation numbers:

a) The oxidation number of an uncombined element is zero.

b) In ionic compounds, the oxidation number of each element is the charge on its ion. In NaCl, the oxidation number of Na $= +1$, and that of Cl $= -1$.

c) The sum of the oxidation numbers of all the elements in a compound is zero. In $AlCl_3$, the oxidation numbers are: Al $= +3$; Cl $= -1$, so that the sum of the oxidation numbers is $+3 + 3(-1) = 0$.

d) The sum of the oxidation numbers of all the elements in an ion is equal to the charge on the ion. In $SO_4{}^{2-}$, the oxidation numbers are S $= +6$, O $= -2$. The sum of the oxidation numbers for all the atoms is $+6 + 4(-2) = -2$, the same as the charge of the $SO_4{}^{2-}$ ion.

e) In a covalent compound, one element must be given a positive oxidation number and the other a negative oxidation number, such that the sum of the oxidation numbers for all the atoms is zero. The following elements always have the same oxidation numbers in all their compounds. A knowledge of their oxidation numbers helps one to assign oxidation numbers to the other elements combined with them:

Na, K $+1$ H $+1$, except in metal hydrides

Mg, Ca $+2$ F -1
Al $+3$ Cl -1, except in compounds with O and F

O -2, except in peroxides and compounds with F

The oxidation number method

A consideration of the changes in oxidation numbers which occur during a redox reaction helps you to decide which reactants have been oxidised and which have been reduced. It can also be very helpful when you need to balance the equation for the reaction. The following two points cover what is involved when you use the oxidation number method to balance the equation for a redox reaction:

a) When an element is oxidised, its oxidation number increases; when an element is reduced, its oxidation number decreases. If x atoms (or ions) of an element A react with y atoms (or ions) of an element B, i.e.

$$x\,A \;+\; y\,B \;\longrightarrow$$

then, if the oxidation number of A changes by a units, and the oxidation number of B changes by b units, you can see that

$$xa \;=\; yb$$

For example, in the reaction between tin(II) and iron(III) ions,

$$Sn^{2+}(aq) \;+\; 2\,Fe^{3+}(aq) \;\longrightarrow\; Sn^{4+}(aq) \;+\; 2\,Fe^{2+}$$

For Sn, no. of ions = 1, change in ox. no. = 2, and product = 2

For Fe, no. of ions = 2, change in ox. no. = 1, and product = 2

b) In a balanced equation

LHS sum of ox. nos. of elements = RHS sum of ox. nos. of elements

In the reaction

$$KIO_3(aq) \;+\; 2\,Na_2SO_3(aq) \;\longrightarrow\; KIO(aq) \;+\; 2\,Na_2SO_4(aq)$$

the elements K, Na and O keep the same oxidation states during the reaction, while I and S change.

Ox. no. of I in $KIO_3 = +5$; in $KIO = +1$

Ox. no. of S in $Na_2SO_3 = +4$; in $Na_2SO_4 = +6$

Sum of ox. nos. on LHS $= (+5) + 2(+4) = +13$

Sum of ox. nos. on RHS $= (+1) + 2(+6) = +13$

When applying the oxidation number method to a reaction between A and B, remember:

$$\left(\begin{array}{l} \text{No. of atoms of A} \times \text{Change} \\ \text{in oxidation number of A} \end{array}\right) = \left(\begin{array}{l} \text{No. of atoms of B} \times \text{Change} \\ \text{in oxidation number of B} \end{array}\right)$$

Sum of ox. nos. on LHS = Sum of ox. nos. on RHS

EXAMPLE 1 What is the oxidation number of germanium in $GeCl_4$?

METHOD Chlorine is one of the elements with a constant oxidation number of -1.

(Oxidation number of Ge) $+ 4(-1) = 0$.

ANSWER Oxidation number of Ge $= +4$.

EXAMPLE 2 What is the oxidation number of manganese in Mn_2O_7?

METHOD   Oxygen is one of the elements with a constant oxidation number of -2.

$2(\text{Oxidation number of Mn}) + 7(-2) = 0.$

ANSWER   Oxidation number of Mn $= +7$.

EXAMPLE 3   What is the oxidation number of iron in $Fe(CN)_6^{3-}$?

METHOD   Since the cyanide ion is CN^-, it has an oxidation number of -1.
(Oxidation number of Fe) $+ 6(-1) = -3.$

ANSWER   Oxidation number of Fe $= +3$.

EXAMPLE 4   Use the oxidation number method to balance the equation

$$MnO_4^-(aq) + H^+(aq) + Fe^{2+}(aq) \longrightarrow Mn^{2+}(aq) + Fe^{3+}(aq) + H_2O(l)$$

METHOD   Hydrogen and oxygen have the same oxidation numbers on both sides of the equation; only manganese and iron need be considered.
In MnO_4^-, the oxidation number of Mn $= +7$
In Mn^{2+}, the oxidation number of Mn $= +2$
Thus, manganese decreases its oxidation number by 5 units, and iron must increase its oxidation number by 5 units.
From Fe^{2+} to Fe^{3+} is an increase of 1 unit; therefore the equation needs $5\,Fe^{2+} \longrightarrow 5\,Fe^{3+}$. This makes the equation

$$MnO_4^-(aq) + H^+(aq) + 5\,Fe^{2+}(aq) \longrightarrow Mn^{2+}(aq) + 5\,Fe^{3+}(aq) + H_2O(l)$$

To combine with 4 oxygen atoms, $8H^+$ are needed:

$$MnO_4^-(aq) + 8H^+(aq) + 5\,Fe^{2+}(aq) \longrightarrow Mn^{2+}(aq) + 5\,Fe^{3+}(aq) + 4H_2O(l)$$

EXERCISE 17   Problems on Oxidation Numbers

1. What is the oxidation number of the named element in the following compounds?

   a) Ba in $BaCl_2$ b) Fe in $Fe(CN)_6^{4-}$ c) Cl in Cl_2
   d) Li in Li_2O e) Fe in $Fe(CN)_6^{3-}$ f) Cl in ClO^-
   g) P in P_2O_3 h) Br in BrO_3^- i) Cl in ClO_3^-
   j) C in CCl_4 k) I in I_2 l) Cl in Cl_2O_7
   m) C in CO   n) I in I^- o) Cl in Cl_2O_3
   p) Cr in CrO_3 q) I in IO_3^- r) O in H_2O_2
   s) Cr in CrO_4^{2-} t) N in NO_2 u) H in LiH
   v) Cr in $Cr_2O_7^{2-}$ w) N in N_2O_4 x) H in HBr
   y) S in SO_3^{2-} z) P in PO_4^{3-}

2. a) Calculate the oxidation numbers of tin and lead on each side of the equation

$$PbO_2(s) + 4H^+(aq) + Sn^{2+}(aq) \longrightarrow Pb^{2+}(aq) + Sn^{4+}(aq) + 2H_2O(l)$$

and state which element has been oxidised and which has been reduced.

b) In the redox reaction

$$2Mn^{2+}(aq) + 5BiO_3^-(aq) + 14H^+(aq) \longrightarrow 2MnO_4^-(aq) + 5Bi^{3+}(aq) + 7H_2O(l)$$

calculate the oxidation numbers of all the elements, and state which have been oxidised and which have been reduced.

c) Calculate the oxidation numbers of arsenic and manganese in each of the species in the reaction:

$$5As_2O_3(s) + 4MnO_4^-(aq) + 12H^+(aq) \longrightarrow 5As_2O_5(s) + 4Mn^{2+}(aq) + 6H_2O(l)$$

State which element has been oxidised and which has been reduced.

3. In each of the following equations, one element is underlined. Calculate its oxidation number in each species, and state whether an oxidation or a reduction has occurred.

a) $2\underline{F}_2(g) + 2OH^-(aq) \longrightarrow \underline{F}_2O(g) + 2\underline{F}^-(aq) + H_2O(l)$

b) $3\underline{Cl}_2(g) + 6OH^-(aq) \longrightarrow \underline{Cl}O_3^-(aq) + 5\underline{Cl}^-(aq) + 3H_2O(l)$

c) $\underline{N}H_4^+\underline{N}O_3^-(s) \longrightarrow \underline{N}_2O(g) + 2H_2O(l)$

d) $\underline{Cr}_2O_7^{2-}(aq) + 14H^+(aq) + 6e^- \longrightarrow 2\underline{Cr}^{3+}(aq) + 7H_2O(l)$

e) $\underline{C}_2O_4^{2-}(aq) \longrightarrow 2\underline{C}O_2(g) + 2e^-$

4. a) Only N and I alter in oxidation number in the reaction

$$N_2H_6O(aq) + IO_3^-(aq) + 2H^+(aq) + Cl^-(aq) \longrightarrow N_2(g) + ICl(aq) + 4H_2O(l)$$

Calculate the oxidation number of N in N_2H_6O.

b) In the reaction below, only S and Br change in oxidation number.

$$Na_2H_{10}S_2O_8(aq) + 4Br_2(aq) \longrightarrow 2H_2SO_4(aq) + 2NaBr(aq) + 6HBr(aq)$$

Calculate the oxidation number of S in $Na_2H_{10}S_2O_8$.

5. Use the oxidation number method to balance the equations

a) $IO_4^-(aq) + I^-(aq) + H^+(aq) \longrightarrow I_2(aq) + 4H_2O(l)$

b) $BrO_3^-(aq) + I^-(aq) + H^+(aq) \longrightarrow Br^-(aq) + I_2(aq) + H_2O(l)$

c) $V^{3+}(aq) + H_2O_2(aq) \longrightarrow VO^{2+}(aq) + H^+(aq)$

d) $SO_2(g) + H_2O(l) + Br_2(aq) \longrightarrow H^+(aq) + SO_4^{2-}(aq) + Br^-(aq)$

e) $NH_3(g) + O_2(g) \longrightarrow N_2(g) + H_2O(g)$

f) $NH_3(g) + O_2(g) \longrightarrow N_2O(g) + H_2O(g)$

g) $NH_3(g) + O_2(g) \longrightarrow NO(g) + H_2O(g)$

h) $Fe^{2+}C_2O_4^{2-}(aq) + Ce^{3+}(aq) \longrightarrow CO_2(g) + Ce^{2+}(aq) + Fe^{3+}(aq)$

6. When potassium dichromate solution reacts with acidified potassium iodide solution, titration shows that 1 mole of potassium dichromate produces 3 moles of iodine. Use the oxidation number method to complete and balance the equation

$$Cr_2O_7^{2-}(aq) \ + \ I^-(aq) \ + \ H^+(aq) \ \longrightarrow \ 3I_2(aq)$$

POTASSIUM MANGANATE(VII) TITRATIONS

When potassium manganate(VII) acts as an oxidising agent in acid solution, it is reduced to a manganese(II) salt:

$$MnO_4^-(aq) \ + \ 8H^+(aq) \ + \ 5e^- \ \longrightarrow \ Mn^{2+}(aq) \ + \ 4H_2O(l)$$

Potassium manganate(VII) is not sufficiently pure to be used as a primary standard, and solutions of the oxidant are standardised by titration against a primary standard such as sodium ethanedioate. This reductant can be obtained in a high state of purity as crystals of formula $Na_2C_2O_4 \cdot 2H_2O$, which are neither deliquescent nor efflorescent, and can be weighed out exactly to make a standard solution.

Once it has been standardised, a solution of potassium manganate(VII) can be used to estimate reducing agents such as iron(II) salts. No indicator is needed as the oxidant changes from purple to colourless at the end point.

EXAMPLE 1   *Standardising potassium manganate(VII) against the primary standard, sodium ethanedioate*
A 25.0 cm³ portion of sodium ethanedioate solution of concentration 0.200 mol dm⁻³ is warmed and titrated against a solution of potassium manganate(VII). If 17.2 cm³ of potassium manganate(VII) are required, what is its concentration?

METHOD   Let M = concentration of the manganate(VII) solution.

Amount (mol) of ethanedioate = $25.0 \times 10^{-3} \times 0.200$ mol

Amount (mol) of manganate(VII) = $17.2 \times 10^{-3} \times M$ mol

The equations for the half-reactions are

$$MnO_4^-(aq) \ + \ 8H^+(aq) \ + \ 5e^- \ \longrightarrow \ Mn^{2+}(aq) \ + \ 4H_2O(l) \quad [1]$$
$$C_2O_4^{2-}(aq) \ \longrightarrow \ 2CO_2(g) \ + \ 2e^- \quad [2]$$

Multiplying [1] by 2 and [2] by 5, and adding the two equations gives

$$2MnO_4^-(aq) + 16H^+(aq) + 5C_2O_4^{2-}(aq) \ \longrightarrow \ 2Mn^{2+}(aq) + 8H_2O(l) + 10CO_2(g)$$

No. of moles of $MnO_4^- = \frac{2}{5} \times$ No. of moles of $C_2O_4^{2-}$.

$$\therefore \quad 17.2 \times 10^{-3} \times M = \tfrac{2}{5} \times 25.0 \times 10^{-3} \times 0.200$$

$$M = \frac{2 \times 25.0 \times 10^{-3} \times 0.200}{5 \times 17.2 \times 10^{-3}} = 0.116 \, \text{mol dm}^{-3}$$

ANSWER The potassium manganate(VII) solution has a concentration of $0.116 \, \text{mol dm}^{-3}$.

EXAMPLE 2 *Oxidising iron(II) compounds*
Ammonium iron(II) sulphate crystals have the following formula: $(NH_4)_2SO_4 \cdot FeSO_4 \cdot nH_2O$. In an experiment to determine n, $8.492 \, \text{g}$ of the salt were dissolved and made up to $250 \, \text{cm}^3$ of solution with distilled water and dilute sulphuric acid. A $25.0 \, \text{cm}^3$ portion of the solution was further acidified and titrated against potassium manganate(VII) solution of concentration $0.0150 \, \text{mol dm}^{-3}$. A volume of $22.5 \, \text{cm}^3$ was required.

METHOD The equations for the two half-reactions are

$$MnO_4^-(aq) + 8H^+(aq) + 5e^- \longrightarrow Mn^{2+}(aq) + 4H_2O(l) \quad [1]$$
$$Fe^{2+}(aq) \longrightarrow Fe^{3+}(aq) + e^- \quad [2]$$

Multiplying [2] by 5 and then adding it to [1] gives

$$MnO_4^-(aq) + 8H^+(aq) + 5Fe^{2+}(aq) \longrightarrow Mn^{2+}(aq) + 5Fe^{3+}(aq) + 4H_2O(l)$$

$$\text{Amount (mol) of manganate(VII)} = 22.5 \times 10^{-3} \times 0.0150$$
$$= 0.338 \times 10^{-3} \, \text{mol}$$

$$\text{No. of moles of iron(II)} = 5 \times \text{No. of moles of manganate(VII)}$$
$$= 5 \times 0.338 \times 10^{-3} = 1.69 \times 10^{-3} \, \text{mol}$$

$$\text{Concn of iron(II)} = \frac{1.69 \times 10^{-3}}{25.0 \times 10^{-3}} = 0.0674 \, \text{mol dm}^{-3}$$

$$\text{Concn of } (NH_4)_2SO_4 \cdot FeSO_4 \cdot nH_2O = \frac{\text{Mass in 1 dm}^3 \text{ of solution}}{\text{Molar mass}}$$

$$= \frac{4 \times 8.492}{\text{Molar mass}}$$

$$0.0674 = \frac{4 \times 8.492}{\text{Molar mass}}$$

$$\text{Molar mass} = 503.9 \, \text{g mol}^{-1}$$

$$\text{Molar mass of } (NH_4)_2SO_4 \cdot FeSO_4 \cdot nH_2O = 284 + 18n = 504 \, \text{g mol}^{-1}$$

$$\therefore \quad n = 12$$

ANSWER The formula of the crystals is $(NH_4)_2SO_4 \cdot 12H_2O$

EXAMPLE 3 *Oxidising hydrogen peroxide*
A solution of hydrogen peroxide was diluted 20.0 times. A 25.0 cm^3 portion of the diluted solution was acidified and titrated against 0.0150 mol dm^{-3} potassium manganate(VII) solution. 45.7 cm^3 of the oxidant were required. Calculate the concentration of the hydrogen peroxide solution a) in mol dm^{-3} and b) the 'volume concentration'. (This means the number of volumes of oxygen obtained from one volume of the solution.)

METHOD The equations for the half-reactions are

$$MnO_4^-(aq) + 8H^+(aq) + 5e^- \longrightarrow Mn^{2+}(aq) + 4H_2O(l) \qquad [1]$$
$$H_2O_2(aq) \longrightarrow O_2(g) + 2H^+(aq) + 2e^- \qquad [2]$$

Multiplying [1] by 2 and [2] by 5, and adding the two equations gives

$$2MnO_4^-(aq) + 6H^+(aq) + 5H_2O_2(aq) \longrightarrow 2Mn^{2+}(aq) + 8H_2O(l) + 5O_2(g)$$

Amount (mol) of $MnO_4^-(aq)$ = $45.7 \times 10^{-3} \times 0.0150$
$$= 0.685 \times 10^{-3} \, mol$$

No. of moles of H_2O_2 = $\frac{5}{2} \times$ No. of moles of MnO_4^-
$$= \tfrac{5}{2} \times 0.685 \times 10^{-3} = 1.71 \times 10^{-3} \, mol$$

Concn of H_2O_2 = $(1.71 \times 10^{-3})/(25.0 \times 10^{-3})$ = $0.0684 \, mol \, dm^{-3}$

Concn of original solution = 20.0×0.0684 = $1.37 \, mol \, dm^{-3}$.

When hydrogen peroxide decomposes,

$$2H_2O_2(aq) \longrightarrow 2H_2O(l) + O_2(g)$$

2 moles of hydrogen peroxide form 1 mole of oxygen. Therefore a solution of hydrogen peroxide of concentration 2 mol dm^{-3} is a 22.4 volume solution (the volume of 1 mole of oxygen).

A solution of H_2O_2 of concentration 1.37 mol dm^{-3} is a $22.4 \times 1.37/2$ = 15.4 volume solution.

ANSWER The concentration of hydrogen peroxide is: a) 1.37 mol dm^{-3}, and b) 15.4 volume.

EXAMPLE 4 *Finding the percentage of iron in ammonium iron(III) sulphate*
Iron(III) ions can be estimated by first reducing them to iron(II) ions, and then, after destroying the excess of reducing agent, oxidising them to iron(III) ions with a standard solution of potassium manganate(VII). Zinc amalgam and sulphuric acid are used as the reducing agent. Note that hydrochloric acid cannot be used, and the reducing agent tin(II) chloride cannot be used as potassium manganate(VII) oxidises chloride ions to chlorine.

7.418 g of ammonium iron(III) sulphate are dissolved and made up to 250 cm^3 after the addition of dilute sulphuric acid, 25.0 cm^3 of the solution are pipetted into a bottle containing zinc amalgam, and shaken until a drop of the solution gives no colour when tested with a solution of a thiocyanate (which turns deep red in the presence of iron(III) ions). The aqueous solution is then separated by decantation from the zinc amalgam. On addition of more dilute sulphuric acid and titration against standard potassium manganate(VII) solution, 18.7 cm^3 of 0.0165 mol dm^{-3} solution are required. Calculate the percentage of iron in ammonium iron(III) sulphate.

METHOD Amount (mol) of manganate(VII) in volume used $= 18.7 \times 10^{-3} \times 0.0165$

$$= 0.0309 \times 10^{-3}\,mol$$

From the equation

$$MnO_4^-(aq) + 8H^+(aq) + 5Fe^{2+}(aq) \longrightarrow Mn^{2+}(aq) + 5Fe^{3+}(aq) + 4H_2O(l)$$

Amount (mol) of Fe^{2+}(aq) in 25.0 cm^3 $= 5 \times 0.309 \times 10^{-3}$

$$= 1.55 \times 10^{-3}\,mol$$

Amount (mol) of Fe^{2+}(aq) in whole solution $= 1.55 \times 10^{-2}\,mol$

Mass of iron in sample $=$ Amount (mol) $\times$ Relative atomic mass

$$= 1.55 \times 10^{-2} \times 55.8 = 0.865\,g$$

ANSWER Percentage of iron $= \dfrac{0.865}{7.418} \times 100 = 11.7\%.$

POTASSIUM DICHROMATE(VI) TITRATIONS

Potassium dichromate(VI) can be obtained in a high state of purity, and its aqueous solutions are stable. It is used as a primary standard. The colour change when chromium(VI) changes to chromium(III) in the reaction

$$Cr_2O_7^{2-}(aq) + 14H^+(aq) + 6e^- \longrightarrow 2Cr^{3+}(aq) + 7H_2O(l)$$

is from orange to green. As it is not possible to see a sharp change in colour, an indicator is used. Barium N-phenylphenylamine-4-sulphonate gives a sharp colour change, from blue-green to violet, when a slight excess of potassium dichromate has been added. Phosphoric(V) acid must be present to form a complex with the Fe^{3+} ions formed during the oxidation reaction; otherwise Fe^{3+} ions affect the colour change of the indicator.

Since dichromate(VI) has a slightly lower redox potential than manganate(VII), it can be used in the presence of chloride ions, without oxidising them to chlorine.

EXAMPLE   *Determination of the percentage of iron in iron wire*
A piece of iron wire of mass 2.225 g was put into a conical flask containing dilute sulphuric acid. The flask was fitted with a bung carrying a Bunsen valve, to allow the hydrogen generated to escape but prevent air from entering. The mixture was warmed to speed up reaction. When all the iron had reacted, the solution was cooled to room temperature and made up to 250 cm^3 in a graduated flask. With all these precautions, iron is converted to Fe^{2+} ions only, and no Fe^{3+} ions are formed. 25.0 cm^3 of the solution were acidified and titrated against a 0.0185 mol dm^{-3} solution of potassium dichromate(VI). The volume required was 31.0 cm^3. Calculate the percentage of iron in the iron wire.

METHOD   Amount (mol) of $Cr_2O_7{}^{2-}$(aq) used $= 31.0 \times 10^{-3} \times 0.0185$

$$= 0.574 \times 10^{-3}\,mol$$

The equations for the two half-reactions are

$$Cr_2O_7{}^{2-}(aq) + 14H^+(aq) + 6e^- \longrightarrow 2Cr^{3+}(aq) + 7H_2O(l) \quad [1]$$
$$Fe^{2+}(aq) \longrightarrow Fe^{3+}(aq) + e^- \quad [2]$$

Multiplying [2] by 6 and adding gives

$$Cr_2O_7{}^{2-}(aq) + 14H^+(aq) + 6Fe^{2+}(aq) \longrightarrow 2Cr^{3+}(aq) + 6Fe^{3+}(aq) + 7H_2O(l)$$

Amount (mol) of Fe^{2+} in 25.0 cm^3 $= 6 \times 0.574 \times 10^{-3}$

$$= 3.45 \times 10^{-3}\,mol$$

Amount (mol) of Fe^{2+} in the whole solution $= 3.45 \times 10^{-2}\,mol$

Mass of Fe in the whole solution $= 3.45 \times 10^{-2} \times 55.8 = 1.93$ g

Percentage of Fe in wire $= \dfrac{1.93}{2.225} \times 100 = 86.7\%$

ANSWER   The wire is 86.7% iron.

SODIUM THIOSULPHATE TITRATIONS

Sodium thiosulphate reduces iodine to iodide ions, and forms sodium tetrathionate, $Na_2S_4O_6$:

$$2S_2O_3{}^{2-}(aq) + I_2(aq) \longrightarrow 2I^-(aq) + S_4O_6{}^{2-}(aq)$$

Sodium thiosulphate, $Na_2S_2O_3 \cdot 5H_2O$, is not used as a primary standard as the water content of the crystals is variable. A solution of sodium thiosulphate can be standardised against a solution of iodine, or a solution of potassium iodate(V) or potassium dichromate or potassium manganate(VII).

EXAMPLE 1   *Standardisation of a sodium thiosulphate solution, using iodine*
Iodine has a limited solubility in water. It dissolves in a solution of potassium iodide because it forms the very soluble complex ion, I_3^-.

$$I_2(s) \; + \; I^-(aq) \; \rightleftharpoons \; I_3^-(aq)$$

An equilibrium is set up between iodine and tri-iodide ions, and if iodine molecules are removed from solution by a reaction, tri-iodide ions dissociate to form more iodine molecules. A solution of iodine in potassium iodide can thus be titrated as though it were a solution of iodine in water.

When sufficient of a solution of thiosulphate is added to a solution of iodine, the colour of iodine fades to a pale yellow. Then $2 \, cm^3$ of starch solution are added to give a blue colour with the iodine. Addition of thiosulphate is continued drop by drop, until the blue colour disappears.

$2.835 \, g$ of iodine and $6 \, g$ of potassium iodide are dissolved in distilled water and made up to $250 \, cm^3$. A $25.0 \, cm^3$ portion titrated against sodium thiosulphate solution required $17.7 \, cm^3$ of the solution. Calculate the concentration of the thiosulphate solution.

METHOD   Molar mass of iodine $= \; 2 \times 127 \; = \; 254 \, g \, mol^{-1}$
Concn of iodine solution $= \; 2.835 \times 4/254 \; = \; 0.0446 \, mol \, dm^{-3}$
Amount (mol) of I_2 in $25.0 \, cm^3 \; = \; 25.0 \times 10^{-3} \times 0.0446$
$$= \; 1.115 \times 10^{-3} \, mol$$

From the equation

$$2S_2O_3^{2-}(aq) \; + \; I_2(aq) \; \longrightarrow \; 2I^-(aq) \; + \; S_4O_6^{2-}(aq)$$

No. of moles of 'thio' $= \; 2 \times$ No. of moles of I_2
Amount (mol) of 'thio' in volume used $= \; 2.23 \times 10^{-3} \, mol$

$$\text{Concn of 'thio'} \; = \; \frac{2.23 \times 10^{-3}}{17.7 \times 10^{-3}} \; = \; 0.126 \, mol \, dm^{-3}$$

ANSWER   The concentration of the thiosulphate solution is $0.126 \, mol \, dm^{-3}$.

EXAMPLE 2   *Standardisation of thiosulphate against potassium iodate (V)*
Potassium iodate(V) is a primary standard. It reacts with iodide ions in the presence of acid to form iodine:

$$IO_3^-(aq) \; + \; 5I^-(aq) \; + \; 6H^+(aq) \; \longrightarrow \; 3I_2(aq) \; + \; 3H_2O(l)$$

A standard solution of iodine can be prepared by weighing out the necessary quantity of potassium iodate(V) and making up to a known volume of solution. When a portion of this solution is added to an excess of potassium iodide in acid solution, a calculated amount of iodine is liberated.

1.015 g of potassium iodate(V) are dissolved and made up to $250\,cm^3$. To a $25.0\,cm^3$ portion are added an excess of potassium iodide and dilute sulphuric acid. The solution is titrated with a solution of sodium thiosulphate, starch solution being added near the end-point. $29.8\,cm^3$ of thiosulphate solution are required. Calculate the concentration of the thiosulphate solution.

METHOD

Molar mass of $KIO_3 = 39.1 + 127 + (3 \times 16.0) = 214\,g\,mol^{-1}$

Concn of KIO_3 solution $= 1.015 \times 4/214 = 0.0189\,mol\,dm^{-3}$

Amount (mol) of KIO_3 in $25\,cm^3 = 25.0 \times 10^{-3} \times 0.0189$
$$= 0.473 \times 10^{-3}\,mol$$

Since
$$IO_3^-(aq) + 5I^-(aq) + 6H^+(aq) \longrightarrow 3I_2(aq) + 3H_2O(l)$$
and
$$2S_2O_3^{2-}(aq) + I_2(aq) \longrightarrow 2I^-(aq) + S_4O_6^{2-}(aq)$$

No. of moles of 'thio' $= 6 \times$ No. of moles of IO_3^-
$$= 6 \times 0.473 \times 10^{-3} = 2.84 \times 10^{-3}\,mol$$

Concn of 'thio' $= (2.84 \times 10^{-3})/(29.8 \times 10^{-3}) = 0.0950\,mol\,dm^{-3}$

ANSWER

The sodium thiosulphate solution has a concentration $0.0950\,mol\,dm^{-3}$.

EXAMPLE 3

Standardisation of thiosulphate solution with potassium dichromate(VI)

A standard solution is made by dissolving 1.015 g of potassium dichromate(VI) and making up to $250\,cm^3$. A $25.0\,cm^3$ portion is added to an excess of potassium iodide and dilute sulphuric acid, and the iodine liberated is titrated with sodium thiosulphate solution. $19.2\,cm^3$ of this solution are needed. Find the concentration of the thiosulphate solution.

METHOD

Molar mass of $K_2Cr_2O_7 = 294\,g\,mol^{-1}$

Concn of dichromate solution $= 1.015 \times 4/294 = 0.0138\,mol\,dm^{-3}$

Amount (mol) of dichromate in $25\,cm^3 = 25.0 \times 10^{-3} \times 0.138\,mol$
$$= 0.345 \times 10^{-3}\,mol$$

The equations for the half-reactions are
$$Cr_2O_7^{2-}(aq) + 14H^+(aq) + 6e^- \longrightarrow 2Cr^{3+}(aq) + 7H_2O(l) \quad [1]$$
$$2I^-(aq) \longrightarrow I_2(aq) + 2e^- \quad [2]$$

Multiplying [2] by 3, and adding to [1] gives the equation
$$Cr_2O_7^{2-}(aq) + 14H^+(aq) + 6I^-(aq) \longrightarrow 2Cr^{3+}(aq) + 7H_2O(l) + 3I_2(aq)$$

No. of moles of $I_2 = 3 \times$ No. of moles of $Cr_2O_7^{2-}$

Amount (mol) of I_2 in $25\ cm^3$ = $3 \times 0.345 \times 10^{-3}$
$$= 1.035 \times 10^{-3}\ mol$$

No. of moles of 'thio' = $2 \times$ No. of moles of I_2 (see Example 1)

Amount (mol) of 'thio' in volume used = $2.07 \times 10^{-3}\ mol$

Concn. of 'thio' = $(2.07 \times 10^{-3})/(19.2 \times 10^{-3})$ = $0.108\ mol\ dm^{-3}$

ANSWER   The concentration of the thiosulphate solution is $0.108\ mol\ dm^{-3}$.

EXAMPLE 4   *Estimation of chlorine*

Chlorine displaces iodine from iodides. The iodine formed can be determined by titration with a standard thiosulphate solution. Chlorate(I) solutions are often used as a source of chlorine as they liberate chlorine readily on reaction with acid:

$$ClO^-(aq) + 2H^+(aq) + Cl^-(aq) \longrightarrow Cl_2(aq) + H_2O(l)$$

The amount of chlorine available in a domestic bleach which contains sodium chlorate(I) can be found by allowing the bleach to react with an iodide solution to form iodine, and then titrating with thiosulphate solution:

$$ClO^-(aq) + 2H^+(aq) + 2I^-(aq) \longrightarrow I_2(aq) + Cl^-(aq) + H_2O(l)$$

A domestic bleach in solution is diluted by pipetting $10.0\ cm^3$ and making this volume up to $250\ cm^3$. A $25.0\ cm^3$ portion of the solution is added to an excess of potassium iodide and ethanoic acid and titrated against sodium thiosulphate solution of concentration $0.0950\ mol\ dm^{-3}$, using starch as an indicator. The volume required is $21.3\ cm^3$. Calculate the percentage of available chlorine in the bleach.

METHOD   Amount (mol) of 'thio' = $21.3 \times 10^{-3} \times 0.0950$ = $2.03 \times 10^{-3}\ mol$

Since $2S_2O_3^{2-}(aq) + I_2(aq) \longrightarrow S_4O_6^{2-}(aq) + 2I^-(aq)$

Amount (mol) of I_2 = $1.015\ mol$

Since iodine is produced in the reaction

$$ClO^-(aq) + 2I^-(aq) + 2H^+(aq) \longrightarrow I_2(aq) + Cl^-(aq) + H_2O(l)$$

Amount (mol) of ClO^- in $25\ cm^3$ of solution = $1.015 \times 10^{-3}\ mol$

Since chlorate(I) liberates chlorine in the reaction

$$ClO^-(aq) + 2H^+(aq) + Cl^-(aq) \longrightarrow Cl_2(aq) + H_2O(l)$$

No. of moles of Cl_2 = No. of moles of chlorate(I)
$$= 1.015 \times 10^{-3}\ mol$$

Mass of chlorine = $71.0 \times 1.015 \times 10^{-3}$ = $0.0720\ g$

This is the mass of chlorine available in 25 cm³ of solution

$$\text{Percentage of available } Cl_2 = \frac{\text{Mass of } Cl_2 \text{ from } 250 \text{ cm}^3 \text{ solution}}{\text{Mass of bleach solution used}} \times 100$$

$$= \frac{0.0720 \times 10}{10} \times 100 = 7.2\%$$

ANSWER   The percentage of available chlorine in bleach is 7.2%.

EXAMPLE 5 *Estimation of copper(II) salts*

Copper(II) ions oxidise iodide ions to iodine. The iodine produced can be titrated with standard thiosulphate solution, and, from the amount of iodine produced, the concentration of copper(II) ions in the solution can be calculated.

A sample of 4.256 g of copper(II) sulphate-5-water is dissolved and made up to 250 cm³. A 25.0 cm³ portion is added to an excess of potassium iodide. The iodine formed required 18.0 cm³ of a 0.0950 mol dm⁻³ solution of sodium thiosulphate for reduction. Calculate the percentage of copper in the crystals.

METHOD   Amount (mol) of 'thio' $= 18.0 \times 10^{-3} \times 0.0950 = 1.71 \times 10^{-3}$ mol

No. of moles of $I_2 = \frac{1}{2} \times$ No. of moles of 'thio' $= 0.855 \times 10^{-3}$ mol

Since $2Cu^{2+}(aq) + 4I^-(aq) \longrightarrow Cu_2I_2(s) + I_2(aq)$

No. of moles of Cu $= 2 \times$ No. of moles of $I_2 = 1.71 \times 10^{-3}$ mol

Mass of Cu $= 63.5 \times 1.71 \times 10^{-3} = 0.109$ g

Mass of Cu in whole solution $= 1.09$ g

$$\text{Percentage of Cu} = \frac{1.09}{4.256} \times 100 = 25.6\%.$$

ANSWER   The percentage of copper in the crystals is 25.6%.

*EXAMPLE 6 *Deriving an equation for the reaction between bromine and thiosulphate ions*

A solution of bromine was prepared and two titrations were performed:

a) 25.0 cm³ of the solution were added to an excess of potassium iodide. The iodine liberated required 19.5 cm³ of a 0.120 mol dm⁻³ solution of sodium thiosulphate.

b) 25.0 cm³ of the bromine solution were titrated directly against the thiosulphate solution. 2.45 cm³ of thiosulphate solution were required.

c) The solution from titration b) was tested for the presence of various anions. Sulphate ions were detected.

Derive an equation for the reaction.

METHOD   In titration a)

Amount (mol) of 'thio' $= 19.5 \times 10^{-3} \times 0.120 = 2.35 \times 10^{-3}\,\text{mol}$

No. of moles of $I_2 = \frac{1}{2} \times$ No. of moles of 'thio' $= 1.18 \times 10^{-3}\,\text{mol}$

Since the reaction is

$$Br_2(aq) + 2I^-(aq) \longrightarrow 2Br^-(aq) + I_2(aq)$$

No. of moles of $Br_2 =$ No. of moles of I_2

Amount (mol) of Br_2 in $25.0\,\text{cm}^3 = 1.18 \times 10^{-3}\,\text{mol}$

In titration b)

Amount (mol) of 'thio' in volume used $= 2.45 \times 10^{-3} \times 0.120$

$$= 0.294\,\text{mol}$$

Moles of Br_2/Moles of $S_2O_3{}^{2-} = 1.18 \times 10^{-3}/0.294 \times 10^{-3} = 4/1$

Bromine is reduced to bromide ions:

$$4Br_2(aq) + 8e^- \longrightarrow 8Br^-(aq)$$

Thiosulphate ions form sulphate ions:

$$S_2O_3{}^{2-}(aq) \longrightarrow 2SO_4{}^{2-}(aq)$$

To balance the equation, H_2O is needed to supply the extra oxygen:

$$S_2O_3{}^{2-}(aq) + 5H_2O(l) \longrightarrow 2SO_4{}^{2-}(aq) + 10H^+(aq) + 8e^-$$

Putting the two half-reactions together gives

ANSWER $4Br_2(aq) + S_2O_3{}^{2-}(aq) + 5H_2O(l) \longrightarrow 8Br^-(aq) + 2SO_4{}^{2-}(aq) + 10H^+(aq)$

EXERCISE 18   Problems on Redox Reactions

1. Write balanced half-reaction equations for the oxidation of each of the following:

   a) NO_2^- to NO_3^- b) $AsO_3{}^{3-}$ to $AsO_4{}^{3-}$ c) $Hg_2{}^{2+}$ to Hg^{2+}
   d) H_2O_2 to O_2 e) V^{3+} to VO^{2+}

2. Write balanced half-reaction equations for the reduction of each of the following in acid solution:

   a) NO_3^- to NO_2 b) NO_3^- to NO c) NO_3^- to NH_4^+
   d) BrO_3^- to Br_2 e) PbO_2 to Pb^{2+}

3. Complete and balance the following ionic equations:

   a) $MnO_4^-(aq) + H_2O_2(aq) + H^+(aq) \longrightarrow$
   b) $MnO_2(s) + H^+(aq) + Cl^-(aq) \longrightarrow$
   c) $MnO_4^-(aq) + C_2O_4{}^{2-}(aq) + H^+ \longrightarrow$
   d) $Cr_2O_7{}^{2-}(aq) + C_2O_4{}^{2-}(aq) + H^+(aq) \longrightarrow$
   e) $Cr_2O_7{}^{2-}(aq) + I^-(aq) + H^+(aq) \longrightarrow$
   f) $H_2O_2(aq) + NO_2^-(aq) \longrightarrow$

4. How many moles of the following reductants will be oxidised by 3.0×10^{-3} mol of potassium manganate(VII) in acid solution?

   a) Fe^{2+} b) Sn^{2+} c) $(CO_2^-)_2$ d) H_2O_2 e) I^-

5. How many moles of the following will be oxidised by 1.0×10^{-4} mol of potassium dichromate(VI)?

   a) Fe^{2+} b) SO_3^{2-} c) Br^- d) $(CO_2^-)_2$ e) Hg_2^{2+}?

6. How many moles of the following will be reduced by 2.0×10^{-3} moles of Sn^{2+}?

   a) $Fe(CN)_6^{3-}$ b) Cl_2 c) Mn^{4+} (to Mn^{2+})
   d) Ce^{4+} (to Ce^{3+})   e) BrO_3^- (to Br^-)?

7. What volumes of the following solutions will be oxidised by $25.0 \, cm^3$ of $0.0200 \, mol \, dm^{-3}$ potassium manganate(VII) in acid solution?

   a) $0.0200 \, mol \, dm^{-3}$ tin(II) nitrate
   b) $0.0100 \, mol \, dm^{-3}$ iron(II) sulphate
   c) $0.250 \, mol \, dm^{-3}$ hydrogen peroxide
   d) $0.200 \, mol \, dm^{-3}$ chromium(II) nitrate
   e) $0.150 \, mol \, dm^{-3}$ sodium ethanedioate

8. What volumes of the following solutions will be oxidised by $20.0 \, cm^3$ of $0.0150 \, mol \, dm^{-3}$ potassium dichromate(VI) in acid solution?

   a) $0.0200 \, mol \, dm^{-3}$ tin(II) chloride
   b) $0.150 \, mol \, dm^{-3}$ iron(II) chloride
   c) $0.125 \, mol \, dm^{-3}$ sodium ethanedioate
   d) $0.300 \, mol \, dm^{-3}$ sodium sulphite (sulphate(IV))
   e) $0.100 \, mol \, dm^{-3}$ mercury(I) nitrate, $Hg_2(NO_3)_2$

9. $25.0 \, cm^3$ of a sodium sulphite solution require $45.0 \, cm^3$ of 0.0200 $mol \, dm^{-3}$ potassium manganate(VII) solution for oxidation. What is the concentration of the sodium sulphite solution?

10. $35.0 \, cm^3$ of potassium manganate(VII) solution are required to oxidise a $0.2145 \, g$ sample of ethanedioic acid-2-water, $H_2C_2O_4 \cdot 2H_2O$. What is the concentration of the potassium manganate(VII) solution?

11. $37.5 \, cm^3$ of cerium(IV) sulphate solution are required to titrate a $0.2245 \, g$ sample of sodium ethanedioate, $Na_2C_2O_4$. What is the concentration of the cerium(IV) sulphate solution?

12. A piece of iron wire weighs $0.2756 \, g$. It is dissolved in acid, reduced to the Fe^{2+} state, and titrated with $40.8 \, cm^3$ of $0.0200 \, mol \, dm^{-3}$ potassium dichromate solution. What is the percentage purity of the iron wire?

13. A piece of limestone weighing $0.1965 \, g$ was allowed to react with an excess of hydrochloric acid. The calcium in it was precipitated as calcium ethanedioate. The precipitate was dissolved in sulphuric acid, and the ethanedioate in the solution needed $35.6 \, cm^3$ of a 0.0200 $mol \, dm^{-3}$ solution of potassium manganate(VII) for titration. Calculate the percentage of $CaCO_3$ in the limestone.

14. A solution of potassium dichromate is standardised by titration with sodium ethanedioate solution. If $47.0\,cm^3$ of the dichromate solution were needed to oxidise $25.0\,cm^3$ of ethanedioate solution of concentration $0.0925\,mol\,dm^{-3}$, what is the concentration of the potassium dichromate solution?

15. $2.4680\,g$ of sodium ethanedioate are dissolved in water and made up to $250\,cm^3$ of solution. When a $25.0\,cm^3$ portion of the solution is titrated against cerium(IV) sulphate, $35.7\,cm^3$ of the cerium(IV) sulphate solution are required. What is its concentration?

16. A $25.0\,cm^3$ portion of a solution containing Fe^{2+} ions and Fe^{3+} ions was acidified and titrated against potassium manganate(VII) solution. $15.0\,cm^3$ of a $0.0200\,mol\,dm^{-3}$ solution of potassium manganate(VII) were required. A second $25.0\,cm^3$ portion was reduced with zinc and titrated against the same manganate(VII) solution. $19.0\,cm^3$ of the oxidant solution were required. Calculate the concentrations of a) Fe^{2+}, and b) Fe^{3+} in the solution.

17. a) What volume of acidified potassium manganate(VII) of concentration $0.0200\,mol\,dm^{-3}$ is decolourised by $100\,cm^3$ of hydrogen peroxide of concentration $0.0100\,mol\,dm^{-3}$?
    b) What volume of oxygen is evolved at s.t.p.?

18. A $0.6125\,g$ sample of potassium iodate(V), KIO_3, is dissolved in water and made up to $250\,cm^3$. A $25.0\,cm^3$ portion of the solution is added to an excess of potassium iodide in acid solution. The iodine formed requires $22.5\,cm^3$ of sodium thiosulphate solution for titration. What is the concentration of the thiosulphate solution?

19. $25.0\,cm^3$ of a solution of X_2O_5 of concentration $0.100\,mol\,dm^{-3}$ is reduced by sulphur dioxide to a lower oxidation state. To reoxidise X to its original oxidation number required $50.0\,cm^3$ of $0.0200\,mol\,dm^{-3}$ potassium manganate(VII) solution. To what oxidation number was X reduced by sulphur dioxide?

20. Manganese(II) sulphate is oxidised to manganese(IV) oxide by potassium manganate(VII) in acid solution. A flocculant is added to settle the solid MnO_2 so that it does not obscure the colour of the manganate(VII). If $25.0\,cm^3$ of manganese(II) sulphate solution require $22.5\,cm^3$ of $0.0200\,mol\,dm^{-3}$ potassium manganate(VII) solution, what is the concentration of $MnSO_4$?

*21. A solution of hydroxylamine hydrochloride contains $0.1240\,g$ of $NH_2OH\cdot HCl$. On boiling, it is oxidised by an excess of acidified iron(III) sulphate. The iron salt formed is titrated against potassium manganate(VII) solution of concentration $0.0160\,mol\,dm^{-3}$. A volume of $44.6\,cm^3$ of the oxidant is required.
    a) Find the ratio of moles NH_2OH : moles Fe^{3+}.
    b) State the change in oxidation number of Fe.
    c) State the oxidation number of N in NH_2OH.

    d) Deduce the oxidation number of N in the product of the reaction.

    e) Decide what compound of nitrogen in this oxidation state is likely to be formed in the reaction.

    f) Write the equation for the reaction.

22. A piece of impure copper was allowed to react with dilute nitric acid. The copper(II) nitrate solution formed liberated iodine from an excess of potassium iodide solution. The iodine was estimated by titration with a solution of sodium thiosulphate. If a 0.877 g sample of copper was used, and the volume required was 23.7 cm^3 of 0.480 mol dm^{-3} thiosulphate solution, what is the percentage of copper in the sample?

23. A household bleach contains sodium chlorate(I), NaOCl. The chlorate(I) ion will react with potassium iodide to give iodine, which can be estimated with a standard thiosulphate solution.

    a) Write the equations for the reaction of ClO$^-$ and I$^-$ to give I$_2$ and for the reaction of iodine and thiosulphate ions.

    b) A 25.0 cm^3 sample of household bleach is diluted to 250 cm^3. A 25.0 cm^3 portion of the solution is added to an excess of potassium iodide solution and titrated against 0.200 mol dm^{-3} sodium thiosulphate solution. The volume required is 18.5 cm^3. What is the concentration of sodium chlorate(I) in the bleach?

COMPLEXOMETRIC TITRATIONS

The complexes formed by a number of metal ions with

bis [bis(carboxymethyl)amino] ethane,

$(HO_2CCH_2)_2NCH_2CH_2N(CH_2CO_2H)_2$,

which is usually referred to as edta (short for its old name) are very stable, and can be used for the estimation of metal ions by titration. The end-point in the titration is shown by an indicator which forms a coloured complex with the metal ion being titrated. If Eriochrome Black T is used as indicator, the metal-indicator colour of red is seen at the beginning of the titration. As the titrant is added, the metal ions are removed from the indicator and complex with edta. At the end-point, the colour of the free indicator, blue, is seen:

Metal-indicator (red) + edta $\longrightarrow$ Metal-edta + Indicator (blue)

EXAMPLE   *Determination of the hardness of tap water*

Hardness in water is caused by the presence of calcium ions and magnesium ions. Both these ions complex strongly with edta. The amounts of temporary hardness and permanent hardness can be determined separately by performing complexometric titrations on tap water and boiled tap water. 100 cm^3 of tap water are measured into a flask. An alkaline buffer and Eriochrome Black T are added, and the solution is titrated against 0.100 mol dm^{-3} edta solution. The volume required is 2.10 cm^3.

A second $100\,cm^3$ of tap water are measured into a $250\,cm^3$ beaker, and boiled for 30 minutes. After cooling, the water is filtered into a $100\,cm^3$ graduated flask, and made up to the mark by the addition of distilled water. On titration as before, the volume of edta needed is $1.25\,cm^3$. Calculate the concentration of calcium and magnesium present as permanent hardness and the concentration of calcium and magnesium present as temporary hardness.

METHOD

Total hardness requires $\qquad$ $2.10\,cm^3$ of $0.100\,mol\,dm^{-3}$ edta

Permanent hardness requires $\qquad$ $1.25\,cm^3$ of $0.100\,mol\,dm^{-3}$ edta

Temporary hardness requires $\qquad$ $0.85\,cm^3$ of $0.100\,mol\,dm^{-3}$ edta

Amount (mol) of metal as permanent hardness

$$= 1.25 \times 10^{-3} \times 0.100\,mol$$

$$= 0.125 \times 10^{-3}\,mol\ \text{in}\ 100\,cm^3\ \text{water}$$

$$= 1.25 \times 10^{-3}\,mol\,dm^{-3}$$

Amount (mol) of metal as temporary hardness

$$= 0.85 \times 10^{-3} \times 0.100\,mol$$

$$= 0.085 \times 10^{-3}\,mol\ \text{in}\ 100\,cm^3\ \text{water}$$

$$= 0.85 \times 10^{-3}\,mol\,dm^{-3}$$

ANSWER

The concentration of calcium and magnesium present as temporary hardness is $8.5 \times 10^{-4}\,mol\,dm^{-3}$; the concentration of calcium and magnesium present as permanent hardness is $1.25 \times 10^{-3}\,mol\,dm^{-3}$.

EXERCISE 19   Problems on Complexometric Titrations

1. Calculate the concentration of a solution of zinc sulphate from the following data. $25.0\,cm^3$ of the solution, when added to an alkaline buffer and Eriochrome Black T indicator, required $22.3\,cm^3$ of a $1.05 \times 10^{-2}\,mol\,dm^{-3}$ solution of edta for titration. The equation for the reaction can be represented as

$$Zn^{2+}(aq)\ +\ edta^{4-}(aq) \longrightarrow Znedta^{2-}(aq)$$

2. To a $50.0\,cm^3$ sample of tap water were added a buffer and a few drops of Eriochrome Black T. On titration against a $0.0100\,mol\,dm^{-3}$ solution of edta, the indicator turned blue after the addition of $9.80\,cm^3$ of the titrant. Calculate the hardness of water in parts per million of calcium, assuming that the hardness is entirely due to the presence of calcium salts. ($1\,p.p.m. = 1\,g\ \text{in}\ 10^6\,g\ \text{water.}$)

3. A $0.2500\,g$ sample of a mixture of magnesium oxide and calcium oxide was dissolved in dilute nitric acid and made up to $1.00\,dm^3$ of solution with distilled water. A $50.0\,cm^3$ portion was buffered and, after addition of indicator, was titrated against $0.0100\,mol\,dm^{-3}$ edta solution. $25.8\,cm^3$ of the titrant were required. Find the percentage by mass of calcium oxide and magnesium oxide in the mixture.

4. Find n in the formula $Al_2(SO_4)_3 \cdot nH_2O$ from the following analysis. 1.000 g of aluminium sulphate hydrate were weighed out and made up to 250 cm³. A 25.0 cm³ portion was allowed to complex with edta by being boiled with 50.0 cm³ of edta solution of concentration 1.00×10^{-2} mol dm⁻³. The excess of edta was determined by adding Eriochrome Black T and titrating against a solution of 1.115×10^{-2} mol dm⁻³ solution of zinc sulphate. 17.9 cm³ of zinc sulphate solution were required to turn the indicator from blue to red. The reactions taking place are

$$Al^{3+}(aq) \;+\; edta^{4-}(aq) \longrightarrow Aledta^{-}(aq)$$
$$Zn^{2+}(aq) \;+\; edta^{4-}(aq) \longrightarrow Znedta^{2-}(aq)$$

PRECIPITATION TITRATIONS

In a precipitation titration, the two solutions react to form a precipitate of an insoluble salt. A solution of a chloride can be estimated by finding the volume of a standard solution of silver nitrate that will precipitate all the chloride ions as insoluble silver chloride:

$$Ag^{+}(aq) \;+\; Cl^{-}(aq) \longrightarrow AgCl(s)$$

Bromides and iodides and thiocyanates can be titrated in the same way. There are various ways of finding out when the end-point has been reached.

EXAMPLE 1 *Determination of chlorides*
25.0 cm³ of a sodium chloride solution required 18.7 cm³ of 0.100 mol dm⁻³ silver nitrate solution for complete precipitation. Calculate the concentration of the sodium chloride solution.

METHOD　Since $\qquad Ag^{+}(aq) \;+\; Cl^{-}(aq) \longrightarrow AgCl(s)$

No. of moles of $AgNO_3$ = No. of moles of Cl^{-}

Amount (mol) of $AgNO_3$ in volume used = $18.7 \times 10^{-3} \times 0.100$

$\qquad\qquad\qquad\qquad\qquad\qquad\qquad = 1.87 \times 10^{-3}$ mol

Amount (mol) of Cl^{-} in 25.0 cm³ = 1.87×10^{-3} mol

Concn of Cl^{-} = $(1.87 \times 10^{-3})/(25.0 \times 10^{-3}) = 7.50 \times 10^{-2}$ mol dm⁻³

ANSWER　The concentration of sodium chloride is 7.50×10^{-2} mol dm⁻³.

EXAMPLE 2 *Determination of a mixture of halides*
2.95 g of a mixture of potassium chloride and potassium bromide is dissolved in water, and the solution is made up to 250 cm³. 25.0 cm³ of this solution required 31.5 cm³ of 0.100 mol dm⁻³ silver nitrate solution. Calculate the percentages of potassium chloride and potassium bromide in the mixture.

METHOD   Let x grams $=$ Mass of potassium chloride

Then $(2.95 - x)$ grams $=$ Mass of potassium bromide

$$\text{Amount (mol) of KCl in } 25.0 \text{ cm}^3 = \frac{x}{74.5} \times \frac{25.0}{250} = \frac{0.1x}{74.5} \text{ mol}$$

The KCl in 25.0 cm^3 requires $\dfrac{x}{74.5}$ dm^3 of 0.100 mol dm^{-3} AgNO$_3$(aq)

$$\text{Amount (mol) of KBr in } 25.0 \text{ cm}^3 = \frac{(2.95 - x)}{119} \times \frac{25.0}{250}$$

$$= \frac{0.100(2.95 - x)}{119}$$

The KBr in 25.0 cm^3 requires $\dfrac{(2.95 - x)}{119}$ dm^3 of 0.100 mol dm^{-3} AgNO$_3$(aq)

$$\text{Therefore, } \frac{x}{74.5} + \frac{(2.95 - x)}{119} = 31.5 \times 10^{-3}$$

$$x = 1.34$$

ANSWER   The mass of potassium chloride is 1.34 g; the mass of potassium bromide is 1.61 g.

EXERCISE 20   Problems on Precipitation Titrations

1. A 25.0 cm^3 portion of a solution of potassium chloride required 18.5 cm^3 of a silver nitrate solution of concentration 0.0200 mol dm^{-3} for titration. What is the concentration of the KCl solution?

2. A solid mixture contains sodium chloride and sodium nitrate. A 0.5800 g sample of the mixture was dissolved in water and made up to 250 cm^3. A 25.0 cm^3 portion was titrated against silver nitrate solution of concentration 0.0180 mol dm^{-3}. The volume required was 27.4 cm^3. Calculate the percentage by mass of sodium chloride in the mixture.

3. 1.2400 g of a mixture of sodium chloride and sodium bromide was dissolved and made up to 1.00 dm^3. A 25.0 cm^3 portion of this solution required 21.7 cm^3 of a silver nitrate solution of concentration 0.0175 mol dm^{-3} for titration. Calculate the percentage composition by mass of the mixture.

4. 25.0 cm^3 of a solution of potassium cyanide required 19.8 cm^3 of a solution of silver nitrate of concentration 0.0350 mol dm^{-3} for titration. Calculate the concentration of the KCN solution.

5. A solution contains sodium chloride and hydrochloric acid. A 25.0 cm³ aliquot required 38.2 cm³ of a 0.0325 mol dm⁻³ solution of silver nitrate for titration. A second 25.0 cm³ aliquot required 7.2 cm³ of a 0.0550 mol dm⁻³ solution of sodium hydroxide for neutralisation. Calculate the concentrations of a) sodium chloride, and b) hydrochloric acid in the solution.

6. Find the percentage by mass of silver in an alloy from the following information. A sample of 1.245 g of the alloy was dissolved in dilute nitric acid and made up to 250 cm³. A 25.0 cm³ portion required 29.8 cm³ of a 0.0214 mol dm⁻³ solution of potassium thiocyanate for titration.

EXERCISE 21   Questions from A-Level Papers

1. Ammonia is produced from its elements on a large scale using the Haber process.
   a) Write an equation for the formation of ammonia from its elements.
   b) The formation of ammonia is an exothermic reaction. In choosing the conditions under which the reaction is to be performed, decisions as to pressure and temperature must be made on economic grounds. State the arguments which influence such decisions.
     i) Argument in favour of using a high pressure
     ii) Argument against using a high pressure
     iii) Argument in favour of using a high temperature
     iv) Argument against using a high temperature.
   c) For reasons of environmental safety the concentration of ammonia in the air downwind of an ammonia production plant was measured by the following procedure.

   A 20 000 litre (measured at s.t.p.) sample of the air was slowly bubbled through an excess of dilute hydrochloric acid. The resulting solution was made alkaline and heated, the ammonia liberated being dissolved in exactly 50 cm³ of 0.1 M hydrochloric acid, which is a large excess. 40.00 cm³ of 0.1 M sodium hydroxide solution were required to neutralise the excess of acid.

   Calculate the concentration of ammonia in the air in units of moles of ammonia per litre of air.   (JMB91)

2. 'Nitrochalk' is a widely used fertiliser which contains a mixture of ammonium nitrate and calcium carbonate.

   A student attempted to determine the percentage by mass of nitrogen in Nitrochalk by the following titrimetric procedure.

   2.00 g of Nitrochalk was heated with 25.0 cm³ of 2.00 mol dm⁻³ sodium hydroxide (that is, excess alkali) until no more ammonia gas was evolved. The mixture was filtered to remove the calcium carbonate. The filtrate was then made up to 250 cm³ by adding distilled water and 25.0 cm³ portions were titrated against 0.100 mol dm⁻³ hydrochloric acid. It was found that 30.0 cm³ of acid was required to neutralise the NaOH left over.

a) Describe briefly a chemical test to show that Nitrochalk contains calcium carbonate.

b) Explain why it is desirable to add the calcium carbonate to the ammonium nitrate.

c) The equation for ammonium nitrate reacting with sodium hydroxide is

$$NH_4NO_3 + NaOH \longrightarrow NH_3 + H_2O + NaNO_3$$

Construct the ionic equation for this reaction and give state symbols.

d) Give *two* reasons why *excess* sodium hydroxide is used.

e) How might the student test to find out when no more ammonia was evolved?

f) How would you determine the end-point of the titration in the experiment?

g) From the data given, calculate the number of moles of ammonium nitrate present in the original sample of fertiliser.

h) Hence calculate:
   i) the percentage by mass of nitrogen in the fertiliser
   ii) the percentage by mass of calcium carbonate in Nitrochalk.

(O90,AS)

3. a) Describe what you would expect to observe on the gradual addition of *excess* dilute sodium hydroxide to separate dilute (approximately 0.1 mol dm⁻³) solutions each containing one of the following cations:

$$Mg^{2+}, Ba^{2+}, Pb^{2+}, Cr^{3+}, Fe^{2+}, Fe^{3+} \text{ and } Cu^{2+}.$$

Give formulae and equations, as appropriate, for the reactions which occur.

b) In acidified aqueous solution iron(II) ions, Fe^{2+}, are oxidised at room temperature by manganate(VII) ions, MnO_4^-. Above 60°C ethanedioate ions, $C_2O_4^{2-}$, are also oxidised by manganate(VII) ions.

Write appropriate ion/electron half equations for:
   i) the reduction of the oxidising agent, MnO_4^-
   ii) the oxidation of the reducing agent, Fe^{2+}
   iii) the oxidation of the reducing agent, $C_2O_4^{2-}$.

By combining (i), (ii) and (iii) write a balanced equation for the redox reaction which occurs between MnO_4^- and iron(II) ethanedioate, FeC_2O_4 (Fe^{2+} and $C_2O_4^{2-}$ ions in acid solution) above 60°C in acid solution. Use this equation to calculate the volume of a manganate(VII) solution of concentration 0.0200 mol dm⁻³, required for complete reaction with $0.2000\,g$ of iron(II) ethanedioate under the above conditions.

$A_r(C) = 12.01$, $A_r(Fe) = 55.85$, $A_r(O) = 16.00$)   (WJEC90)

4. a) Write down the oxidation state of iodine in the IO_3^- anion.

   b) Write down the two ion/electron half equations for the reaction of the IO_3^- anion with iodide ion (I^-) in acidic solution. Hence write down the stoichiometric equation for the overall reaction.

   c) An unknown mass of KIO_3 was treated in aqueous solution with excess of iodide ion and acidified. The resulting solution, on titration against $0.2000\,mol\,dm^{-3}$ sodium thiosulphate solution, required $53.70\,cm^3$ thereof for complete reaction. Find the mass of KIO_3 used.

   $(A_r(K) = 39.10$, $A_r(I) = 126.90$, $A_r(O) = 16.00)$ (WJEC91,p)

5. Moss in lawns is treated with a mixture consisting of sand and ammonium iron(II) sulphate-6-water (*ferrous ammonium sulphate*), $(NH_4)_2SO_4 \cdot FeSO_4 \cdot 6H_2O$. When $3.000\,g$ of the mixture were shaken with dilute sulphuric acid and the resulting mixture titrated with $0.02\,M\,KMnO_4$ solution, $25.00\,cm^3$ of $KMnO_4$ were decolourised.

   a) Describe briefly the essential steps necessary to isolate, from the solid mixture, a pure sample of ammonium iron(II) sulphate-6-water.

   b) By stating necessary reagent(s) and observation(s), give *one* test which would show the presence of sulphate ions in the pure sample. Give an equation for the reaction involved in the test.

   c) Construct an ionic equation for the reaction between Fe^{2+} ions and MnO_4^- ions in acid solution by writing two half-equations (one for Fe^{2+} and one for MnO_4^- with H^+) and combining them to give the overall equation.

   d) Using the information given at the start of the question, calculate the percentage by mass of Fe^{2+} ions in the mixture.   (JMB91)

6. Iodine monochloride, ICl, is used to determine the degree of unsaturation in oils. The ICl adds rapidly to the carbon–carbon double bonds present. In an experiment, $0.127\,g$ of an unsaturated oil was treated with $25.0\,cm^3$ of $0.100\,M$ iodine monochloride solution. The mixture was kept in the dark until the reaction was complete. The unreacted ICl was then treated with an excess of aqueous potassium iodide, forming I_2. The liberated iodine was found to react with $40.0\,cm^3$ of $0.100\,M$ sodium thiosulphate.

   a) Suggest why it is necessary to keep the mixture of oil and iodine monochloride in the dark.

   b) Write an equation for the reaction between iodine monochloride and potassium iodide.

   c) Calculate the number of moles of sodium thiosulphate which were used in the titration.

d) Calculate the number of moles of iodine liberated, given that iodine reacts with sodium thiosulphate according to the equation

$$I_2 \ + \ 2Na_2S_2O_3 \ \longrightarrow \ 2NaI \ + \ Na_2S_4O_6$$

Hence, calculate the number of moles of unreacted iodine monochloride.

e) Calculate the number of moles of iodine monochloride which reacted with the 0.127 g of unsaturated oil.

f) Direct addition of iodine to an unsaturated oil is slow. However, unsaturation is quoted as the *iodine number*. The iodine number is the number of grams of iodine which in theory can be added to 100 g of oil. Calculate the iodine number of this oil, given that 1 mole of ICl is equivalent to 1 mole of I_2.   (JMB91)

7. a) Indicate simple test-tube experiments you could do to show the formation of one complex of copper(II) and one complex of cobalt(II) starting from solutions containing their usual hydrated ions. State what you would observe, and give the formulae of the complex ions produced.

b) Brass is a mixture of copper and zinc. It dissolves in nitric acid to give a mixture of $Cu^{2+}(aq)$ and $Zn^{2+}(aq)$ ions. For example

$$3Cu(s) \ + \ 2NO_3^-(aq) \ + \ 8H^+(aq) \ \longrightarrow \ 3Cu^{2+}(aq) \ + \ 2NO(g) \\ + \ 4H_2O(l)$$

The copper ions may be analysed by means of iodide and sodium thiosulphate. The zinc ions do not react during this analysis.

1.00 g of brass was dissolved in nitric acid and, after boiling off oxides of nitrogen and neutralisation, excess potassium iodide was added

$$2Cu^{2+}(aq) \ + \ 4I^-(aq) \ \longrightarrow \ 2CuI(s) \ + \ I_2(aq)$$

The iodine reacted with 0.0100 moles of sodium thiosulphate

$$I_2(aq) \ + \ 2S_2O_3^{2-}(aq) \ \longrightarrow \ 2I^-(aq) \ + \ S_4O_6^{2-}(aq)$$

Calculate the percentage by mass of copper in the brass. (O90, AS)

*8. a) Sodium peroxide, Na_2O_2, reacts with chlorine dioxide, ClO_2, to give a single solid compound, *P*, containing only sodium, chlorine and oxygen. An aqueous solution of *P* did not give a white precipitate on treatment with aqueous silver nitrate. When 1.00 g of *P* was heated for an hour at 260 °C there was no change in mass. The resulting solid was dissolved in 100 cm^3 of water and titrated with aqueous silver nitrate of concentration 0.100 mol dm^{-3}. After the addition of 36.80 cm^3 of the aqueous silver nitrate no more silver chloride was precipitated.

The mixture was heated to boiling, then treated with sulphur dioxide to reduce all chlorine-containing species to chloride ion. Subsequent titration with the silver nitrate solution required a further 73.60 cm^3 to precipitate all the chloride ion present. Deduce the formula of P and give an equation for the reaction which occurs when P is heated at $260\,°C$.

(Relative atomic masses: $Cl = 35.5$, $Na = 23.0$, $O = 16.0$.)

b) In the complete absence of air 10.00 cm^3 of an aqueous solution of sodium nitrite ($NaNO_2$) of concentration $0.100 \text{ mol dm}^{-3}$ were added to the same volume of acidified potassium iodide solution of concentration $0.500 \text{ mol dm}^{-3}$. The liberated iodine was titrated with sodium thiosulphate solution ($Na_2S_2O_3$) of concentration $0.100 \text{ mol dm}^{-3}$, 10.00 cm^3 being required to reach the end-point.

In a second experiment, using the same volume of the aqueous solution of sodium nitrite but with free access of air, 30.00 cm^3 of sodium thiosulphate solution were required to reach the end-point. Obtain an equation for the reaction between nitrate ions and iodide ions and explain the reasons for the differences between the first and second experiments. (L92,S)

9. a) Give the electronic structures of:
   i) a zinc atom
   ii) a zinc ion.

   b) Zinc is an element in the d-block which forms *colourless* ions in aqueous solution.
   i) Explain why zinc ions are colourless whereas those of many d-block metals are coloured.
   ii) Give the formula of the ion of another metal in the d-block which is colourless in aqueous solution.

   c) State with a reason whether you would expect ruthenium (Ru) compounds to be coloured or colourless in solution.

   d) 25.0 cm^3 of an aqueous solution containing $0.050 \text{ mol dm}^{-3}$ of an ion $M^{3+}(aq)$ was reduced using excess zinc, and the unreacted zinc removed. The resulting solution required 5.0 cm^3 of an aqueous solution of potassium manganate(VII) of the same molar concentration to restore M to its original +3 oxidation state. To what oxidation state was M^{3+} reduced by the zinc? Show your working.

   The manganate(VII) ion is reduced according to the equation

   $$MnO_4^-(aq) \ + \ 8H^+(aq) \ + \ 5e^- \ \longrightarrow \ Mn^{2+}(aq) \ + \ 4H_2O(l)$$

   (O90,AS)

*10. Consider the following information about the yellow-orange paramagnetic gas, A, which is an oxide of chlorine.

   1. A given volume, v, of A is decomposed into its elements by sparking and the resulting gases occupy a volume $\dfrac{3v}{2}$, measured under the same conditions. When the chlorine gas is absorbed in potassium iodide solution, the residual gas occupies a volume, v, under the same conditions as before.

2. When 0.1250 g of A reacts under slightly alkaline conditions with a small excess of hydrogen peroxide, H_2O_2, solution, 20.76 cm^3 of oxygen gas is evolved, measured at 273 K and 1.01×10^5 Pa.

3. When the solution resulting from. 2. above was boiled to destroy excess H_2O_2 and reacted under slightly acidic conditions with excess potassium iodide solution, iodine was liberated which required 74.12 cm^3 of 0.1000 mol dm^{-3} sodium thiosulphate ($Na_2S_2O_3$) solution for a complete reaction.

4. When the solution resulting from 3. above was boiled with nitric acid (to remove excess iodide ion as volatile iodine) and titrated against 0.0500 mol dm^{-3} silver nitrate ($AgNO_3$) solution, 37.06 cm^3 thereof was required to precipitate all the chloride ion present. (*No chlorine containing compounds remained in solution thereafter.*)

5. When a further 0.1250 g of A is allowed to react with water at 0 °C in the dark, a slow reaction ensues requiring some weeks for completion. Thereafter the resulting solution is found to require 6.18 cm^3 of 0.0500 mol dm^{-3} silver nitrate solution to precipitate all the chloride ion present. The silver salt of the other product of reaction is water soluble and when the filtrate from the above was acidified and treated with excess potassium iodide solution, the liberated iodine required 92.65 cm^3 of 0.1000 mol dm^{-3} sodium thiosulphate for complete reaction.

a) Use the information in 1. to obtain an empirical formula for A and to deduce a probable identity. Use the information in 4. to confirm your conclusion.

b) Use your conclusions in a) and the information in 2., 3., 4. and 5. to suggest identities for all the oxochloro species which are formed and react therein.

c) i) Write balanced stoichiometric equations for all the redox processes in 2., 3., 4. and 5. above. State the changes in oxidation state undergone by chlorine and by the other species involved. Write balanced pairs of ion/electron half equations as appropriate.

   ii) Demonstrate the consistency of *all* the quantitative information with your conclusions.

   (1 mole of a gas occupies 2.24×10^4 cm^3 at 273 K and 1.01×10^5 Pa, $A_r(O) = 16.00$, $A_r(Cl) = 35.45$, one mole of I_2 reacts with two moles of $S_2O_3^{2-}$.

   Hint For the formulation of ion/electron half equations, the addition of species such as H^+, H_2O, OH^- to one side or other of the equation is frequently of assistance.)   (WJEC92,S)

11. a) When an acidified solution of potassium dichromate(VI) is added to a solution of an iron(II) compound, the Fe^{2+} ions are oxidised to Fe^{3+} ions.

   i) Write an equation to show the ionic half reaction involving the iron(II).

   ii) Give the formula and colour of the chromium ion formed in the redox reaction.

b) A sample of steel weighing 0.200 g is dissolved in dilute aqueous sulphuric acid. The resulting solution requires 34.0 cm^3 of potassium manganate(VII) of concentration 0.02 mol dm^{-3} in a titration. The reactions which take place are represented by the equations

$$Fe(s) \ + \ 2H^+(aq) \longrightarrow Fe^{2+}(aq) \ + \ H_2(g)$$

$$5Fe^{2+}(aq) \ + \ MnO_4^-(aq) \ + \ 8H^+(aq) \longrightarrow 5Fe^{3+}(aq) \ + \ Mn^{2+}(aq)$$
$$+ \ 4H_2O(l)$$

   i) Explain why no indicator is required in this titration.
   ii) Calculate the percentage of iron in the steel.

c) i) Give a simple test and its result to show that a sample of a mineral contains the element calcium.
   ii) When aqueous sodium hydroxide is added to a solution of an iron(III) compound, iron(III) hydroxide is formed. Classify the type of reaction which takes place. Write an equation to represent the reaction.

d) Discuss some of the environmental problems that can arise in the quarrying of minerals.   (AEB90,AS)

12. This question concerns the determination of the amount of preservative, sodium sulphite (Na_2SO_3), in a sample of beefburgers. In an experiment 1 kg of meat was boiled with an excess of dilute hydrochloric acid (Step 1). The sulpher dioxide gas released was completely absorbed in an excess of dilute aqueous sodium hydroxide (Step 2). The resulting solution was then acidified with dilute sulphuric acid and titrated with 0.02 M KMnO$_4$ solution (Step 3); 30.00 cm^3 were required to reach the end-point.

Use the following equations to answer the questions below.

Step 1
$$Na_2SO_3 \ + \ 2HCl \longrightarrow 2NaCl \ + \ SO_2 \ + \ H_2O$$

Step 2
$$SO_2 \ + \ 2OH^- \longrightarrow H_2O \ + \ SO_3^{2-}$$

Step 3
$$5SO_3^{2-} \ + \ 2MnO_4^- \ + \ 6H^+ \longrightarrow 5SO_4^{2-} \ + \ 2Mn^{2+} \ + \ 3H_2O$$

a) i) How many moles of Na_2SO_3 are equivalent to 1 mol of MnO_4^-?
   ii) How many moles of MnO_4^- were used in the titration?
   iii) How many moles of Na_2SO_3 were present in 1 kg of the meat?
   iv) Government chemists often express the amount of Na_2SO_3 in meat as parts per million (1 ppm = 1 g of Na_2SO_3 in 10^6 g of meat). Express the amount of Na_2SO_3 in the meat as ppm.

b) i) In Step 1, why is it necessary to use an excess of dilute hydrochloric acid and to boil the solution?

ii) In Step 3, why is it essential not to use dilute hydrochloric acid to acidify the solution?

iii) In Step 3, what colour change would you observe at the end-point? (JMB92)

13. a) i) State the conditions under which magnesium and calcium will react with water, and write balanced equations for the reactions.

ii) Explain any differences between the two reactions in terms of the atomic properties of the two metals.

b) Compare the chemistries of magnesium and calcium with reference to the following:
  i) the solubilities of their sulphates in water
  ii) the thermal stabilities of their carbonates
  iii) the reactions of their oxides with water.

c) A mineral, which can be represented by the formula $Mg_x Ba_y (CO_3)_z$, was analysed as described below.

From the results, calculate the formula of the mineral.

A sample of the mineral was dissolved in excess hydrochloric acid and the solution made up to $100 \, cm^3$ with water. During the process $48 \, cm^3$ of carbon dioxide, measured at $25\,°C$ and 1 atmosphere pressure, were evolved.

A $25.0 \, cm^3$ portion of the resulting solution required $25.0 \, cm^3$ of edta solution of concentration $0.02 \, mol \, dm^{-3}$ to reach an end-point. A further $25.0 \, cm^3$ portion gave a precipitate of barium sulphate of mass $0.058 \, g$ on treatment with excess dilute sulphuric acid. You may assume that Group 2 metal ions form 1:1 complexes with edta.

(Molar volume of any gas at $25\,°C$ and 1 atmosphere pressure = $24 \, dm^3$.) (AEB90)

14. Zinc sulphate can be used as a dietary supplement in cases of suspected zinc deficiency. The compound crystallises as a hydrated salt, and is readily water-soluble.

a) In a simple experiment to determine the extent of hydration, a technician carefully heated $3.715 \, g$ of the crystals to a moderate temperature until no further loss in mass occurred.

The anhydrous salt had a mass of $2.086 \, g$.

Relative atomic masses are included in the Periodic Table on pp. 294–5.
  i) How many moles of zinc sulphate are there in $2.086 \, g$ of anhydrous zinc sulphate?
  ii) How many moles of water were lost?
  iii) What is the value of x in the formula $ZnSO_4 \cdot xH_2O$?

b) The daily recommended intake of zinc in the USA is 15 mg.
  i) What mass of zinc sulphate crystals would need to be taken to obtain this intake?

ii) If this is taken via a 5 cm³ dose of aqueous zinc sulphate, calculate the concentration of this solution in $mol\,dm^{-3}$ of the hydrated salt.

c) The organic ligand ethylenediamine tetra-acetic acid (edta) can be used to titrate Zn^{2+} in solution. The edta complexes the zinc, releasing hydrogen ions: edta works best in this experiment at about pH 10.

  i) 5.932 g of hydrated zinc sulphate was dissolved in water and made up to 200 cm³ of solution. 20.00 cm³ aliquots of this solution required 20.65 cm³ of 0.1000 $mol\,dm^{-3}$ edta for complete reaction. In what mole ratio do Zn^{2+} and edta react?

  ii) Explain why this titration is done in the presence of aqueous ammonia containing dissolved ammonium chloride.

  iii) Why would an attempt to determine the concentration of zinc ions in solution by precipitating zinc hydroxide with excess sodium hydroxide and weighing it be an unsatisfactory method?

(L92)

*15. This question concerns a hydrated potassium manganese ethanedioate complex, and experiments to determine its composition.

Stage 1

5.000 g of a pure sample of the complex were dissolved in dilute sulphuric acid to make 100 cm³ of solution. 10.0 cm³ of this solution were placed in a conical flask, warmed to 60°C, and titrated with 0.0500 $mol\,dm^{-3}$ potassium manganate(VII) until a persistent faint pink colour was observed in the clear solution. This was repeated, and a consistent titre of 23.5 cm³ of the potassium manganate(VII) solution was obtained.

Stage 2

A 50 cm³ portion of the same solution of the complex was treated with excess sodium carbonate solution, and the brown precipitate formed was filtered off. After washing and drying, the precipitate was heated strongly in air until there was no further change in mass; the brownish-black residue of manganese(IV) oxide weighed 0.425 g.

Stage 3

Another pure sample of the complex weighing 5.000 g was dehydrated carefully by gentle warming; the residue weighed 4.648 g.

a) i) Write a balanced equation for the reaction of ethanedioate ions with manganate(VII) ions in acidic solution in Stage 1, given the half equation

$$C_2O_4^{2-}(aq) \longrightarrow 2CO_2(g) + 2e^-$$

  ii) Calculate the mass of ethanedioate ions in the original sample of the complex.

b) i) Explain the chemistry of the steps in Stage 2.

  ii) Calculate the mass of manganese in the original sample of the complex.

c) Using your answers to a) and b), and the results from Stage 3, calculate the formula of the complex.

d) i) What is the oxidation number of manganese in the complex?

ii) Draw a diagram of the displayed structure of the anhydrous complex.   (L92,N,S)

16. a) Describe what you would expect to *observe* on the gradual addition of excess of the following reagents to separate samples of a solution containing copper(II) ions, Cu^{2+}: i) aqueous ammonia; ii) potassium iodide solution. Give the formulae for all the species produced.

Explain the nature of the reaction with potassium iodide solution, giving also a balanced overall chemical equation. *Briefly* indicate also how this reaction could be used as a basis for the volumetric estimation of copper.

b) i) Describe the nature of the bonding in lead(IV) chloride, $PbCl_4$, and compare it with that in lead(II) chloride, $PbCl_2$. Similarly, compare these two compounds with respect to 1) physical properties, 2) thermal stability, and 3) behaviour on treatment with water. Give balanced chemical equations wherever appropriate.

ii) Show clearly how your answers in b) i) 2) and 3) above are consistent with the following numerical data.

1) Gentle heating of 0.5000 g of lead(IV) chloride, allowing volatile products to escape, results in a weight loss of 0.1016 g.

2) Treatment of a further 0.5000 g of lead(IV) chloride with water yields a solid product and a solution which requires 28.65 cm^3 of 0.2000 mol dm^{-3} silver nitrate solution for complete reaction with the chloride ion present. (Here $Ag^+(aq) + Cl^-(aq) \longrightarrow AgCl(s)$.)

($A_r(Pb) = 207.21$, $A_r(Cl) = 35.45$.)

c) You are provided with a solution which is known to contain *either* a chloride *or* a bromide *or* an iodide. Describe and explain *one* test only which would allow you unambiguously to identify the halide present.   (WJEC92)

17. Hydrogen peroxide (H_2O_2) may be prepared in the laboratory by treating barium peroxide (BaO_2) with dilute sulphuric acid. Pure hydrogen peroxide is a pale blue syrupy liquid m.pt. $-1\,^{\circ}C$ and b.pt. $150\,^{\circ}C$.

a) i) Write a balanced equation for the reaction of barium peroxide with sulphuric acid.

ii) How would you remove the barium sulphate produced in the reaction?

b) Hydrogen peroxide decomposes on heating. The dilute solution of hydrogen peroxide, obtained from the reaction of barium peroxide with sulphuric acid, may be concentrated using the apparatus below.

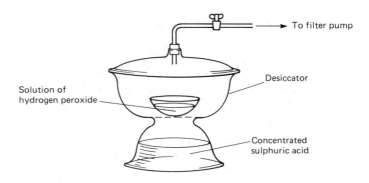

i) Explain why a filter pump is connected to the desiccator.
ii) What is the purpose of the concentrated sulphuric acid?

c) The structure of hydrogen peroxide may be represented by either of the following structures.

$$H-O-O-H$$

$$\begin{array}{c} \\ H \end{array} O-O \begin{array}{c} H \\ \end{array}$$

I   II

i) Using the outer electrons draw a dot and cross diagram for hydrogen peroxide.

ii) Using the dot and cross diagram suggest which of the structures I or II is more likely for hydrogen peroxide. Explain your reasoning.

d) The hydrogen peroxide sold in shops is usually described as '20-volume'. This means that $1 \, cm^3$ of hydrogen peroxide solution produces $20 \, cm^3$ of oxygen. Hydrogen peroxide decomposes on standing and faster with a catalyst.

$$2H_2O_2 \longrightarrow 2H_2O + O_2$$

i) Calculate the number of moles of hydrogen peroxide in $1 \, dm^3$ of a 20-volume solution. (One mole of a gas occupies $24 \, dm^3$ at room temperature and pressure.)

ii) Hydrogen peroxide liberates iodine quantitatively from solutions of acidified potassium iodide.

$$H_2O_2 + 2I^- + 2H^+ \longrightarrow I_2 + 2H_2O$$

The liberated iodine may be titrated with sodium thiosulphate solution.

$$2S_2O_3^{2-} + I_2 \longrightarrow S_4O_6^{2-} + 2I^-$$

A bottle, labelled '20-volume hydrogen peroxide', had been standing on the shelf of a pharmacy for some time. 25.0 cm^3 of this solution were diluted to a total volume of 250 cm^3. A 25.0 cm^3 portion of the diluted solution was acidified and excess potassium iodide added. This treated 25.0 cm^3 portion was titrated against 0.1 M sodium thiosulphate solution and gave an end point of 34.0 cm^3.

Calculate the actual strength, by 'volume', of the solution labelled 20-volume hydrogen peroxide.

e) Describe how a solution of hydrogen peroxide may be used to identify the presence of Mn^{2+} in aqueous solution. State the result expected.   (NI90,p)

9 The Atom

MASS SPECTROMETRY

In a mass spectrometer, an element or compound is vaporised and then ionised. The ions are accelerated, collimated into a beam and deflected by a magnetic field. The amount of the deflection depends on the ratio of mass/charge of the ions, as well as the values of the accelerating voltage and the magnetic field. The magnetic field is kept constant while the accelerating voltage is varied continuously to focus each species in turn into the ion detector. The detector records each species as a peak on a trace. From the value of the voltage associated with a particular peak the ratio of mass/charge for that ionic species can be found. Since each ion has a charge of $+1$, the ratio mass/charge is equal to the mass of the ion. The mass spectrometer can be calibrated to read out ionic masses directly. The heights of the peaks are proportional to the relative abundance of the different ions.

EXAMPLE 1 The mass spectrum of borom shows two peaks, one at 10.0 u, and the other at 11.0 u. The heights of the peaks are in the ratio 18.7% : 81.3%. Calculate the relative atomic mass of boron.

METHOD The relative heights of the peaks show that the relative abundance of ^{10}B and ^{11}B is 18.7% ^{10}B : 81.3% ^{11}B.

In 1000 atoms, there are 187 of mass 10.0 u = 1870 u

and 813 of mass 11.0 u = 8943 u

The mass of 1000 atoms = 10 813 u

The average atomic mass = 10.8 u

ANSWER The relative atomic mass of boron is 10.8.

EXAMPLE 2 The mass spectrum of neon shows three peaks, corresponding to masses of 22, 21 and 20 u. The heights of the peaks are in the ratio 11.2 : 0.2 : 114. Calculate the average atomic mass of neon.

METHOD Multiplying the relative abundance (the height of the peak) by the mass to find the total mass of each isotope present gives

Mass of neon-22 = 11.2×22.0 = 246.4 u

Mass of neon-21 = 0.2×21.0 = 4.2 u

Mass of neon-20 = 114×20.0 = 2280 u

Totals are 125.4 = 2530.6 u

Average mass of neon atom = 2530.6/125.4 = 20.18 u.

ANSWER The average atomic mass of neon is 20.2 u.

EXERCISE 22   Problems on Mass Spectrometry

1. The mass spectrum of rubidium consists of a peak at mass 85 and a peak at mass 87 u. The relative abundance of the isotopes is 72 : 28. Calculate the mean atomic mass of rubidium.

2. If ^{69}Ga and ^{71}Ga occur in the proportions 60 : 40, calculate the average atomic mass of gallium.

3. Fig. 9.1 shows the mass spectrum of magnesium. The heights of the three peaks and the mass numbers of the isotopes are shown in Fig. 9.1. Calculate the relative atomic mass of magnesium.

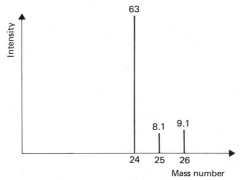

Fig. 9.1   Mass spectrum of magnesium

4. The mass spectrum of chlorine shows peaks at masses 70, 72 and 74 u. The heights of the peaks are in the ratio of 9 to 6 to 1. What is the relative abundance of ^{35}Cl and ^{37}Cl? What is the average atomic mass of chlorine?

5. Calculate the relative atomic mass of lithium, which consists of 7.4% of ^{6}Li and 92.6% of ^{7}Li.

6. A sample of water containing ^{1}H, ^{2}H and ^{16}O was analysed in a mass spectrometer. The trace showed peaks at mass numbers 1, 2, 3, 4, 17, 18, 19 and 20. Suggest which ions are responsible for these peaks.

7. Calculate the average atomic mass of potassium, which consists of 93% ^{39}K and 7.0% ^{41}K.

8. Fig. 9.2 shows a mass spectrometer trace for copper nitrate. Each of the eight peaks is produced by a different species of ion. Suggest what these ions are.

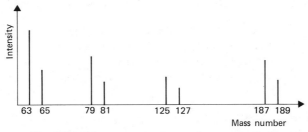

Fig. 9.2   The mass spectrum of copper nitrate

EXERCISE 23   Questions from A-level Papers

1. a)   i) State the main natural source of alkanes.
         ii) State *two* important uses of alkanes.

    b)   i) State the conditions under which methane reacts with chlorine.
         ii) Name the type of reaction which occurs in this case.
         iii) Write the mechanism of this reaction (in b) i)) excluding steps which produce $CHCl_3$ or CCl_4.

    c) The main peaks in the mass spectrum of an alkane, a ketone, and a carboxylic acid are listed below but it is not known which is which.
         i) Assign each molecule to its correct mass spectrum and also write its structural formula. (If isomers exist, only one formula is wanted.) ($A_r(H) = 1$, $A_r(C) = 12$, $A_r(O) = 16$.)

Mass spectrum	m/e	Type of molecule	Structural formula
A	60, 45, 15		
B	58, 43, 28, 15		
C	58, 43, 29, 15		

         ii) Give reasons in *each* case for your assignments in c) i).

    d) The boiling points of the alkane, ketone and carboxylic acid in c) are 0, 56 and 118 °C respectively. Explain why these differ despite the fact that the molecules have very similar relative molecular masses.   (WJEC92)

2. Two isomeric aromatic compounds X and Y have the following percentage composition by mass: $C = 66.4\%$; $H = 5.5\%$; $Cl = 28.1\%$.

   The relative molecular mass of the compounds is 126.5.

   a) Show that the molecular formula for X and Y is C_7H_7Cl.

   b) i) One of the compounds, X, yields a white precipitate when warmed with aqueous silver nitrate, whereas Y does not. What can you deduce from this?
       ii) Suggest a structure for X and give *one* of the possible structures for Y.

   c) Compounds related to Y are used as insecticides. What property leads to these compounds causing environmental damage?

d) Sketch a possible mass spectrum for compound Y on a copy of the axes below and explain, using your sketch, how it would enable you to distinguish between compound X and compound Y (there is no need to draw a possible mass spectrum for X). (L91)

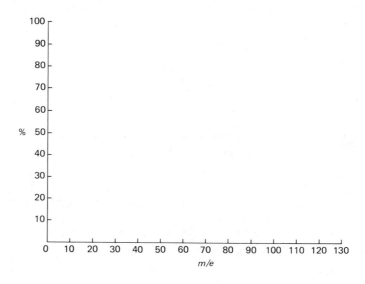

3. a) Identify, and give the main characteristics of, the particles contained in atomic nuclei.

b) Relative atomic and molecular masses are measured on a scale in which $^{12}C = 12$ exactly. Explain what this means and indicate why relative atomic mass is a more useful concept than atomic mass.

c) Chlorine is essentially a mixture of two isotopes, ^{35}Cl and ^{37}Cl. By reference to chlorine explain what is meant by the term *isotope.*

d) The following is the mass spectrum of chlorine, Cl_2.

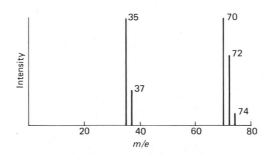

i) Identify each peak in the spectrum.
ii) Suggest a reason why the peaks are not all the same height.

e) Boron B, consists of a mixture of 20% ^{10}B and 80% ^{11}B. On a copy
    of the axes, sketch the mass spectrum of boron.   (L91)

*4. a) Outline the principles of mass spectrometry and indicate some of
        of its applications to modern chemistry.

    b) Bromine consists of two isotopes with mass numbers 79 and 81
        which have percentage abundance of 50.5 and 49.5 respectively.
        i) Calculate the relative atomic mass of bromine.
        ii) What are the m/e values for the Br_2^+ ions in the mass spectrum
            of bromine and what are the relative heights of the corres-
            ponding peaks?

    c) Both ^{14}C and ^{40}K are radioactive and decay by β-emission.

        Briefly state how ^{14}C is used to date archaeological items of animal
        or vegetable origin. Comment on the fact that the age of certain
        minerals may be inferred from the ^{40}K/^{40}Ar ratio.   (AEB91,S)

        (For part c), see Chapter 14.)

5. a) How are the procedures of i) refluxing and ii) distillation carried
        out in the laboratory? Explain the purpose of carrying out the
        procedures by reference to two different chemical reactions.

    b) What do you understand by the term *relative atomic mass*?

    c) The mass spectrum of an organic compound which can be obtained
        by the oxidation of an alcohol is shown below.

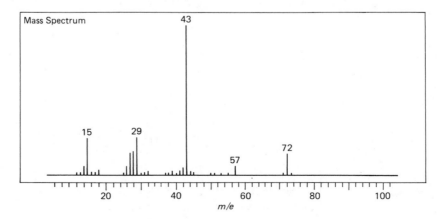

The compound has the following composition by mass:

$$C = 66.7\%$$
$$H = 11.1\%$$
$$O = 22.2\%$$

Calculate the empirical formula of the compound and by interpreting the labelled peaks on the mass spectrum determine the molecular formula of the compound. (L92,AS)

*6. a) Give a detailed account of the design and operation of a mass spectrometer.

b) The figure below shows part of the mass spectrum of CCl_4.

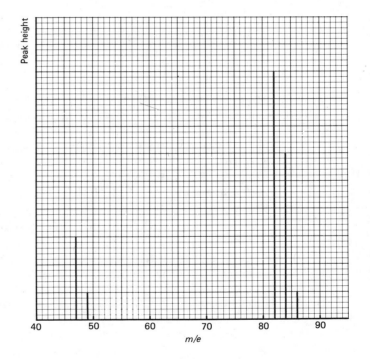

i) Assign the peaks shown to the appropriate ions.
ii) Explain the observed relative peak heights *within* the groups of m/e 82-84-86 and 47-49.
iii) Very small additional peaks (not shown) include those at m/e 83 and 48 which are about 1% of the height of the 82 and 47 peaks respectively. Suggest a reason for these peaks.
iv) A small peak is also seen at m/e 41; suggest a reason for this.

N.B. The peaks in b) iii) and b) iv) are not caused by impurities.
$(A_r(C) = 12.01, \ A_r(Cl) = 35.45.)$

c) If methane is leaked into the mass spectrometer and the electron beam energy slowly increased from zero, the ionisation curves shown in the figure below are obtained.

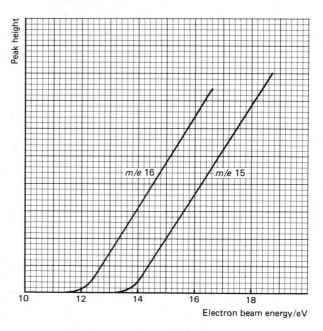

i) Explain the origin and shape of the curves.

ii) Given that the ionisation energy of the methyl radical ($CH_3\cdot \longrightarrow CH_3^+ + e$) is 9.8 eV, use the figure to calculate the bond dissociation energy $D(H_3C-H)$ in methane.

N.B. 1. 1 electron volt (eV), the energy gained by an electron when accelerated through 1 volt, is equivalent to 96.5 kJ mol^{-1}.

2. Only single events occur in the mass spectrometer; electrons do not strike a species twice.

d) High resolution mass spectrometry measures m/e values to several places of decimals. Since isotopic masses differ from whole numbers by different amounts, it is possible to find the molecular formula of a fragment without knowing which elements are present.

The molecular ion of a solid organic molecule, important in living systems, has an m/e value of 75.032 015 and contains all of the isotopes whose exact relative atomic masses are listed below. Deduce the molecular formula and suggest a structure.

N.B. The m/e value given is corrected for the loss of mass of the electron.

$(A_r(^1H) = 1.007\,825, \; A_r(^{12}C) = 12.000\,000,$
$A_r(^{14}N) = 14.003\,07, \; A_r(^{16}O) = 15.994\,91.)$ (WJEC92,S)

NUCLEAR REACTIONS

In a nuclear reaction, a rearrangement of the protons and neutrons in the nuclei of the atoms takes place, and new elements are formed.

The atomic number or proton number, Z, of an element is the number of protons in the nucleus of an atom of the element. The mass number or nucleon number, A, is the number of protons and neutrons in the nucleus of an atom. Isotopes of an element differ in mass number but have the same atomic number. Isotopes are represented as ^A_Z Symbol, e.g. $^{12}_6C$. Protons are represented as ^1_1H, electrons (β-particles) as $^0_{-1}e$, neutrons as ^1_0n, and α-particles as ^4_2He. In the equation for a nuclear reaction, the sum of the mass numbers is the same on both sides, and the sum of the atomic numbers is the same on both sides of the equation.

For practice in balancing nuclear equations, study the following examples.

EXAMPLE 1　Complete the equation

$$^{16}_7N \longrightarrow {}^a_bO + {}^0_{-1}e$$

METHOD　Consider mass numbers:　$16 = a + 0 \quad \therefore a = 16$

Consider atomic numbers:　$7 = b + (-1) \quad \therefore b = 8$

ANSWER　$^{16}_7N \longrightarrow {}^{16}_8O + {}^0_{-1}e$

EXAMPLE 2　Find the values of a and b in the equation

$$^{27}_{13}Al + {}^1_0n \longrightarrow {}^4_2He + {}^a_bX$$

METHOD　Consider mass numbers:　$27 + 1 = 4 + a \quad \therefore a = 24$

Consider atomic numbers:　$13 + 0 = 2 + b \quad \therefore b = 11$

ANSWER　$a = 24$ and $b = 11$.

EXERCISE 24　　Problems on Nuclear Reactions

Complete the following equations, supplying values for the missing mass numbers (nucleon numbers) and atomic numbers (proton numbers).

1. a) $^9_4Be + \gamma \longrightarrow {}^8_4Be + {}^a_bX$

   b) $^{14}_7N + {}^4_2He \longrightarrow {}^1_1H + {}^a_bY$

   c) $^9_4Be + {}^1_1H \longrightarrow {}^6_3Li + {}^a_bZ$

   d) $^{209}_{83}Bi + {}^2_1D \longrightarrow {}^a_bX + {}^1_1H$

   e) $^{16}_8O + {}^1_0n \longrightarrow {}^{13}_6C + {}^a_bY$

   f) $^{10}_5B + {}^a_bY \longrightarrow {}^{13}_7N + {}^1_0n$

   g) $^{14}_7N + {}^1_0n \longrightarrow {}^a_bQ + {}^1_1H$

h) $^{19}_{9}F + ^{1}_{0}n \longrightarrow ^{a}_{b}Z + ^{4}_{2}He$

i) $^{207}_{82}Pb \longrightarrow ^{a}_{b}X + ^{0}_{-1}e$

j) $^{27}_{13}Al + ^{1}_{0}n \longrightarrow ^{24}_{b}Y + ^{a}_{2}Z$

k) $^{35}_{17}Cl + ^{p}_{q}X \longrightarrow ^{r}_{16}S + ^{1}_{1}H$

l) $^{6}_{3}Li + ^{1}_{0}n \longrightarrow ^{4}_{2}He + ^{c}_{d}X$

EXERCISE 25   Questions from A-level Papers

1. a) Complete the following table:
      (m_p = mass of a proton, e = charge of a proton.)

	Mass relative to m_p	Charge relative to e
Alpha particle		
Beta particle		

b) Complete the following table which relates to the changes in the mass number and atomic number of an atomic nucleus when it emits
   i) an alpha particle
   ii) a beta particle.

Particle emitted	Change in mass number	Change in atomic number
Alpha particle		
Beta particle		

c) An isotope of the element uranium, $^{235}_{92}U$, emits successively seven alpha particles and four beta particles to form a stable isotope of another element X. Deduce:
   i) the mass number of X
   ii) the atomic number of X
   iii) the identity of X.   (O90)

2. a)   i) Distinguish clearly between *mass number* and *relative atomic mass*.

ii) Calculate the relative atomic mass of the element magnesium from its mass spectrum below.

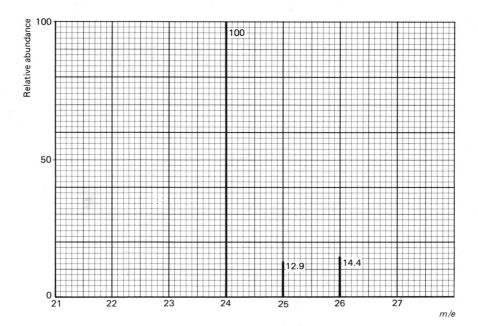

b)   i) Particles lost in radioactive decay are of two types; alpha particles (helium nuclei) and beta particles (high energy electrons). Identify Q and R in the following equations.

$$^{238}U \longrightarrow Q + \text{alpha particle}$$
$$^{40}K \longrightarrow R + \text{beta particle}$$

ii) ^{40}K may lose an electron by a completely different process. Name this process and write an equation to represent the change.   (AEB,90)

3. a) The relative atomic mass of antimony is 121.75. Explain carefully the meaning of this statement.

b) A radioactive isotope of the element thorium $^{232}_{90}Th$ decays according to the following scheme:

$$^{232}_{90}Th \xrightarrow[\text{emission}]{\alpha\text{-particle}} X \xrightarrow[\text{emission}]{\beta\text{-particle}} Y \xrightarrow[\text{emission}]{\beta\text{-particle}} Z$$

Deduce the mass numbers and atomic numbers of X, Y and Z.

	X	Y	Z
Mass number			
Atomic number			

c) A sample of carbon dioxide was prepared from carbon (assume 100% ^{12}C) and isotopically enriched oxygen containing $^{16}O_2$ and $^{18}O_2$ in the molar ratio $4:1$. This sample gave a mass spectrum containing three peaks associated with singly charged molecular ions.

Deduce the relative molecular masses associated with these peaks and their relative intensities.   (O91)

10 Gases

THE GAS LAWS

The behaviour of gases is described by the Gas Laws: Boyle's law, Charles' law, the equation of state for an ideal gas, Graham's law of diffusion, Gay-Lussac's law, Avogadro's law, Dalton's law of partial pressures and the ideal gas equation. We look at each of these in turn. equation. We look at each of these in turn.

BOYLE'S LAW

Boyle's law states that the pressure of a fixed mass of gas at a constant temperature is inversely proportional to its volume:

$$PV = \text{Constant}$$

where P = pressure, V = volume.

CHARLES' LAW

Charles' law states that the volume of a fixed mass of gas at constant pressure is directly proportional to its temperature on the Kelvin scale:

$$\frac{V}{T} = \text{Constant}$$

where T = temperature in kelvins.

Temperature on the Kelvin scale (called the absolute temperature) is obtained by adding 273 to the temperature on the Celsius scale.

$$\text{Temperature (K)} = \text{Temperature (}^{\circ}\text{C)} + 273$$

$$273\,\text{K} = 0\,^{\circ}\text{C}$$

THE EQUATION OF STATE FOR AN IDEAL GAS

Gases which obey Boyle's law and Charles' law are called *ideal* gases. By combining these two laws, the following equation can be obtained. It is called the *equation of state for an ideal gas*:

$$\boxed{\frac{P_1 V_1}{T_1} = \frac{P_2 V_2}{T_2}}$$

A gas has a volume of V_1 at a temperature T_1 and pressure P_1. If the conditions are changed to a pressure P_2 and a temperature T_2, the new volume can be calculated from the equation. It is usual to compare gas volumes at 0 °C and 1 atmosphere (abbreviated as 1 atm). These conditions are referred to as standard temperature and pressure (s.t.p.) or normal temperature and pressure (n.t.p.). Some authors calculate volumes at room temperature (20 °C) and 1 atm. The SI unit of pressure is the pascal.

$$1\,atm \ = \ 1.01 \times 10^5 \, pascals \, (Pa) \ = \ 101 \, kilopascals \, (kPa)$$

$$= \ 1.01 \times 10^5 \, newtons \, per \, square \, metre \, (N\,m^{-2})$$

$$= \ 760 \, mm \, mercury$$

Volumes can be measured in the SI unit, the cubic metre (m^3) or in cubic decimetres (dm^3) or cubic centimetres (cm^3).

$$10^3 \, cm^3 \ = \ 1 \, dm^3 \ = \ 10^{-3} \, m^3$$

Temperatures must be in kelvins.

EXAMPLE   A volume of gas, 265 cm^3, is collected at 70 °C and $1.05 \times 10^5 \, N\,m^{-2}$. What volume would the gas occupy at s.t.p.?

METHOD   The experimental conditions are

$$P_1 \ = \ 1.05 \times 10^5 \, N\,m^{-2}$$

$$T_1 \ = \ 273 + 70 \ = \ 343 \, K$$

$$V_1 \ = \ 265 \, cm^3$$

Standard conditions are

$$P_2 \ = \ 1.01 \times 10^5 \, N\,m^{-2}$$

$$T_2 \ = \ 273 \, K$$

$$\frac{P_1 V_1}{T_1} \ = \ \frac{P_2 V_2}{T_2}$$

$$\frac{1.05 \times 10^5 \times 265}{343} \ = \ \frac{1.01 \times 10^5 \times V_2}{273}$$

$$V_2 \ = \ 219 \, cm^3$$

ANSWER   The volume of gas at s.t.p. would be 219 cm^3.

(*Note* that the pressure and volume are in the same units on both sides of the equation.)

EXERCISE 26   Problems on Gas Volumes

1. Correct the following gas volumes to s.t.p.:
   a) 205 cm^3 at 27 °C and 1.01 × 10^5 N m^{-2}
   b) 355 cm^3 at 310 K and 1.25 × 10^5 N m^{-2}
   c) 5.60 dm^3 at 425 K and 1.75 × 10^5 N m^{-2}
   d) 750 cm^3 at 308 K and 2.00 × 10^4 N m^{-2}
   e) 1.25 dm^3 at 25 °C and 2.14 × 10^5 N m^{-2}

2. A certain mass of an ideal gas has a volume of 3.25 dm^3 at 25 °C and 1.01 × 10^5 N m^{-2}. What pressure is required to compress it to 1.88 dm^3 at the same temperature?

3. An ideal gas occupies a volume 2.00 dm^3 at 25 °C and 1.01 × 10^5 N m^{-2}? What will the volume of gas become at 40 °C and 2.25 × 10^5 N m^{-2}?

4. An ideal gas occupies 2.75 dm^3 at 290 K and 8.70 × 10^4 N m^{-2}. At what temperature will it occupy 3.95 dm^3 at 1.01 × 10^5 N m^{-2}?

5. An ideal gas occupies 365 cm^3 at 298 K and 1.56 × 10^5 N m^{-2}. What will be its volume at 310 K and 1.01 × 10^5 N m^{-2}?

6. Correct the following gas volumes to s.t.p.:
   a) 256 cm^3 of an ideal gas measured at 50 °C and 650 mm Hg
   b) 47.2 cm^3 of an ideal gas measured at 62 °C and 726 mm Hg
   c) 10.0 dm^3 of an ideal gas measured at 200 °C and 850 mm Hg
   d) 4.25 dm^3 of an ideal gas measured at 370 °C and 2.12 atm
   e) 600 cm^3 of an ideal gas measured at 95 °C and 0.98 atm

GRAHAM'S LAW OF DIFFUSION

At constant temperature and pressure, the rate of diffusion of a gas is inversely proportional to the square root of its density:

$$r \propto \frac{1}{\sqrt{\rho}}$$

where r = rate of diffusion and ρ = density.

Comparing the rates of diffusion of two gases A and B gives

$$\frac{r_A}{r_B} = \sqrt{\frac{\rho_B}{\rho_A}}$$

This expression applies to rates of effusion (passage through a small aperture) as well as to diffusion (passage from a region of high concentration to a region of low concentration). It provides a method of measuring molar masses. The molar mass of a gas is proportional to its density (see p. 81: Density = Molar mass/Gas molar volume).

Graham's law can therefore be written as

$$\frac{r_A}{r_B} = \sqrt{\frac{M_B}{M_A}}$$

where M_A and M_B are the molar masses of A and B.

EXAMPLE 1 A gas, A, diffuses through a porous plug at a rate of $1.43\, cm^3 s^{-1}$. Carbon dioxide diffuses through the plug at a rate of $0.43\, cm^3 s^{-1}$. Calculate the molar mass of A.

METHOD   Molar mass of carbon dioxide $= 44.0\, g\, mol^{-1}$

$$\frac{r_{CO_2}}{r_A} = \sqrt{\frac{M_A}{M_{CO_2}}}$$

$$\frac{0.43}{1.43} = \sqrt{\frac{M_A}{44.0}}$$

$$M_A = 4.0$$

ANSWER   The molar mass of A is $4.0\, g\, mol^{-1}$.

EXAMPLE 2 It takes 54.4 seconds for $100\, cm^3$ of a gas, X, to effuse through an aperture, and 36.5 seconds for $100\, cm^3$ of oxygen to effuse through the same aperture. What is the molar mass of X?

METHOD   Since

$$\frac{r_{O_2}}{r_X} = \sqrt{\frac{M_X}{M_{O_2}}}$$

$$\frac{100/36.5}{100/54.4} = \sqrt{\frac{M_X}{32}}$$

$$M_X = 71$$

ANSWER   The molar mass of X is $71\, g\, mol^{-1}$.

EXERCISE 27   Problems on Diffusion and Effusion

1. A certain volume of hydrogen takes $2\, min\, 10\, s$ to diffuse through a porous plug, and an oxide of nitrogen takes $10\, min\, 23\, s$. What is:
   a) the molar mass, b) the formula of the oxide of nitrogen?

2. Plugs of cotton wool, one soaked in concentrated ammonia solution and the other soaked in concentrated hydrochloric acid, are inserted into opposite ends of a horizontal glass tube. A disc of solid ammonium chloride forms in the tube. If the tube is 1 m long, how far from the ammonia plug is the solid deposit?

3. A certain volume of sulphur dioxide diffuses through a porous plug in 10.0 min, and the same volume of a second gas takes 15.8 min. Calculate the relative molecular mass of the second gas.

4. Nickel forms a carbonyl, $Ni(CO)_n$. Deduce the value of n from the fact that carbon monoxide diffuses 2.46 times faster than the carbonyl compound.

5. A certain volume of oxygen diffuses through an apparatus in 60.0 seconds. The same volumes of gases A and B, in the same apparatus under the same conditions, diffuse in 15.0 and 73.5 seconds respectively. Gas A is flammable and gas B turns starch-iodide paper blue. Identify A and B.

6. $25\ cm^3$ of ethane effuses through a small aperture in 40 s. What time is taken by $25\ cm^3$ of carbon dioxide?

7. Xenon diffuses through a pin-hole at a rate of $2.00\ cm^3\ min^{-1}$. At what rate will hydrogen effuse through the same hole at the same temperature and pressure?

8. In 3.00 minutes, $7.50\ cm^3$ of carbon dioxide effuse through a pinhole. What volume of helium would effuse through the same hole under the same conditions in the same time?

9. A mixture of carbon monoxide and carbon dioxide diffuses through a porous diaphragm in one half of the time taken for the same volume of bromine vapour. What is the composition by volume of the mixture?

10. In 4.00 minutes, $16.2\ cm^3$ of water vapour effuse through a small hole. In the same time, $8.1\ cm^3$ of a mixture of NO_2 and N_2O_4 effuse through the same hole. Calculate the percentage by volume of NO_2 in the mixture.

THE GAS MOLAR VOLUME

Avogadro's law (see Chapter 7) states that equal volumes of gases, measured at the same temperature and pressure, contain equal numbers of molecules. It follows that the volume occupied by a mole of gas is the same for all gases. It is called the *gas molar volume* and measures $22.4\ dm^3$ at s.t.p. ($24.0\ dm^3$ at $20\ °C$ and 1 atm).

If the volume occupied by a known mass of gas is known, the molar mass of the gas can be calculated.

EXAMPLE   11.0 g of a gas occupy 5.60 dm³ at s.t.p. What is the molar mass of the gas?

METHOD   Mass of 5.60 dm³ of gas = 11.0 g
Mass of 22.4 dm³ of gas = 11.0 × 22.4/5.60 = 44.0 g

ANSWER   The molar mass of the gas is 44.0 g mol⁻¹.

EXERCISE 28   Problems on Gas Molar Volume

Use $R = 8.314 \, J \, K^{-1} mol^{-1}$; GMV $= 22.41 \, dm^3$ at s.t.p.

1. Calculate the molar mass of a gas which has a density of $1.798 \, g \, dm^{-3}$ at 298 K and 101 kN m⁻².

2. At 273 K and $1.01 \times 10^5 \, N \, m^{-2}$, 2.965 g of argon occupy 1.67 dm³. Calculate the molar mass of the gas.

3. Calculate the volume occupied by 0.250 mol of an ideal gas at $1.01 \times 10^5 \, N \, m^{-2}$ and 20 °C.

4. A volume, 500 cm³ of krypton, measured at 0 °C and $9.8 \times 10^4 \, N \, m^{-2}$, has a mass of 1.809 g. Calculate the molar mass of krypton.

5. What amount (number of moles) of an ideal gas occupies 5.80 dm³ at $2.50 \times 10^5 \, N \, m^{-2}$ and 300 K?

6. Propane has a density of $1.655 \, g \, dm^{-3}$ at 323 K and $1.01 \times 10^5 \, N \, m^{-2}$. Calculate its molar mass.

7. What volume is occupied by 0.250 mole of an ideal gas at 373 K and $1.25 \times 10^5 \, N \, m^{-2}$?

8. An ideal gas occupies 1.50 dm³ at 300 K and $1.25 \times 10^5 \, N \, m^{-2}$. What is the amount (in moles) of gas present?

DALTON'S LAW OF PARTIAL PRESSURES

In a mixture of gases, the total pressure is the sum of the pressures that each of the gases would exert if it alone occupied the same volume as the mixture. The contribution that each gas makes to the total pressure is called the *partial pressure.*

EXAMPLE   3.0 dm³ of carbon dioxide at a pressure of 200 kPa and 1.0 dm³ of nitrogen at a pressure of 300 kPa are introduced into a 1.5 dm³ vessel. What is the total pressure in the vessel?

METHOD   When the carbon dioxide contracts from $3.0\,dm^3$ to $1.5\,dm^3$, the pressure increases from 200 to $200 \times \dfrac{3.0}{1.5}\,kPa$, i.e. $400\,kPa$. The partial pressure of carbon dioxide in the vessel is $400\,kPa$.

When the nitrogen expands from $1.0\,dm^3$ to $1.5\,dm^3$, the pressure decreases from 300 to $300 \times 1.0/1.5 = 200\,kPa$. The partial pressure of nitrogen is $200\,kPa$.

$$\text{Total pressure} = P_{CO_2} + P_{N_2}$$
$$= 400 + 200 = 600\,kPa$$

ANSWER   The total pressure is $600\,kPa$.

EXERCISE 29   Problems on Partial Pressures of Gases

1. Use the following values of the vapour pressure of water at various temperatures.

Temperature	Vapour Pressure/$N\,m^{-2}$
15 °C	1.70×10^3
20 °C	2.33×10^3
25 °C	3.16×10^3
30 °C	4.23×10^3

   a) $200\,cm^3$ of oxygen are collected over water at an atmospheric pressure of $9.80 \times 10^4\,N\,m^{-2}$ and a temperature of $20\,°C$. What is the partial pressure of the oxygen? What will be its volume at s.t.p.?

   b) $250\,cm^3$ of gas are collected over water at an atmospheric pressure of $9.70 \times 10^4\,N\,m^{-2}$ and a temperature of $30\,°C$. What is the partial pressure of the gas? Correct its volume to s.t.p.

   c) What is the volume of 1 mole of nitrogen measured over water at an atmospheric pressure of $9.70 \times 10^4\,N\,m^{-2}$ and a temperature of $25\,°C$?

2. $2.00\,dm^3$ of nitrogen at a pressure of $1.01 \times 10^5\,N\,m^{-2}$ and $5.00\,dm^3$ of hydrogen at a pressure of $5.05 \times 10^5\,N\,m^{-2}$ are injected into a $10.0\,dm^3$ vessel. What is the pressure of the mixture of gases?

3. A mixture of gases at a pressure of $1.01 \times 10^5\,N\,m^{-2}$ contains 25.0% by volume of oxygen. What is the partial pressure of oxygen in the mixture?

4. Into a $5.00\,dm^3$ vessel are introduced $2.50\,dm^3$ of methane at a pressure of $1.01 \times 10^5\,N\,m^{-2}$, $7.50\,dm^3$ of ethane at a pressure of $2.525 \times 10^5\,N\,m^{-2}$ and $0.500\,dm^3$ of propane at a pressure of $2.02 \times 10^5\,N\,m^{-2}$. What is the resulting pressure of the mixture?

5. A mixture of gases at a pressure $7.50 \times 10^4\,N\,m^{-2}$ has the volume composition 40% N_2; 35% O_2; 25% CO_2.
   a) What is the partial pressure of each gas?
   b) What will the partial pressures of nitrogen and oxygen be if the carbon dioxide is removed by the introduction of some sodium hydroxide pellets?

6. A mixture of gases at $1.50 \times 10^5\,N\,m^{-2}$ has the composition 40% NH_3; 25% H_2; 35% N_2 by volume.
   a) What is the partial pressure of each gas?
   b) What will the partial pressures of the other gases become if the ammonia is removed by the addition of some solid phosphorus(V) oxide?

THE IDEAL GAS EQUATION

Gases which obey Boyle's law and Charles' law are called *ideal gases*. Combining these two laws gives the equation:

$$\frac{P \times V}{T} = \text{Constant for a given mass of gas}$$

It follows from Avogadro's law that, if a mole of gas is considered, the constant will be the same for all gases. It is called the universal gas constant, and given the symbol R, so that the equation becomes

$$PV = RT$$

This equation is called the *ideal gas equation*. For n moles of gas, the equation becomes

$$\boxed{PV = nRT}$$

The value of the constant R can be calculated. Consider 1 mole of gas at s.t.p. Its volume is $22.414\,dm^3$. Inserting values of P, V and T in SI units into the ideal gas equation.

$P = 1.0132 \times 10^5\,N\,m^{-2}$ $T = 273.15\,K$
$V = 22.414 \times 10^{-3}\,m^3$ $n = 1\,mol$

gives $1.0132 \times 10^5 \times 22.414 \times 10^{-3} = 1 \times 273.15 \times R$

$$R = 8.314$$

The units of R are PV/nT, i.e.

$$\frac{N\,m^{-2}\,m^3}{mol\,K} = N\,m\,mol^{-1}\,K^{-1} = J\,K^{-1}\,mol^{-1}$$

Thus, $R = 8.314\,J\,K^{-1}\,mol^{-1}$ (joules per kelvin per mole).

THE KINETIC THEORY OF GASES

To explain the gas laws, the kinetic theory of gases was put forward. The kinetic theory considers that the molecules of gas are in constant motion in straight lines. The pressure which the gas exerts results from the bombardment of the walls of the container by the molecules.

The kinetic energy of a molecule $= \frac{1}{2}mc^2$ (m = mass, c = velocity).

The kinetic energy of the gas $= \frac{1}{2}mN\overline{c^2}$ (N = number of molecules, $\overline{c^2}$ = average value of the square of the velocity for all the molecules; $\sqrt{\overline{c^2}}$ = root mean square velocity).

From the kinetic theory can be derived the equation

$$PV = \frac{1}{3}mN\overline{c^2}$$

Since the kinetic energy of the molecules is proportional to T (kelvins)

$$PV = \text{Constant} \times T$$

This is the ideal gas equation. The agreement between theory and experimental results is good support for the kinetic theory.

The kinetic theory can be used to calculate the root mean square velocity of gas molecules.

EXAMPLE Calculate the root mean square velocity of hydrogen molecules at s.t.p.

METHOD 1 Use $M(H_2) = 2.02\,\text{g mol}^{-1}$. In the equation $PV = \frac{1}{3}mN\overline{c^2}$, substitute $PV = RT$ for 1 mole of gas, and $mN = M$, the molar mass of gas in kg. Substituting $mN = 2.02 \times 10^{-3}\,\text{kg mol}^{-1}$ in $\frac{1}{3}mN\overline{c^2} = RT$, gives

$$\overline{c^2} = 3 \times 8.31 \times 273/(2.02 \times 10^{-3})$$
$$\sqrt{\overline{c^2}} = 1.84 \times 10^3\,\text{m s}^{-1}$$

ANSWER The root mean square velocity of hydrogen molecules at s.t.p. is $1.84 \times 10^3\,\text{m s}^{-1}$.

METHOD 2 Use the density of hydrogen ($9.00 \times 10^{-2}\,\text{kg m}^{-3}$ at s.t.p.). Since $mN/V = \rho$, the density of the gas, substituting in $P = \frac{1}{3}\rho\overline{c^2}$ gives

$$\sqrt{\overline{c^2}} = \sqrt{\frac{3 \times 1.01 \times 10^5}{9.00 \times 10^{-2}}} = 1.84 \times 10^3\,\text{m s}^{-1}$$

ANSWER As before, the root mean square velocity is $1.84 \times 10^3\,\text{m s}^{-1}$.

EXERCISE 30   Problems on the Kinetic Theory and the Ideas Gas Equation

Use $R = 8.314\,\text{J K}^{-1}\text{mol}^{-1}$.

1. Krypton has a density of $3.44\,\text{g dm}^{-3}$ at 25 °C and $1.01 \times 10^5\,\text{N m}^{-2}$. Calculate its molar mass.

2. The density of hydrogen at 273 K and $1.01 \times 10^5 \, N\,m^{-2}$ is $8.96 \times 10^{-2} \, g\,dm^{-3}$. Calculate the root mean square velocity of the hydrogen molecules under these conditions.

3. Using the equation $PV = \frac{1}{3}mN\bar{c^2}$ calculate the kinetic energy of the molecules in one mole of an ideal gas at 0 °C.

4. Calculate the root mean square velocity for argon at s.t.p. ($M_r(Ar) = 40.0$).

5. A volume of $1.00 \, dm^3$ is occupied by 1.798 g of a gas at 298 K and 101 kPa. Calculate the molar mass of the gas.

6. Calculate the ratio of the root mean square velocities of oxygen and xenon molecules at 27 °C. ($A_r(O) = 16.0$, $A_r(Xe) = 131.$)

7. Calculate the root mean square velocity of hydrogen iodide molecules at 27 °C. ($A_r(I) = 127.$)

8. a) Calculate the ratio of the root mean square velocity of hydrogen molecules to the root mean square velocity of argon molecules at the same temperature.

   b) At what temperature will argon molecules have the same root mean square velocity as hydrogen molecules at 0 °C? ($A_r(Ar) = 40.0$.)

EXERCISE 31   Questions from A-level Papers

1. On decay one atom of the radium isotope $^{226}_{88}Ra$ emits one α particle which forms an atom of helium gas. A sample of $^{226}_{88}Ra$ produced $4.48 \times 10^{-6} \, dm^3$ of helium measured at 273 K and 1 atm ($1.01 \times 10^5 \, Pa$) pressure, in a given time.

   Calculate the mass of $^{226}_{88}Ra$ which decayed in that time if 1 mol of helium occupies $22.4 \, dm^3$ at 273 K and 1 atm pressure.   (WJEC90,p)

2. a) State the ideal gas equation.

   b) Gas X has a density of $0.714 \, g\,dm^{-3}$ at 273 K and 101.3 kPa, and diffuses twice as quickly as gas Y under identical conditions.
      i) Calculate the relative molecular mass of gas X.
      ii) Name and state the law which relates the rate of diffusion to relative molecular mass.
      iii) Calculate the relative molecular mass of Y.

   c) i) State van der Waals' equation for a real gas.
      ii) Give *two* properties of real gases for which van der Waals' equation attempts to compensate, stating which term in the equation is responsible for the compensation.   (AEB90)

3. a) The ideal gas equation can be written as $pV = nRT$.

   Use this equation to calculate the volume occupied by one mole of an ideal gas at 300 K and 100 kPa pressure.

b) An organic compound, X, contains carbon, hydrogen and oxygen only. When vaporised at 101 kPa and 373 K, 0.100 g of X occupied a volume of 66.7 cm³. Calculate the relative molecular mass of X.

c) On combustion in excess oxygen, 1 mol of X produced 2 mol of carbon dioxide and 3 mol of water.
   i) What is the molecular formula of X?
   ii) Write structures for *two* compounds with this molecular formula.
   iii) Write a balanced equation for the complete combustion of X in oxygen.

d) X is a liquid at room temperature. When X is treated with metallic sodium, hydrogen is evolved.
   i) Use this information to deduce the structure of X.
   ii) Write a balanced equation for the reaction of X with metallic sodium. (AEB91)

4. a) Calculate the value of the product pV (where p is the pressure and V is the volume of gas at a fixed temperature T) for one mole of an ideal gas at 300 K. State the units of this product.

b) The graph shows experimental values of the product pV for a mass m of a certain gas G at a fixed temperature T (the units have been omitted from the graph).

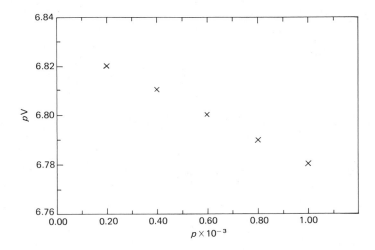

   i) Give *two* properties of molecules which could lead to the product pV for one mole of a real gas being different from that for an ideal gas.
   ii) Which one of these properties results in pV decreasing as p increases as shown in the graph above?
   iii) Using the variables, p, V and T, write an expression for the relative molecular mass M_r which applies to a mass m of an ideal gas. Use the data in the graph above to calculate the

relative molecular mass of the gas G, given that the experimental results were obtained with a sample of 100 g of G at 300 K and that the pressure and volume were measured in kPa and m^3, respectively.   (JMB91)

5. A gaseous hydrocarbon has the composition, by mass: C, 85.7%; H, 14.3%.

a) Calculate the empirical formula of the hydrocarbon.

b) A 0.25 g sample of the hydrocarbon has a volume of 100 cm^3 at s.t.p. Calculate the relative molecular mass and the molecular formula of the hydrocarbon.

c) i) Draw *three* possible full structural formulae for the hydrocarbon.
    At least one of the structures in c) i) is that of an alkene.

   ii) Draw two repeat units of the addition polymer that could be obtained from one of the alkenes.   (C91)

11 Liquids

DETERMINATION OF MOLAR MASS

The gas syringe method

The gas syringe method can be used to find the molar mass of a liquid with a low boiling point. A small weighed quantity of liquid is injected into a gas syringe. The volume of vapour formed is measured, and its temperature and pressure are noted. From the values of mass and volume, the molar mass can be calculated.

EXAMPLE 1 A gas syringe contains 18.4 cm^3 of air at 57 °C. 0.187 g of a volatile liquid is injected into the syringe. The volume of gas in the syringe is then 54.6 cm^3 at 57 °C and 1.01 × 10^5 Pa. Calculate the molar mass of the liquid.

METHOD Using the values
$$P = 1.01 \times 10^5 \, \text{Pa}$$
$$V = 36.2 \, \text{cm}^3 = 36.2 \times 10^{-6} \, \text{m}^3$$
$$T = 273 + 57 = 330 \, \text{K}$$
$$R = 8.314 \, \text{J K}^{-1} \text{mol}^{-1}$$

in the equation $PV = \dfrac{m}{M} RT$ gives

$$1.01 \times 10^5 \times 36.2 \times 10^{-6} = \frac{0.187}{M} \times 8.314 \times 330$$

$$M = 140$$

ANSWER The molar mass is 140 g mol^{-1}.

The values of molar mass obtained by this method are not very accurate. A knowledge of the empirical formula enables the value to be corrected. For example, if the compound has the empirical formula CH_2O and an experimental value of 57 g mol^{-1} for the molar mass, one can see that $C_2H_4O_2$ is the molecular formula, and 60 g mol^{-1} is the correct molar mass.

ANOMALOUS RESULTS FROM MEASUREMENTS OF MOLAR MASS

Sometimes, an unexpectedly low result for molar mass is obtained. This happens when the molecules of the vapour on which measurements are being made dissociate, causing an increase in the actual number of particles present. If 1 mole of molecules of XY dissociate partially into X and Y, and α is the degree of dissociation, then

Species: XY $\rightleftharpoons$ X + Y

Number of moles: $(1 - \alpha)$ α α Total = $(1 + \alpha)$

$(1 - \alpha)$ moles of XY remain, and α moles of X and α moles of Y are formed.

Thus $\dfrac{\text{Actual number of moles}}{\text{Expected number of moles}} = \dfrac{1 + \alpha}{1}$

Since the volume occupied by a gas is proportional to the number of moles of gas,

$$\frac{\text{Actual volume of gas}}{\text{Expected volume of gas}} = \frac{1 + \alpha}{1}$$

Since we are finding molar mass from the equation, given on p. 80,

$$PV = nRT = \frac{m}{M} RT$$

where m = mass of substance, and M = its molar mass, if the volume, V, is greater than expected, M, the molar mass, is less than expected.

Thus

$$\frac{\text{Actual volume}}{\text{Expected volume}} = \frac{\text{Molar mass calculated from formula}}{\text{Measured molar mass}} = \frac{1 + \alpha}{1}$$

If the volume is kept constant, the pressure increases instead of the gas expanding and

$$\frac{\text{Calculated molar mass}}{\text{Measured molar mass}} = \frac{\text{Measured pressure}}{\text{Calculated pressure}} = \frac{1 + \alpha}{1}$$

If one molecule dissociates into n particles, the expression becomes:

$$\frac{\text{Calculated molar mass}}{\text{Measured molar mass}} = \frac{\text{Measured pressure}}{\text{Calculated pressure}} = \frac{1 + (n - 1)\alpha}{1}$$

EXAMPLE 1   The molar mass of phosphorus(V) chloride at 140 °C is 166. Calculate the degree of dissociation.

METHOD   The molar mass of PCl_5 = $31.0 + (5 \times 31.5)$ = 208.5 g mol^{-1}
The dissociation which occurs is

$$PCl_5(g) \rightleftharpoons PCl_3(g) + Cl_2(g)$$

$$\frac{\text{Calculated molar mass}}{\text{Measured molar mass}} = \frac{1 + \alpha}{1}$$

Thus, $n = 2$, and

$$\frac{\text{Calculated molar mass}}{\text{Measured molar mass}} = \frac{208.5}{166} = \frac{1 + \alpha}{1}$$

Therefore α = 0.26 (26%).

ANSWER   The degree of dissociation is 0.26.

EXAMPLE 2 When $1.00\,g$ of iodine is heated at $1200\,°C$ in a $500\,cm^3$ vessel a pressure of $1.51 \times 10^2\,kPa$ develops. Calculate the degree of dissociation.

METHOD   Since

$$PV = nRT,$$

$$P = \frac{1.00}{254} \times \frac{8.314 \times 1473}{500 \times 10^{-6}}$$

$$= 9.64 \times 10^4\,Pa$$

$$\frac{\text{Observed pressure}}{\text{Calculated pressure}} = \frac{1.51 \times 10^5}{9.64 \times 10^4} = \frac{1 + (n-1)\alpha}{1}$$

Since the dissociation is

$$I_2(g) \rightleftharpoons 2I(g)$$

$$n = 2 \quad \text{and} \quad \frac{1.51 \times 10^5}{9.60 \times 10^4} = 1 + \alpha$$

Solving this equation gives $\alpha = 0.58$ (or 58%).

ANSWER   The degree of dissociation is 0.58.

A measurement of molar mass higher than the value calculated from the formula is a sign that molecules are associated. In 1 mole of A, if 2 molecules of A form a dimer, and if the degree of dimerisation is α,

Species: $2A \rightleftharpoons A_2$

No. of moles: $(1 - \alpha)$ $\alpha/2$ Total $= (1 - \alpha/2)$

$$\frac{\text{Actual no. of moles}}{\text{Expected no. of moles}} = \frac{1 - \alpha/2}{1}$$

$$\frac{\text{Actual volume}}{\text{Calculated volume}} = 1 - \alpha/2$$

$$\frac{\text{Calculated molar mass}}{\text{Measured molar mass}} = 1 - \alpha/2$$

In general, if n molecules associate,

$$\frac{\text{Calculated molar mass}}{\text{Measured molar mass}} = 1 - \frac{(n-1)\alpha}{n}$$

EXAMPLE 3 A value of 200 is obtained for the molar mass of aluminium chloride. Calculate the degree of dimerisation of aluminium chloride at the temperature at which the measurement was made.

METHOD   Calculated molar mass $= 133.5\,\mathrm{g\,mol^{-1}}$

$$\frac{\text{Calculated molar mass}}{\text{Measured molar mass}} = 1 - \frac{\alpha}{2}$$

$$133.5/200 = 1 - \frac{\alpha}{2}$$

$$\alpha = 0.67$$

ANSWER   The degree of association is 0.67.

EXERCISE 32   Problems on Molar Masses of Volatile Substances

The gas constant, $R = 8.314\,\mathrm{J\,K^{-1}\,mol^{-1}}$.
$1\,\mathrm{atm} = 1.01 \times 10^5\,\mathrm{N\,m^{-2}} = 1.01 \times 10^5\,\mathrm{Pa} = 101\,\mathrm{kPa}$.

1. Calculate the molar mass of a liquid B, given that 0.850 g of B produced 55.5 cm^3 of vapour (corrected to s.t.p).

2. A compound of phosphorus and fluorine contains 24.6% by mass of phosphorus. 1.000 g of this compound has a volume of 178 cm^3 at s.t.p. Deduce the molecular formula of the compound.

3. 0.110 g of a liquid produced 42.0 cm^3 of vapour, measured at 147 °C and $1.01 \times 10^5\,\mathrm{N\,m^{-2}}$. What is the molar mass of the liquid?

4. 0.228 g of liquid was injected into a gas syringe. The volume of vapour formed was 84.0 cm^3 at 17 °C and $1.01 \times 10^5\,\mathrm{N\,m^{-2}}$. Calculate the molar mass of the substance.

5. 0.452 g of a volatile solid displaced 82.0 cm^3 of air, collected at 20 °C and $1.023 \times 10^5\,\mathrm{N\,m^{-2}}$. If the saturated vapour pressure of water at 20 °C is $2.39 \times 10^3\,\mathrm{N\,m^{-2}}$, calculate the molar mass of the solid.

6. Fig. 11.1 shows the results of gas syringe measurements on ethanol ($\circ$), propanone ($\square$) and ethoxyethane ($\bullet$), all at 80 °C and 1 atm. For each liquid, several measurements of the volume of vapour formed after the injection of a known mass of liquid were made.

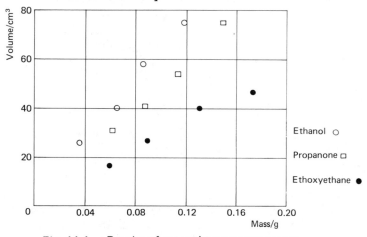

Fig. 11.1   Results of gas syringe measurements

Trace the results on to a piece of paper, and plot the best straight line through the points for each vapour. Find the slope of each line.

The reciprocal of the slope, mass/volume, is the density of the vapour, ρ. From the equation

$$PV = \frac{m}{M}RT$$

$$\left(\text{since } \frac{m}{V} = \rho\right)$$

$$M = \rho\frac{RT}{P}$$

a) Insert the value you have obtained for the density of ethanol vapour into the equation, and find the molar mass of ethanol. Do the same for b) propanone, and c) ethoxyethane.

EXERCISE 33   Problems on Association and Dissociation

1. 20.85 g of phosphorus(V) chloride are allowed to vaporise in a 5.00 dm³ vessel at 175 °C. A pressure of 1.04×10^5 N m⁻² develops. Calculate the degree of dissociation of PCl_5 into PCl_3 and Cl_2.

2. 10.32 g of aluminium chloride are allowed to vaporise in a 1.00 dm³ vessel at 80 °C. A pressure of 1.70×10^5 N m⁻² develops. What is the degree of association of $AlCl_3$ into Al_2Cl_6 molecules?

3. Nitrogen dioxide exists in an equilibrium mixture:
$$N_2O_4(g) \rightleftharpoons 2NO_2(g)$$
The relative molar mass of nitrogen dioxide at 25 °C is 80.0. What percentage of the molecules in the mixture is N_2O_4?

4. A sample of iodine of mass 25.4 g is vaporised in a 2.00 dm³ vessel at 800 K. A pressure of 4.32×10^5 N m⁻² develops. Calculate the degree of dissociation of iodine molecules into atoms.

5. The molar mass of iron(III) chloride measured at 900 K is 246 g mol⁻¹. Calculate the degree of dimerisation of $FeCl_3$ molecules.

VAPOUR PRESSURE

In a liquid, the molecules are in constant motion. Some molecules, those with energy considerably above average, will have enough energy to escape from the liquid into the vapour state. If a liquid is introduced into a closed container, some of the liquid will evaporate. The molecules in the vapour state will exert a pressure. When equilibrium is reached between the liquid state and the vapour state, the pressure exerted by the vapour is called the *vapour pressure* of the liquid. To be correct, one should call it the *saturated vapour pressure* or the *equilibrium vapour pressure*. The magnitude of the vapour pressure depends on the identity of the liquid and on the temperature: it does not depend on the amount of liquid present.

EXAMPLE   The saturated vapour pressure of water at $65\,°C$ is $25.05\,kN\,m^{-2}$. What mass of water will be present in the vapour phase if $10.0\,cm^3$ of water are injected into a $1.000\,dm^3$ vessel?

METHOD   Use the ideal gas equation, $PV = nRT$, and substitute

$$P = 25.05 \times 10^3\,N\,m^{-2} \qquad R = 8.314\,J\,K^{-1}\,mol^{-1}$$

$$T = 338\,K \qquad V = 1.000\,dm^3 = 1.000 \times 10^{-3}\,m^3$$

giving $25.05 \times 10^3 \times 1.000 \times 10^{-3} = n \times 8.314 \times 338$

Amount (mol) of water, $n = 8.92 \times 10^{-3}\,mol$

Mass of water $= 18.0 \times 8.92 \times 10^{-3} = 0.161\,g$

ANSWER   The mass of water that evaporates is $0.161\,g$.

EXERCISE 34   Problems on Vapour Pressures of Liquids

1. $10.0\,cm^3$ of ethyl ethanoate are introduced into an evacuated $10.0\,dm^3$ vessel at $25\,°C$. What mass of ethyl ethanoate will vaporise? The saturated vapour pressure of ethyl ethanoate at $25\,°C$ is $9.55 \times 10^3\,N\,m^{-2}$.

2. At $95\,°C$, the saturated vapour pressure of bromobenzene is $1.54 \times 10^4\,N\,m^{-2}$. What mass of bromobenzene will vaporise when a small amount of liquid bromobenzene is introduced into a $2.50\,dm^3$ flask at $95\,°C$?

3. At $0\,°C$, the saturated vapour pressure of water is $6.10 \times 10^2\,N\,m^{-2}$. How many molecules of water vapour will be present in each cm^3 of air in a vessel containing ice at $0\,°C$?

4. If analysis shows that $0.0230\,g$ of water are present in $1.00\,dm^3$ of air at $25\,°C$, what is the saturated vapour pressure of water at $25\,°C$?

SOLUTIONS OF SOLIDS IN LIQUIDS: OSMOTIC PRESSURE

A semipermeable membrane is a film of material which can be penetrated by a solvent but not by a solute. When two solutions are separated by a semipermeable membrane, solvent passes from the more dilute to the more concentrated. This phenomenon is called *osmosis*. The pressure which must be applied to a solution to prevent the solvent from diffusing in is called the osmotic pressure of the solution. There is an analogy with gas pressure. One mole of a solid, A, when vaporised, occupies a volume of $22.4\,dm^3$ at $0\,°C$ and $1.01 \times 10^5\,N\,m^{-2}$. One mole of A dissolved in $22.4\,dm^3$ of solvent at $0\,°C$ exerts an osmotic pressure of $1.01 \times 10^5\,N\,m^{-2}$.

> 1 mole of solute in 22.4 dm³ of solvent at 0 °C has an osmotic
> pressure of $1.01 \times 10^5 \, N \, m^{-2}$ (1 atmosphere).

The expression which relates osmotic pressure to concentration and
temperature is similar to the Ideal Gas Equation,

$$\pi V = nRT$$

where π is the osmotic pressure, V is the volume, T is the temperature
(Kelvin), n is the amount (in mol) of solute, and R is a constant which
has the same value as the gas constant, $8.314 \, J \, K^{-1} \, mol^{-1}$. This
equation is obeyed by ideal solutions.

The osmotic pressure of a solution depends on the concentration of
solute present: it is a colligative property. Measurements of osmotic
pressure can be used to give the molar masses of solutes.

EXAMPLE   Calculate the molar mass of a solute, given that 35.0 g of the solute
in 1.00 dm³ water have an osmotic pressure of $5.15 \times 10^5 \, N \, m^{-2}$ at
20 °C.

METHOD $$\pi V = nRT$$

where $\pi = 5.15 \times 10^5 \, N \, m^{-2}$, $V = 1.00 \times 10^{-3} \, m^3$,
$R = 8.314 \, J \, K^{-1} \, mol^{-1}$, and $T = 293 \, K$.

$\therefore \qquad 5.15 \times 10^5 \times 1.00 \times 10^3 = n \times 8.314 \times 293$

$$n = 0.211$$

$$n = \frac{35.0}{M}$$

$$M = 165 \, g \, mol^{-1}.$$

ANSWER   The solute has a molar mass of $166 \, g \, mol^{-1}$.

EXERCISE 35   Problems on Osmotic Pressure

1. Find the osmotic pressure of the following aqueous solutions at 25 °C:
   a) a sucrose solution of concentration 0.213 mol dm⁻³
   b) a solution containing 144 g dm⁻³ of glucose
   c) a solution containing 12.0 g of urea in 200 cm³ of solution.

2. Find the molar masses of the following solutes:
   a) 1.50 g of A in 200 cm³ of aqueous solution, having an osmotic
   pressure of $2.66 \times 10^5 \, N \, m^{-2}$ at 20 °C

b) 20.0 g of B in 100 cm^3 of aqueous solution, having an osmotic pressure of 3.00×10^6 N m^{-2} at 27 °C

c) 5.00 g of C in 200 cm^3 of solution, having an osmotic pressure of 2.39×10^5 N m^{-2} at 25 °C.

3. A polysaccharide has the formula $(C_{12}H_{22}O_{11})_n$. A solution containing 5.00 g dm^{-3} of the sugar has an osmotic pressure of 7.12×10^2 N m^{-2} at 20 °C. Find n in the formula.

4. A solution of PVC $(CH_2CHCl)_n$, in dioxan has a concentration of 4.00 g dm^{-3} and an osmotic pressure of 65 N m^{-2} at 20 °C. Calculate the value of n.

5. Calculate the osmotic pressure of an aqueous solution containing 25.0 g dm^{-3} of a protein of relative molecular mass 5.00×10^4 at 27 °C.

6. A solution of 2.00 g of a polymer in 1 dm^3 of water has an osmotic pressure of 300 N m^{-2} at 20 °C. Calculate the molar mass of the polymer.

7. The osmotic pressure of blood is 7 atm at 37 °C. What is the concentration of the sodium chloride solution which has the same osmotic pressure as that of blood at normal body temperature (37 °C)?

8. At 20 °C the osmotic pressure of an aqueous solution containing 3.221×10^{-3} g cm^{-3} of an enzyme was found to be 5.637×10^2 N m^{-2}. What is the relative molecular mass of the enzyme?

SOLUTIONS OF LIQUIDS IN LIQUIDS

How do you express the composition of a liquid–liquid mixture? One way is by stating the mole fraction of each constituent:

$$\text{Mole fraction of A in A–B mixture} = \frac{\text{No. of moles of A}}{\text{Total no. of moles}}$$

$$= \frac{n_A}{n_A + n_B}$$

The vapour above a mixture of the liquids A and B will contain both A and B.

Raoult's law states that the saturated vapour pressure of each component in the mixture is equal to the product of the mole fraction of that component and the saturated vapour pressure of that component when pure, at the same temperature.

If p_A = vapour pressure of A,

p_A^0 = saturated vapour pressure of pure A,

and x_A = mole fraction of A, then

$$p_A = x_A \times p_A^0 \qquad p_B = x_B \times p_B^0$$

Raoult's law is obeyed by mixtures of similar compounds. They are said to form *ideal solutions*. The vapour above a mixture of liquids does not have the same composition as the mixture. It is richer in the more volatile component. The mole fractions of A and B in the vapour phase are in the ratio of their mole fractions in the liquid phase multiplied by the ratio of the saturated vapour pressures of the two liquids. If x_A' and x_B' are the mole fractions of A and B in the vapour phase,

$$\frac{x_A'}{x_B'} = \frac{x_A}{x_B} \times \frac{p_A^0}{p_B^0}$$

EXAMPLE 1 Calculate the vapour pressure of a solution containing 50.0 g heptane and 38.0 g octane at 20 °C. The vapour pressures of the pure liquids at 20 °C are heptane 473 Pa; octane 140 Pa.

METHOD Amount (mol) of heptane = 50/100 = 0.50 mol

Amount (mol) of octane = 38/114 = 0.33 mol

Mole fraction of heptane = 0.50/0.83

Mole fraction of octane = 0.33/0.83

$$p\,(\text{heptane}) = p^0(\text{heptane}) \times x(\text{heptane})$$

$$= 473 \times 0.50/0.83 = 284.0$$

$$p\,(\text{octane}) = p^0(\text{octane}) \times x(\text{octane})$$

$$= 140 \times 0.33/0.83 = 55.9$$

ANSWER Total vapour pressure = 284.0 + 55.9 = 340 Pa.

EXAMPLE 2 Two pure liquids A and B have vapour pressures $1.50 \times 10^4\,\text{N m}^{-2}$ and $3.50 \times 10^4\,\text{N m}^{-2}$ at 20 °C. If a mixture of A and B obeys Raoult's law, calculate the mole fraction of A in a mixture of A and B which has a total vapour pressure of $2.90 \times 10^4\,\text{N m}^{-2}$ at 20 °C.

METHOD If n_A is the mole fraction of A, $(1 - n_A)$ is the mole fraction of B. Then, $(n_A \times 1.50 \times 10^4) + (1 - n_A)(3.50 \times 10^4) = 2.90 \times 10^4$

$$1.50n_A + 3.50 - 3.50n_A = 2.90$$

$$2.00n_A = 0.60$$

$$n_A = 0.30$$

ANSWER The mole fraction of A is 0.30.

EXERCISE 36   Problems on Vapour Pressures of Solutions of Two Liquids

1. Two pure liquids A and B have vapour pressures of respectively 17000 and 35000 N m^{-2} at 25 °C. An equimolar mixture of A and B has a vapour pressure of 26 000 N m^{-2} at 25 °C. Calculate the vapour pressure of a mixture containing four moles of A and one mole of B at 25 °C.

2. Hexane and heptane are totally miscible and form an ideal two component system. If the vapour pressures of the pure liquids are 56000 and 24000 N m^{-2} at 50 °C calculate:
   a) the total vapour pressure, and
   b) the mole fraction of heptane in the vapour above an equimolar mixture of hexane and heptane.

3. The vapour pressure of water at 298 K is 3.19×10^3 Pa. What are the partial vapour pressures of water in mixtures of:
   a) 27 g water and 69 g ethanol
   b) 9.0 g of water and 92 g of ethanol
   at this temperature?

4. A and B are two miscible liquids which form an ideal solution. The vapour pressures at 20 °C are: A, 40 kPa, B, 32 kPa. Calculate the total pressure of the vapour in equilibrium with mixtures of:
   a) 3 moles of A and 1 mole of B at 20 °C
   b) 1 mole of A and 4 moles of B at 20 °C.

IMMISCIBLE LIQUIDS: SUM OF VAPOUR PRESSURES

Steam distillation

In a system of immiscible liquids, each liquid exerts its own vapour pressure independently of the other. The vapour pressure of the system is equal to the sum of the vapour pressures of the pure components. This is the basis for steam distillation. Phenylamine will distil over in steam at 98 °C, although its boiling point is 184 °C. At 98 °C, the sum of the vapour pressures of phenylamine and water is equal to atmospheric pressure. The ratio of the amounts of the two liquids in the distillate is equal to the ratio of their vapour pressures:

$$\frac{n_A}{n_W} = \frac{P_A}{P_W}$$

where n_A and n_W are the amounts of phenylamine and water in the distillate, and p_A and p_W are the vapour pressures of phenylamine and water at 98 °C.

Since $n = m/M$ (where m = mass, M = molar mass)

$$\frac{m_A}{M_A} \times \frac{M_W}{m_W} = \frac{p_A}{p_W}$$

This equation can be used to find m_A/m_W, the ratio of masses of amine and water in the distillate. Steam distillation has been used as a method of determining molar masses. In this case, the masses of the liquid and water in the distillate must be measured and inserted into the equation to give the unknown molar mass.

EXAMPLE   Bromobenzene distils in steam at 95 °C. The vapour pressures of bromobenzene and water at 95 °C are $1.59 \times 10^4 \, N\,m^{-2}$ and $8.50 \times 10^4 \, N\,m^{-2}$. Calculate the percentage by mass of bromobenzene in the distillate.

METHOD   Let the percentage of bromobenzene $= y$.

In the equation $\dfrac{n_{C_6H_5Br}}{n_{H_2O}} = \dfrac{p_{C_6H_5Br}}{p_{H_2O}}$

$$\frac{y/157}{(100-y)/18} = \frac{1.59 \times 10^4}{8.50 \times 10^4}$$

$$y = 62.0$$

ANSWER   The distillate contains 62.0% by mass of bromobenzene.

EXERCISE 37   Problems on Steam Distillation

1. The liquid A distils in steam. At the boiling point, the partial pressures of the two liquids are $A = 6.59 \times 10^3 \, N\,m^{-2}$; $H_2O = 9.44 \times 10^4 \, N\,m^{-2}$. If the molar mass of A is $95 \, g\,mol^{-1}$, what is the percentage by mass of A in the distillate?

2. Phenylamine, $C_6H_5NH_2$, distils in steam at 98 °C and $1.01 \times 10^5 \, N\,m^{-2}$. If the saturation vapour pressure of water is $9.40 \times 10^4 \, N\,m^{-2}$, what is the percentage by mass of phenylamine in the distillate?

3. Naphthalene, $C_{10}H_8$, distils in steam at 98 °C and $1.01 \times 10^5 \, N\,m^{-2}$. If the vapour pressure of water is $9.50 \times 10^4 \, N\,m^{-2}$, calculate the mass of distillate that contains 10.0 g of naphthalene.

DISTRIBUTION OF A SOLUTE BETWEEN TWO IMMISCIBLE SOLVENTS

Consider a solid which is appreciably soluble in both of a pair of immiscible liquids. When the solid is shaken with the two liquids, it distributes itself between the two layers. It is found that the ratio of the solute concentrations in the two layers is always the same. If c_U and c_L are the concentrations in the upper and lower layers, then

$$c_U/c_L = k$$

The constant k is called the *partition coefficient* or *distribution coefficient*, and is constant for a particular temperature.

EXAMPLE 1   The partition coefficient for iodine between water and carbon disulphide at 20 °C is 2.43×10^{-3}. A $100\,cm^3$ sample of a solution of iodine in water, of concentration $1.00 \times 10^{-3}\,mol\,I_2\,dm^{-3}$ is shaken with $10.0\,cm^3$ of carbon disulphide. What fraction of the iodine is extracted by carbon disulphide?

METHOD   Let x be the number of moles of I_2 extracted by CS_2.

No. of moles of I_2 (total) $= 100 \times 10^{-3} \times 10^{-3} = 1.00 \times 10^{-4}$

Use $c_U/c_L = k$ (c_U for water layer, c_L for CS_2 layer):

Then, $$\frac{(1.00 \times 10^{-4}) - x}{100} \bigg/ \frac{x}{10.0} = 2.43 \times 10^{-3}$$

$$(1.00 \times 10^{-4}) - x = 2.43 \times 10^{-2} x$$

giving $$x = 9.76 \times 10^{-5}$$

ANSWER   Fraction of iodine extracted $= (9.76 \times 10^{-5})/(1.00 \times 10^{-4})$
$= 0.98.$

EXAMPLE 2   The partition coefficient of X between ether and water is 25.0 at 20 °C. Calculate the mass of X extracted from a solution containing $10.0\,g$ of X in $1.00\,dm^3$ of water by a) $100\,cm^3$ of ether, b) two successive portions of $50.0\,cm^3$ of ether.

METHOD   Let the mass of X extracted by $100\,cm^3$ of ether be m_1.

Use $c_U/c_L = k$ (c_U for ether layer, c_L for water layer)

Concn of X in ether $= m_1/100\,g\,cm^{-3}$

Concn of X in water $= (10.0 - m_1)/1000\,g\,cm^{-3}$

$$\frac{m_1}{100} \bigg/ \frac{(10.0 - m_1)}{1000} = 25.0$$

giving $m_1 = 7.14\,g$.

Let $m_2 =$ mass of X extracted by the first $50.0\,cm^3$ of ether, and $m_3 =$ mass of X extracted by the second $50.0\,cm^3$ of ether.

Then

$$\frac{m_2}{50.0} \bigg/ \frac{(10.0 - m_2)}{1\,000} = 25.0$$

giving $m_2 = 5.55\,\text{g}$.

If $5.55\,\text{g}$ of X are extracted by ether, $4.45\,\text{g}$ remain in the aqueous solution.

$$\therefore \qquad \frac{m_3}{50} \bigg/ \frac{(4.45 - m_3)}{1\,000} = 25.0$$

giving $m_3 = 2.47\,\text{g}$.

ANSWER Total mass of X extracted by ether in two portions $= 5.55\,\text{g} + 2.47\,\text{g}$ $= 8.02\,\text{g}$.

(*Note* that this is greater than the value of $7.14\,\text{g}$ calculated for the mass of X extracted by using all the ether at once.)

Partition can be used to investigate an equilibrium in aqueous solution between a covalent species and an ionic species, for example, the equilibrium

$$I_2(aq) \;+\; I^-(aq) \;\rightleftharpoons\; I_3^-(aq)$$

Only the covalent I_2 molecules will dissolve in an organic solvent. If an aqueous solution of iodine in iodide ions is shaken with an organic solvent, the concentration of iodine in the solvent can be measured and divided by the partition coefficient to give the concentration of iodine molecules in the aqueous layer. The concentration of iodine combined as I_3^- ions is obtained by subtracting the free iodine from the total iodine concentration. The concentration of I^- ions is obtained by subtracting $[I_3^-]$ from the original concentration of I^- ions.

EXAMPLE 3 Iodine is dissolved in water containing $0.160\,\text{mol dm}^{-3}$ of potassium iodide, and the solution is shaken with tetrachloromethane. The concentration of iodine in the aqueous layer was found to be $0.080\,\text{mol dm}^{-3}$; that in the organic layer $0.100\,\text{mol dm}^{-3}$. The partition coefficient for iodine between tetrachloromethane and water is 85. Calculate the equilibrium constant for the reaction:

$$I_2(aq) \;+\; I^-(aq) \;\rightleftharpoons\; I_3^-(aq).$$

METHOD Since $[I_2]$ in $CCl_4 = 0.100\,\text{mol dm}^{-3}$

$[I_2]$ free in water $= 0.100/85 = 0.001\,18\,\text{mol dm}^{-3}$

$[I_2]$ total $= 0.080\,\text{mol dm}^{-3}$

$[I_2]$ combined as $I_3^- = 0.080 - 0.001\,18 = 0.0788\,\text{mol dm}^{-3}$

$[I^-]$ total $= 0.160\,\text{mol dm}^{-3}$

$[I^-]$ free $= 0.160 - 0.0788 = 0.0812\,\text{mol dm}^{-3}$

Putting these values into the expression

$$\frac{[I_3^-]}{[I_2][I^-]} = K$$

gives

$$K = \frac{0.0788}{0.001\,18 \times 0.0812} = 822\,\text{mol}^{-1}\,\text{dm}^3$$

ANSWER   The equilibrium constant is $820\,\text{mol}^{-1}\,\text{dm}^3$.

EXERCISE 38   Problems on Partition

1. X is 12.0 times more soluble in trichloromethane than in water. What mass of X will be extracted from $1.00\,\text{dm}^3$ of an aqueous solution containing 25.0 g by shaking with $100\,\text{cm}^3$ of trichloromethane?

2. The partition coefficient of Y between ethoxyethane (ether) and water is 80. If $200\,\text{cm}^3$ of an aqueous solution containing 5.00 g of Y is shaken with $50.0\,\text{cm}^3$ of ethoxyethane, what mass of Y is extracted from the solution?

3. Z is allowed to reach an equilibrium distribution between the liquids ethoxyethane and water. The ether layer is $50.0\,\text{cm}^3$ in volume and contains 4.00 g of Z. The aqueous layer is $250\,\text{cm}^3$ in volume and contains 1.00 g of Z. What is the partition coefficient of Z between ethoxyethane and water?

4. $500\,\text{cm}^3$ of an aqueous solution of concentration $0.120\,\text{mol dm}^{-3}$ is shaken with $50.0\,\text{cm}^3$ of ethoxyethane. The partition coefficient of the solute between ethoxyethane and water is 60.0. Calculate the amount (in mol) of solute which will be extracted by ethoxyethane.

5. The distribution coefficient of A between ethoxyethane and water is 90. An aqueous solution of A with a volume of $500\,\text{cm}^3$ contains 5.00 g. What mass of A will be extracted by:
   a) $100\,\text{cm}^3$ of ethoxyethane, and
   b) two successive portions of $50.0\,\text{cm}^3$ of ethoxyethane?

6. An organic acid is allowed to reach an equilibrium distribution in a separating funnel containing $50.0\,\text{cm}^3$ of ethoxyethane and $500\,\text{cm}^3$ of water. On titration, $25.0\,\text{cm}^3$ of the ethoxyethane layer required $22.5\,\text{cm}^3$ of $1.00\,\text{mol dm}^{-3}$ sodium hydroxide solution, and $25.0\,\text{cm}^3$ of the aqueous layer required $9.0\,\text{cm}^3$ of $0.100\,\text{mol dm}^{-3}$ sodium hydroxide solution. Calculate the partition coefficient for the acid between ethoxyethane and water.

7. A small amount of iodine is shaken in a separating funnel containing 50.0 cm³ of tetrachloromethane and 500 cm³ of water. On titration, 25.0 cm³ of the aqueous layer require 6.7 cm³ of a 0.0550 mol dm⁻³ solution of sodium thiosulphate. 25.0 cm³ of the organic solvent require 27.2 cm³ of 1.15 mol dm⁻³ sodium thiosulphate solution. Calculate the distribution coefficient for iodine between water and tetrachloromethane.

8. A solid A is three times as soluble in solvent X as in solvent Y. A has the same relative molecular mass in both solvents. Calculate the mass of A that would be extracted from a solution of 4 g of A in 12 cm³ of Y by extracting it with: a) 12 cm³ of X, and b) three successive portions of 4 cm³ of X.

9. a) The partition coefficient of T between ethoxyethane and water at room temperature is 19. A solution of 5.0 g of T in 100 cm³ of water is extracted with 100 cm³ of ethoxyethane at room temperature. What mass of T will be present in the ethoxyethane layer?

   b) What would be the total mass of T extracted if the ethoxyethane were used in two separate 50 cm³ portions, instead of the single 100 cm³ portion?

EXERCISE 39   Questions from A-level Papers

1. Hexane and heptane have the following properties

	Relative molar mass	Boiling point/°C	Vapour pressure at 25°C/Pa
Hexane	86.2	68.7	20 180
Heptane	100.2	98.1	6018

   a) State Raoult's Law as it applies to hexane in an ideal mixture of hexane and heptane.

   b) Suggest an experiment which you could use to investigate the variation in boiling point with liquid composition for a mixture of two miscible, flammable liquids such as hexane and heptane. You should give a diagram of the apparatus and the quantities of liquids you would use. State how the mixtures would be made up, give a careful description of how accurate measurements would be taken, and give any safety procedures involved.

   c) In a liquid mixture containing 23.0 g hexane and 42.0 g of heptane at 25°C calculate
      i) the mole fraction of hexane in the liquid
      ii) the partial vapour pressure of each component in the vapour
      iii) the total vapour pressure above the mixture

iv) the mole fraction of hexane in the vapour

v) the difference between your answers in i) and iv) and state what technique depends on this difference.

d) Mixtures of ethoxyethane, $(C_2H_5)_2O$ and trichloromethane, $CHCl_3$, do not obey Raoult's Law; they have higher boiling points than expected. Explain this observation in terms of the intermolecular forces present in the pure liquids and in the mixture.   (O&C91)

2. a) Explain the principle involved in the process of *solvent extraction.* State why the process is useful.

b) During nuclear fuel reprocessing, solvent extraction with an organic solvent is used to recover uranium salts from aqueous solution. If the partition coefficient of a uranium salt between the organic solvent and acidified water has a value of 50, calculate the ratio of the total number of moles of uranium in the organic solvent layer to that in the water layer after $500 \, dm^3$ of an aqueous solution of uranium salt has been extracted with $200 \, dm^3$ of organic solvent.

(*Hint*   Number of moles = Concentration $\times$ Volume.) (WJEC92,p)

*3. a)   i) Describe the basic assumptions of the kinetic theory of ideal gases.

ii) Suggest the experimental conditions under which these assumptions are no longer likely to be valid. Give your reasons.

b)   i) A known mass of an organic liquid A was allowed to evaporate completely and the volume of vapour produced was recorded. The experiment was repeated with different masses of A. The following data were produced at $100°C$ and 1 atmosphere pressure $(1.01 \times 10^5 \, N \, m^{-2})$.

Mass of liquid/g	Volume of vapour/cm^3
0.012	5
0.052	21
0.080	33
0.120	50

Using a graphical method, determine the molar mass of the unknown liquid, stating two assumptions that you make.

ii) The quantitative analysis of A gives C, 64.9%; H, 13.5%; O, 21.6%. Determine the empirical and hence the molecular formula of A.

iii) When A is passed over heated aluminium oxide a gas is formed which on hydrogenation produces an alkane. Substance B, an isomer of A, when treated overall in a similar way produces the same alkane. Suggest what types of molecule A and B are likely to be, giving your reasons.

iv) When molecules such as A and B are passed through a mass spectrometer, they invariably break up into smaller fragments.

These fragments produce appropriate peaks in a mass spectrum. For example, a fragment such as $C_2H_5^+$ produces a peak at 29 m/e ratio, and $C_3H_7^+$ at 43 m/e ratio.

Substance A produces a mass spectrum which includes peaks at 31 and 43 m/e ratio. However, the isomeric compound B has peaks at 29, 45 and 59 m/e ratio.

Deduce the species responsible for the peaks in the mass spectra and hence suggest suitable molecular structures for A and B.   (O&C91,S)

4. a) 0.18 g of an organic liquid occupies 75 cm^3 at 373 K and 1.00×10^5 Pa when vaporised. Use the ideal gas equation to calculate the relative molecular mass of the liquid.

   b) Describe an experiment, which could be carried out in a school laboratory, to determine the relative molecular mass of a volatile liquid such as that given in a).

   c) Outline the essential assumptions underlying the simple kinetic theory for ideal gases. How does the behaviour of real gases deviate from that of ideal gases? Explain how the ideal gas equation was modified by van der Waals to account for these deviations. Under what conditions is a gas most nearly ideal?   (AEB90)

5. a) Describe an experiment to measure the relative molecular mass of a volatile liquid. State how the results obtained are used to calculate the relative molecular mass.

   b) The density of hydrogen cyanide gas at 20 °C and 101.3 kPa (1.0 atm) pressure is 1.21×10^{-3} kg l^{-1} (1.21 g l^{-1}). Calculate the apparent relative molecular mass of the gas and account for your answer.

   c) i) State the basic assumptions of the kinetic theory of gases.

      ii) Explain, using the expression $pV = \frac{1}{3}Nm\overline{c^2}$, how the relative molecular mass of a gas can be determined from experimental data on relative rates of diffusion.   (JMB92)

6. a) Use the following data to plot, on graph paper, the variation of vapour pressure against temperature for methanol and for water. Plot both graphs on the same paper using the same axes.

Temperature/°C	0	20	40	60	70	80
Saturated vapour pressure of methanol/kPa	4	12	34	82	123	180
Saturated vapour pressure of water/kPa	1	2	7	20	31	47

   b) Discuss how molecular motion leads to gas pressure and explain why the vapour pressure of methanol varies with temperature in the manner shown by your graph.

c) Use your graph to predict the normal boiling point of methanol. Explain, in terms of the behaviour of molecules, the difference between boiling and evaporation.

d) Explain, in molecular terms, why the vapour pressure of methanol at a given temperature is different from that of water.

e) With the aid of the ideal gas equation calculate the mass of methanol vapour which occupies a volume of $0.001\,00\,m^3$ at $25\,°C$, assuming that the vapour is saturated and that it behaves as an ideal gas.   (JMB90, AS)

7. If a solute, X, is shaken with two immiscible solvents, Y and Z, at constant temperature, it is found that

$$\frac{\text{(Concentration of } X \text{ in } Y)}{\text{(Concentration of } X \text{ in } Z)} = \text{a constant}$$

a) What name is given to the constant?

b) i) For $X =$ phenylamine, $Y =$ water and $Z =$ ethoxyethane the value of the constant is 0.20. Calculate the mass of phenylamine extracted into the ethoxyethane layer when $100\,cm^3$ of water containing $20\,g$ of phenylamine per dm^3 is treated with $100\,cm^3$ of ethoxyethane.

   ii) How could the efficiency of extraction of the phenylamine have been improved using the same total volume of ethoxyethane?

c) i) Give *two* reasons why ethoxyethane is frequently used for solvent extraction in preparative organic chemistry.

   ii) State *one* drawback of ethoxyethane for this purpose. (AEB90)

8. The vapour pressures, at a certain temperature, for the miscible liquids propan-2-ol and benzene as a function of the mole fraction of propan-2-ol appear below.

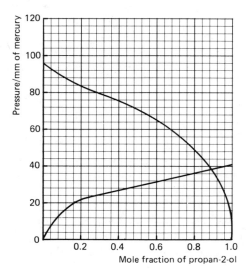

a) i) On a copy of the graph, plot accurately, on the same axes, a line showing the variation of total vapour pressure with composition for these two liquids. Label this line X.

   ii) Draw a line on the same axes to show the variation in total vapour pressure with composition if mixtures of propan-2-ol and benzene obeyed Raoult's Law. Label this line Y.

b) What would be the vapour pressure of benzene above a mixture containing a mole fraction of 0.6 of propan-2-ol?

c) A mixture contains 7.20 g of propan-2-ol and 2.34 g of benzene. What would be the observed vapour pressure above this mixture? (Relative atomic masses: H = 1, C = 12, O = 16.)

d) i) Would a mixture of propan-2-ol and benzene show a positive or a negative deviation from Raoult's Law?

   ii) Explain clearly, in terms of the intermolecular forces involved, how this deviation from Raoult's Law arises.

   iii) Does this deviation lead to a minimum or a maximum boiling point?

   iv) Name one liquid likely to form a mixture with benzene which obeys Raoult's Law and explain briefly why it would do so.

(L92)

9. State Raoult's Law as it applies to mixtures of methanol (b.pt. 64 °C) and ethanol (b.pt. 78 °C) which behave ideally, and explain the reasons for this ideal behaviour.

Give a fully labelled diagram showing the relationship between boiling temperature and composition for mixtures of methanol and ethanol.

Give full practical details for the fractional distillation in the laboratory of a mixture of methanol and ethanol in which the mole fraction of methanol is 0.2 and, by reference to your temperature–composition diagram, explain the principles of the process.

At a particular temperature, the vapour pressures of pure methanol and pure ethanol are 81 mmHg and 45 mmHg, respectively. Calculate the partial pressure of each component above a mixture of 64 g of methanol and 46 g of ethanol at this temperature.

Mixtures of benzene (b.pt. 80 °C) and ethanol show a negative deviation from Raoult's Law. Give a fully labelled temperature–composition diagram for such mixtures and state and explain what happens when benzene is added to ethanol.

(Relative atomic masses: H = 1, C = 12, O = 16.) (L90)

10. This question is concerned with the extraction of caffeine from tea leaves. Tea leaves contain between 3% and 5% by mass of caffeine. The caffeine can be extracted initially with hot water, in which it is fairly soluble (18 g/100 g water at 80 °C; 2.2 g/100 g water at 20 °C). Coloured impurities such as tannic acids can be removed as calcium salts by adding calcium carbonate. After filtering, the caffeine in the filtrate is extracted by shaking with a number of successive portions of

dichloromethane to produce a caffeine solution in dichloromethane. Most of the solvent is distilled off to produce a concentrated solution of caffeine in the organic solvent, and this solution is then evaporated over a water bath to yield the crude product.

The product is purified by dissolving it in hot methylbenzene ($10\,cm^3$ is correct for about $30\,g$ of tea as starting material), adding 15–$20\,cm^3$ of hexane, filtering hot and allowing the filtrate to cool. The crystals are often greenish in colour.

a) Using the basic description set out above, describe the extraction of caffeine from tea leaves in sufficient detail to allow an A-level student to carry out the experiment. Begin with $30\,g$ or so of tea leaves. Safety considerations should be stressed where appropriate.

b) What name is given to the process where methylbenzene and hexane are used? Explain briefly how this process removes both soluble and insoluble impurities.

c) When a substance X, dissolved in a particular solvent, is shaken up with another solvent which is immiscible with the first, an equilibrium of the form

$$[X]_{\text{solvent 1}} \rightleftharpoons [X]_{\text{solvent 2}}$$

is established. Use the equilibrium law to derive a relationship between the concentrations of X in solvent 1 and solvent 2 when equilibrium is attained.

Starting with $100\,cm^3$ of an aqueous solution containing $10\,g$ of X, and given that the equilibrium constant is 1, show that it is more efficient to extract the caffeine, X, with two portions of dichloromethane of volume $50\,cm^3$ rather than with one portion of volume $100\,cm^3$. (Note that one $100\,cm^3$ portion will distribute the $10\,g$ as $5\,g$ into each solvent.)

d) i) Tea leaves contain many other organic compounds. Explain why this method is suitable for the isolation of one specific compound.

ii) Explain how the calcium carbonate allows the removal of tannic acids.   (L91)

11. Anhydrous iron(III) chloride sublimes at $315\,°C$. It is obtained as dark green crystals by passing dry chlorine over heated iron wire. The figure shows an incomplete diagram of the apparatus used for this preparation.

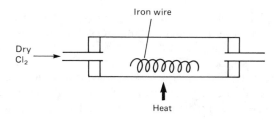

a) Write an equation for the reaction.

b) Copy and complete the diagram by showing how the iron(III) chloride may be collected. Include a means of keeping the whole of the apparatus dry.

c) At 400 °C and 1.00 atmosphere pressure, 3.00 g of anhydrous iron(III) chloride occupies a volume of 510 cm³. Work out the apparent relative molecular mass of iron(III) chloride at this temperature and hence suggest a formula for the molecular species present.

   (1 mole of any gas under these conditions occupies a volume of 55.2 dm³.)

d) A solution of iron(III) chloride may be reduced to iron(II) chloride by passing hydrogen sulphide through it.

$$2FeCl_3(aq) + H_2S(g) \longrightarrow 2FeCl_2(aq) + 2HCl(aq) + S(s)$$

   For this reaction write;
   i) a half-equation for the reduction process
   ii) a half-equation for the oxidation process.

e) State *two* characteristics of a transition metal, such as iron.

                                                                    (O90,AS)

12. a) Name the best method for:
       i) the separation of ammonium chloride from a mixture of this salt with potassium chloride.
       ii) the separation of argon from liquid air on a commercial scale.

   b) i) State the partition law for a solute X between two immiscible solvents P and Q.
      ii) A solute X is soluble without change in water and ethoxy-ethane, being 4 times more soluble in the latter. Calculate the quantity of X extracted from 1 dm³ of an aqueous solution of X, concentration 5 mol dm⁻³, when the aqueous solution is shaken with 1) 500 cm³ of ethoxyethane, 2) two successive 250 cm³ portions of ethoxyethane, and comment on the results.

   c) Give a qualitative account of:
      i) gas–liquid chromatography
      ii) paper chromatography.

   In each case outline the experimental procedure *and* indicate the underlying principles.   (O90)

12 Electrochemistry

ELECTROLYSIS

Electrovalent compounds, when molten or in solution, conduct electricity. The conductors which connect the melt or solution with the applied voltage are called the *electrodes*. The positive electrode is called the *anode*; the negative electrode is the cathode. Chemical reactions occur at the electrodes, and elements are deposited as solids or evolved as gases. These reactions are called *electrolysis*.

If the compound is a salt of a metal low in the electrochemical series, the metal ions are discharged, and a layer of metal is deposited on the cathode.

The electrolysis of a solution of a silver salt to deposit a layer of silver on the cathode is carried out as shown in Fig. 12.1.

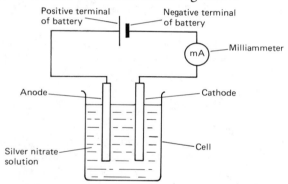

Fig. 12.1   Electrolysis of silver nitrate solution

The cathode process is

$$Ag^+(aq) + e^- \longrightarrow Ag(s)$$

According to this equation, one mole of silver ions accepts one mole of electrons to form one mole of silver atoms.

By weighing the cathode before and after the passage of the current, the mass of silver deposited can be found. By measuring the current with a milliammeter and timing the electrolysis, one can work out the quantity of electric charge which has passed. Electric charge is measured in coulombs (C).

$$\frac{\text{One coulomb}}{\text{of charge}} = \text{One ampere of current flowing for one second}$$

Charge in coulombs (C) = Current in amperes (A)
                          × Time in seconds (s)

From the measurements of mass and charge, one can calculate the amount of electric charge needed to deposit one mole of silver. This quantity is 96 500 coulombs. This quantity of electric charge must be the charge on one mole of electrons. The ratio 96 500 coulombs per mole ($C\,mol^{-1}$) is called the *Faraday constant* after the famous electrochemist Michael Faraday.

Consider the deposition of copper when a copper(II) salt is electrolysed.

$$Cu^{2+}(aq)\ +\ 2e^-\ \longrightarrow\ Cu(s)$$

Since one mole of copper ions needs two moles of electrons to become one mole of copper atoms, the charge needed to deposit one mole of copper is $2 \times 96\,500$ coulombs.

When gold is deposited during the electrolysis of gold(III) salts, the electrode process is

$$Au^{3+}(aq)\ +\ 3e^-\ \longrightarrow\ Au(s)$$

Since one mole of gold ions needs three moles of electrons to form one mole of gold atoms, the charge needed to deposit one mole of gold atoms is $3 \times 96\,500$ coulombs.

Calculations in electrochemistry are as easy as one, two, three, once you have worked out whether:

1 mole of electrons discharge 1 mole of the element, e.g. silver or
2 moles of electrons discharge 1 mole of the element, e.g. copper or
3 moles of electrons discharge 1 mole of the element, e.g. gold.

EXAMPLE 1 A direct current of 10.0 mA flows for 4.00 hours through three cells in series. They contain solutions of silver nitrate, copper(II) sulphate, and gold(III) nitrate. Calculate the mass of metal deposited in each.

METHOD Coulombs = Amperes $\times$ Seconds = $0.0100 \times 4.00 \times 60 \times 60$
$$= 144\,C$$

Electrical charge passed = 144/96 500 moles of electrons
No. of moles of Ag deposited = 144/96 500 mol
Mass of Ag deposited = $108 \times 144/96\,500 = 0.161\,g$
No. of moles of Cu deposited = $\frac{1}{2} \times 144/96\,500$ mol
Mass of Cu deposited = $63.5 \times \frac{1}{2} \times 144/96\,500 = 0.0474\,g$
No. of moles of Au deposited = $\frac{1}{3} \times 144/96\,500$ mol
Mass of Au deposited = $197 \times \frac{1}{3} \times 144/96\,500 = 0.0980\,g$

ANSWER Deposited are: 0.161 g silver; 0.0474 g copper; 0.0980 g gold.

EXAMPLE 2 A metal of relative atomic mass 27 is deposited by electrolysis. If 0.176 g of the metal is deposited on the cathode when 0.15 A flows for $3\frac{1}{2}$ hours, what is the charge on the cations of this metal?

METHOD   Coulombs = Amperes × seconds = $0.15 \times 3\frac{1}{2} \times 60 \times 60 = 1890\,C$

If $1890\,C$ deposit $0.176\,g$ of metal,

then $96\,500\,C$ deposit $\dfrac{96\,500 \times 0.176}{1890}g = 8.98\,g$

1 mole of metal = $27\,g$

Since $8.98\,g$ of metal are discharged by 1 mole of electrons,

$27\,g$ are discharged by $27/8.98 = 3$ moles of electrons

ANSWER   If 1 mole of metal needs 3 moles of electrons, the charge on a metal ion must be $+3$.

EXAMPLE 3   A current of 1.00 A flowing for 1 hour 50 minutes deposits 2.15 g of copper from an aqueous solution of copper(II) sulphate. If the Avogadro constant is $6.02 \times 10^{23}\,mol^{-1}$, calculate the charge on an electron.

METHOD   Coulombs = Amperes × Seconds = $1.00 \times 110 \times 60 = 6600\,C$

No. of moles of Cu = $2.15/63.5$

No. of atoms of Cu = $6.02 \times 10^{23} \times 2.15/63.5$

No. of electrons = $2 \times 6.02 \times 10^{23} \times 2.15/63.5 = 0.406 \times 10^{23}$

$\dfrac{\text{No. of coulombs}}{\text{No. of electrons}} = 6600/(0.406 \times 10^{23}) = 1.63 \times 10^{-19}\,C\,electron^{-1}$

ANSWER   The charge on an electron is $1.63 \times 10^{-19}\,C$.

EVOLUTION OF GASES

Solutions of salts of metals high in the electrochemical series evolve hydrogen at the cathode on electrolysis. The cathode process is:

$$H^+(aq) + e^- \longrightarrow H(g)$$

followed by $$2H(g) \longrightarrow H_2(g)$$

Thus, 2 moles of electrons are needed to evolve 1 mole of hydrogen molecules; that is, 2 g of hydrogen (the molar mass) or $22.4\,dm^3$ at s.t.p. (the molar volume).

At the anode, solutions of halides evolve the halogen, and other salts evolve oxygen. When chlorine is evolved, the anode process is

$$Cl^-(aq) \longrightarrow Cl(g) + e^-$$

followed by $$2Cl(g) \longrightarrow Cl_2(g)$$

Thus, 2 moles of electrons are required for the evolution of 1 mole of chlorine molecules, 71.0 g or $22.4\,dm^3$ at s.t.p.

When oxygen is evolved, the anode process is the discharge of hydroxide ions, derived from the water in the solution:

$$OH^-(aq) \longrightarrow OH(aq) + e^-$$

followed by $4OH(aq) \longrightarrow O_2(g) + 2H_2O(l)$

Thus, 4 moles of electrons are required for the evolution of 1 mole of oxygen molecules ($22.4\,dm^3$ oxygen at s.t.p.).

EXAMPLE 1   State the names and calculate the volumes of gases formed at the cathode and anode at s.t.p. when 0.0250 A of current are passed for 4.00 hours through a solution of sulphuric acid.

METHOD   At the anode, oxygen is evolved, and at the cathode hydrogen.
Coulombs = Amperes × Seconds = $0.0250 × 4.00 × 60 × 60 = 360\,C$
No. of moles of electrons = 360/96 500
No. of moles of H_2 = $\frac{1}{2} × 360/96\,500$
Vol. of H_2 = $22.4 × \frac{1}{2} × 360/96\,500\,dm^3$ = $0.0417\,dm^3$ = $41.7\,cm^3$
Vol. of O_2 = $\frac{1}{2} × 41.7$ = $20.9\,cm^3$

ANSWER $41.7\,cm^3$ hydrogen is formed at the cathode and $20.9\,cm^3$ of oxygen at the anode.

EXERCISE 40   Problems on Electrolysis

In the following problems use Faraday constant = $96\,500\,C\,mol^{-1}$.

SECTION 1

1. Which of these expressions shows the mass of aluminium liberated by a current of 0.1 ampere flowing for 6 minutes through molten alumina?

   a) $\dfrac{27 × 6 × 60 × 0.1}{3 × 96\,500}$ b) $\dfrac{3 × 6 × 60 × 0.1}{27 × 96\,500}$

   c) $\dfrac{27 × 6 × 60 × 0.1}{96\,500}$ d) $\dfrac{27 × 6 × 60}{3 × 96\,500 × 0.1}$

2. A current of 0.15 ampere flowing for 4 hours deposits 0.71 g copper on a cathode. Which expression gives the charge on a copper ion?

   a) $\dfrac{0.15 × 4 × 60 × 60 × 63.5}{96\,500 × 0.71}$ b) $\dfrac{0.15 × 4 × 60 × 60 × 0.71}{96\,500 × 63.5}$

   c) $\dfrac{63.5 × 0.15 × 4 × 60 × 60 × 0.71}{96\,500}$ d) $\dfrac{96\,500 × 0.71}{0.15 × 4 × 60 × 60 × 63.5}$

3. A current of 0.1 ampere passes for 3 hours through a solution of dilute sulphuric acid. Which of the following represents the volume of oxygen, in dm^3 at s.t.p., evolved at the anode?

a) $\dfrac{0.1 \times 3 \times 60 \times 60 \times 4}{96\,500 \times 22.4}$

b) $\dfrac{0.1 \times 3 \times 60 \times 60 \times 22.4}{96\,500 \times 4}$

c) $\dfrac{0.1 \times 3 \times 60 \times 60 \times 22.4}{96\,500}$

d) $\dfrac{0.1 \times 3 \times 60 \times 60}{22.4 \times 96\,500 \times 4}$

4. The mass of lead deposited at the cathode by a current of 0.15 ampere flowing for three hours through a solution of lead(II) nitrate is given by one of the following expressions. Which is correct?

a) $\dfrac{207 \times 0.15 \times 3 \times 60 \times 60}{96\,500}$

b) $\dfrac{96\,500 \times 207 \times 2}{0.15 \times 3 \times 60 \times 60}$

c) $\dfrac{207 \times 0.15 \times 3 \times 60 \times 60}{96\,500}$

d) $\dfrac{207 \times 0.15 \times 3 \times 60 \times 60}{2 \times 96\,500}$

5. A current of 0.2 ampere passing for 5 hours through a solution of gold ions deposits a mass of 2.45 g of gold on the cathode. Which of these expressions gives the charge on a gold ion?

a) $\dfrac{2.45 \times 0.2 \times 5 \times 60 \times 60}{197 \times 96\,500}$

b) $\dfrac{0.2 \times 5 \times 60 \times 60 \times 197}{96\,500 \times 2.45}$

c) $\dfrac{2.45 \times 96\,500}{197 \times 0.2 \times 5 \times 60 \times 60}$

d) $\dfrac{197 \times 0.2 \times 5 \times 60 \times 96\,500}{2.45}$

SECTION 2

1. Calculate the mass of copper that would be deposited on a copper cathode from an aqueous solution of copper(II) sulphate, if the same current, passed for the same time, liberated 0.900 g of silver from an aqueous solution of silver nitrate, $AgNO_3$.

2. A current of 100 mA was passed for 1.00 h through an aqueous solution of $1.00 \, mol \, dm^{-3}$ silver nitrate. A silver cathode was used. Calculate the increase in mass of the cathode. State how the change in mass would be affected by:

a) passing a current of 200 mA
b) passing a current of 100 mA for 2.00 h
c) using a $2.00 \, mol \, dm^{-3}$ solution of silver nitrate.

3. An electric current of 5.00 A was passed through molten anhydrous calcium chloride, $CaCl_2$, for 20.0 minutes between graphite electrodes. Calculate the mass of each product liberated.

4. A current of electricity liberated 6.22×10^{-3} mol of silver from a silver nitrate solution. What mass in grams, of aluminium, would be liberated from a suitable aluminium compound, using the same quantity of electricity?

5. A current is passed for 45 minutes through three solutions in series, using platinum electrodes. In one, 0.203 g of silver is deposited from silver nitrate solution. In a second, hydrogen is evolved from a solution of dilute sulphuric acid, and, in a third, lead is deposited from a solution of lead nitrate. Calculate: a) the current passed, b) the volume of hydrogen collected at $0.983 \times 10^5 \, \mathrm{N \, m^{-2}}$ and 18 °C, c) the mass of lead deposited.

6. A current of 0.750 A passes through 250 cm^3 solution of 0.250 mol dm^{-3} copper(II) sulphate solution. How long (in minutes) will it take to deposit all the copper on the cathode?

7. What current is needed to deposit 0.500 g of nickel from a nickel(II) sulphate solution in 1.00 hour?

8. A current of 1.25 A passes for 5.00 h between platinum electrodes in 500 cm^3 of copper(II) sulphate solution of concentration 2.00 mol dm^{-3}. What will be the concentration of copper(II) sulphate at the end of the time?

9. How long (in hours) will it take a current of 0.100 A to deposit 1.00 kg of silver from an unlimited source of silver ions?

10. A current of 2.05 A passes through a solution of sulphuric acid for 5.00 h. Calculate the volumes of hydrogen and oxygen produced.

11. A steady current was passed through a solution of copper(II) sulphate until 6.05 g of metallic copper were deposited on the cathode. How many coulombs of electricity had been used?

12. In the electrolysis of a solution of potassium chloride, 100 cm^3 of chlorine are produced at 20 °C and $9.9 \times 10^4 \, \mathrm{N \, m^{-2}}$. How many seconds has a current of 0.750 A been flowing through the solution to effect this?

ELECTROLYTIC CONDUCTIVITY

Solutions of electrolytes obey Ohm's law, from which it follows that

$$R = V/I$$

where R = Resistance, V = Potential difference, I = Current.

Resistivity is defined by the equation

$$R = \rho \times l/A$$

where l = Length and A = Cross-sectional area of the conductor.

If R is in ohms (Ω), l in metres (m), and A in square metres (m²), then ρ has the dimensions Ω m.

The reciprocal of resistance is conductance; the reciprocal of resistivity ρ is conductivity κ.

$$1/\rho = \kappa$$

κ has the dimensions $\Omega^{-1}m^{-1}$.

Molar conductivity Λ is defined by the equation

$$\Lambda = \kappa/c$$

---

where c = Molar concentration of solute. If κ is in $\Omega^{-1}m^{-1}$ and c is in $mol\,m^{-3}$, then $\Lambda = \Omega^{-1}m^2\,mol^{-1}$, and Λ is numerically equal to the conductivity of 1 mole of the electrolyte.

The molar conductivity, Λ, of the solute increases as the concentration of the solution decreases, until further dilution has no further effect. The value of molar conductivity at this dilution is called the molar conductivity at infinite dilution, and is represented as Λ_0 or Λ_∞.

MOLAR CONDUCTIVITY AND CONCENTRATION

The molar conductivity of a solute depends on its concentration. When values of molar conductivity Λ are plotted against the volume of solution containing 1 mole of solute (called the dilution), the graphs obtained for different electrolytes fall into two categories. These are shown in Fig. 12.1.

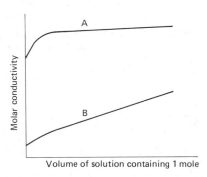

Fig. 12.1   Plots of molar conductivity against dilution

Strong electrolytes are those which are completely ionised in solution. They give graphs of shape A, swiftly rising to a maximum. Weak electrolytes, which are incompletely ionised, give graphs of shape B.

They do not reach a maximum at dilutions for which conductivities can be measured in practice. Arrhenius explained graphs of shape B as arising from an increase in the ionisation of the solute with dilution. The fraction of the solute which is ionised, α, gradually increases with dilution. If measurements could be made in more and more dilute solutions, α would increase up to a value of 1, and Λ would reach its limiting value of Λ_0. This is called the molar conductivity at infinite dilution since further dilution results in no further increase in Λ. In more concentrated solutions,

$$\alpha = \Lambda/\Lambda_0$$

Strong electrolytes show the steep increase in Λ with dilution illustrated in curve A. It seems to indicate that the strong electrolytes are completely ionised except at high concentrations. According to more modern views, strong electrolytes are completely ionised at all concentrations. Another explanation of the low value of Λ at high concentration is proposed. It is attributed to the attraction of oppositely charged ions slowing down the ions in a concentrated solution. The value of α given by Λ/Λ_0 is referred to as the *apparent degree of ionisation* or the *conductivity ratio*.

You can see from Fig. 12.1 that Λ_0 for weak electrolytes cannot be found experimentally. Kohlrausch's law helps here. It enables Λ_0 to be found for weak electrolytes. Kohlrausch's law can be expressed as

$$\Lambda_0 = \Lambda_0(\text{cation}) + \Lambda_0(\text{anion})$$

For every cation and anion, values of Λ_0 can be looked up in tables and added to give Λ_0 for the salt. Alternatively, measurements on strong electrolytes can be used to give values of Λ_0 which can be combined to give Λ_0 for the required weak electrolyte.

EXAMPLE   Calculate the molar conductivity at infinite dilution at 25 °C of ethanoic acid, given that the molar conductivities at infinite dilution of hydrochloric acid, sodium chloride and sodium ethanoate are 4.26×10^{-2}, 1.26×10^{-2} and $9.1 \times 10^{-3}\,\Omega^{-1}\,m^2\,mol^{-1}$ respectively.

METHOD   According to Kohlrausch's law,

$$\Lambda_0(CH_3CO_2H) = \Lambda_0(CH_3CO_2^-) + \Lambda_0(H^+)$$

$$\Lambda_0(HCl) = \Lambda_0(H^+) + \Lambda_0(Cl^-)$$

$$\Lambda_0(NaCl) = \Lambda_0(Na^+) + \Lambda_0(Cl^-)$$

$$\Lambda_0(CH_3CO_2Na) = \Lambda_0(CH_3CO_2^-) + \Lambda_0(Na^+)$$

Thus,

$$\Lambda_0(CH_3CO_2H) = \Lambda_0(HCl) + \Lambda_0(CH_3CO_2Na) - \Lambda_0(NaCl)$$

ANSWER $$\Lambda_0(CH_3CO_2H) = (4.26 \times 10^{-2}) + (9.00 \times 10^{-3}) - (1.26 \times 10^{-2})$$

$$= 3.91 \times 10^{-2}\,\Omega^{-1}\,m^2\,mol^{-1}$$

CALCULATION OF SOLUBILITY FROM CONDUCTIVITY

The solubility of a sparingly soluble salt can be found by measuring the conductance of a saturated solution of the salt. The conductance depends on the concentration of the dissolved ions. From it, the conductivity can be calculated. For a sparingly soluble salt, the concentration is so low, even in a saturated solution, that the salt can be assumed to be completely ionised. The conductivity is compared with the molar conductivity at infinite dilution of the salt. This is calculated as the sum of the molar conductivities at infinite dilution of the cation and anion. Then

$$\Lambda_0 = \kappa/c$$

where Λ_0 = Molar conductivity at infinite dilution, κ = Conductivity, and c = Molar concentration of solute.

Hence c can be calculated. Since the solution is saturated, the concentration is equal to the solubility of the salt.

EXAMPLE   After making allowance for the conductivity of water, the conductivity of a saturated solution of silver chloride at $25\,^{\circ}C = 1.50 \times 10^{-4}\,\Omega^{-1}\,m^{-1}$. If $\Lambda_0(Ag^+) = 6.2 \times 10^{-3}$ and $\Lambda_0(Cl^-) = 7.6 \times 10^{-3}\,\Omega^{-1}\,m^2\,mol^{-1}$, what is the solubility of silver chloride in $mol\,dm^{-3}$?

METHOD   The molar conductivity at infinite dilution, Λ_0, is the sum of those of the ions.

$$\Lambda_0 = \Lambda_0(Ag^+) + \Lambda_0(Cl^-)$$
$$= (6.2 \times 10^{-3}) + (7.7 \times 10^{-3}) = 1.38 \times 10^{-2}$$
$$\Lambda_0 = \kappa/c$$
$$c = \kappa/\Lambda_0$$
$$= (1.50 \times 10^{-4})/(1.38 \times 10^{-2})$$
$$= 1.09 \times 10^{-2}\,mol\,m^{-3} = 1.09 \times 10^{-5}\,mol\,dm^{-3}$$

ANSWER   The solubility of silver chloride is $1.09 \times 10^{-5}\,mol\,dm^{-3}$.

IONIC EQUILIBRIA; WEAK ELECTROLYTES

Electrolytes are ionic compounds. Strong electrolytes consist entirely of ions. Weak electrolytes consist of molecules, some of which dissociate to form ions. The fraction of molecules which dissociate is called the degree of ionisation or degree of dissociation. As the concentration of a solution of a weak electrolyte decreases, the degree of ionisation increases. *Ostwald's dilution law* gives a relationship between the degree of ionisation, α, of a weak electrolyte and its concentration c. Consider a weak acid, HA, in a solution of concentration c. An equilibrium is set up between the undissociated molecules, HA, and the ions H_3O^+ and A^-. H_3O^+ is the oxonium ion, a complex ion formed by the combination of a hydrogen ion and a molecule of water. It can also be written as $H^+(aq)$. Since the degree of ionisation is α, the equilibrium concentrations of the species are: $[HA] = (1-\alpha)c$; $[H_3O^+] = [A^-] = \alpha c$ (where the square brackets stand for concentration). The equilibrium can be represented:

$$HA + H_2O \rightleftharpoons H_3O^+ + A^-$$
$$c(1-\alpha) \qquad\qquad c\alpha \quad c\alpha$$

The ratio $\dfrac{[H_3O^+][A^-]}{[HA][H_2O]}$ is a constant. (See Chapter 11 on Equilibria.)

The value of $[H_2O]$ is constant as the concentration of water is not significantly altered by the ionisation of HA.

The ratio $\dfrac{[H_3O^+][A^-]}{[HA]}$ is called the dissociation constant, K_a, of the acid:

$$K_a = \frac{[H_3O^+][A^-]}{[HA]} = \frac{c\alpha \times c\alpha}{c(1-\alpha)} = \frac{c\alpha^2}{1-\alpha}$$

The Ostwald dilution law is embodied in the equation

$$K_a = \frac{c\alpha^2}{1-\alpha}$$

For some weak electrolytes, α is so small that the error involved in putting $(1-\alpha)$ equal to 1 is negligible. In this case,

$$K_a = c\alpha^2$$
$$\alpha = \sqrt{K_a/c}$$

Ostwald's dilution law can be applied to weak bases. In the case of a weak base B which is partially ionised in solution,

$$B + H_2O \rightleftharpoons BH^+ + OH^-$$

The concentration of water is not significantly altered by the dissociation: $[H_2O] = $ constant.

The base dissociation constant K_b is given by the equation

$$K_b = \frac{[BH^+][OH^-]}{[B]}$$

$$K_b = \frac{c\alpha^2}{1-\alpha}$$

If $\alpha \ll 1$, the expression approximates to $K_b = c\alpha^2$.

CALCULATION OF THE DEGREE OF DISSOCIATION AND THE DISSOCIATION CONSTANT OF A WEAK ELECTROLYTE FROM CONDUCTANCE MEASUREMENTS

From conductance measurements on a solution of known concentration, the molar conductivity Λ is found. The relationship,

$$\alpha = \Lambda/\Lambda_0$$

enables the degree of ionisation α to be found. The value of α can be substituted in the Ostwald equation to give the ionisation constant of the electrolyte.

EXAMPLE 1 a) Calculate the degree of dissociation of ethanoic acid of concentration $1.00 \times 10^{-3}\,mol\,dm^{-3}$. The conductivity of the solution is $4.85 \times 10^{-3}\,\Omega^{-1}\,m^{-1}$ and the values of Λ_0 for ethanoic acid is $3.91 \times 10^{-2}\,\Omega^{-1}\,m^2\,mol^{-1}$.

b) Calculate the dissociation constant of ethanoic acid at the temperature at which the measurements were made.

METHOD a) Conductivity, $\kappa = 4.85 \times 10^{-3}\,\Omega^{-1}\,m^{-1}$

Concentration, $c = 1.00 \times 10^{-3}\,mol\,dm^{-3} = 1.00\,mol\,m^{-3}$

$$\Lambda = \kappa/c = (4.85 \times 10^{-3})/1.00 = 4.85 \times 10^{-3}\,\Omega^{-1}\,m^2\,mol^{-1}$$

$$\Lambda_0(CH_3CO_2H) = 3.91 \times 10^{-2}\,\Omega^{-1}\,m^2\,mol^{-1}$$

$$\alpha = \Lambda/\Lambda_0$$

$$= (4.85 \times 10^{-3})/(3.91 \times 10^{-2}) = 0.124$$

ANSWER The degree of dissociation of ethanoic acid is 0.124.

METHOD b) From the Ostwald equation,

$$K_a = \frac{c\alpha^2}{1-\alpha}$$

With a value of 0.124 for α, $(1-\alpha)$ cannot be put $= 1$

$$K_a = \frac{10^{-3} \times (0.124)^2}{0.876} = 1.76 \times 10^{-5}\,mol\,dm^{-3}$$

ANSWER The dissociation constant of ethanoic acid is $1.76 \times 10^{-5}\,mol\,dm^{-3}$.

EXAMPLE 2 Calculate the degree of dissociation of hydrogen cyanide in a solution of concentration $0.100\,mol\,dm^{-3}$. $K_a = 7.24 \times 10^{-10}\,mol\,dm^{-3}$.

METHOD
$$K_a = \frac{c\alpha^2}{(1-\alpha)}$$

$$7.24 \times 10^{-10} = \frac{0.100\alpha^2}{(1-\alpha)} \qquad \therefore \ 7.24 \times 10^{-9} = \frac{\alpha^2}{1-\alpha}$$

Inspection shows that α is of the order of 10^{-5}, and $(1-\alpha)$ can be put equal to 1.

Then $\qquad\qquad \alpha = \sqrt{7.24 \times 10^{-9}} = 8.5 \times 10^{-5}$

ANSWER   The degree of dissociation is 8.5×10^{-5}.

EXERCISE 41   Problems on Molar Conductivity

1. The molar conductivity of a solution of a weak acid of concentration $0.0400 \, mol \, dm^{-3}$ at $25 \, °C$ is $8.25 \times 10^{-4} \, \Omega^{-1} \, m^2 \, mol^{-1}$. If the molar conductivity at infinite dilution is $3.98 \times 10^{-2} \, \Omega^{-1} \, m^2 \, mol^{-1}$ at $25 \, °C$, calculate the dissociation constant of the acid at $25 \, °C$.

2. A solution of a weak monobasic acid has a concentration of 0.0250 $mol \, dm^{-3}$. The conductivity at $25 \, °C$ is $2.64 \times 10^{-2} \, \Omega^{-1} \, m^{-1}$. The molar conductivity at infinite dilution is $3.91 \times 10^{-2} \, \Omega^{-1} \, m^2 \, mol^{-1}$. Calculate: a) the degree of dissociation, b) the concentration of hydrogen ions in solution, and c) the dissociation constant of the acid.

3. A solution of a weak monobasic acid of concentration $0.0200 \, mol$ dm^{-3} has conductivity $0.184 \, \Omega^{-1} \, m^{-1}$ at $25 \, °C$. The molar conductivity at infinite dilution is $4.00 \times 10^{-2} \, \Omega^{-1} \, m^2 \, mol^{-1}$. Calculate: a) the degree of dissociation, and b) the dissociation constant of the acid at $25 \, °C$.

4. A solution of methanoic acid of concentration $0.100 \, mol \, dm^{-3}$ has conductivity $0.166 \, \Omega^{-1} \, m^{-1}$ at $25 \, °C$. The molar conductivity at infinite dilution is $4.04 \times 10^{-2} \, \Omega^{-1} \, m^2 \, mol^{-1}$. Calculate the dissociation constant of methanoic acid at $25 \, °C$.

5. A solution of an organic acid, RCO_2H, has a concentration of $0.0300 \, mol \, dm^{-3}$ and a molar conductivity of $1.00 \times 10^{-3} \, \Omega^{-1} \, m^2$ mol^{-1} at $25 \, °C$. The molar conductivity at infinite dilution at $25 \, °C$ is $39.0 \times 10^{-3} \, \Omega^{-1} \, m^2 \, mol^{-1}$. Calculate: a) the degree of dissociation, and b) the dissociation constant for the acid at $25 \, °C$.

CALCULATION OF pH, pOH and pK

Acids react with water to give hydrogen ions. These are not simple H^+ ions; they are complex ions of formula H_3O^+, formed by combination of a molecule of water and an H^+ ion. Although they are properly called *oxonium* ions, in this text, H_3O^+ ions will be referred to as hydrogen ions. The hydrogen ion concentration of a solution can be indicated by means of a number on the pH scale.

The pH of a solution is the negative logarithm to the base 10 of the hydrogen ion concentration. pOH, pK_a and pK_b are defined below.

$$pH = -\lg[H_3O^+/\text{mol dm}^{-3}] \qquad pK_a = -\lg(K_a/\text{mol dm}^{-3})$$
$$pOH = -\lg[OH^-/\text{mol dm}^{-3}] \qquad pK_b = -\lg(K_b/\text{mol dm}^{-3})$$

Strong acids and bases

The pH of a solution of a strong acid or base is simply calculated. If the concentration of hydrochloric acid is $0.1\,\text{mol dm}^{-3}$,

$$[H_3O^+] = 0.1\,\text{mol dm}^{-3} = 10^{-1}\,\text{mol dm}^{-3}$$
$$\lg[H_3O^+] = -1$$
$$pH = 1$$

If the concentration of a solution of sodium hydroxide is $0.01\,\text{mol dm}^{-3}$,

$$[OH^-] = 0.01 = 10^{-2}\,\text{mol dm}^{-3}$$

The product of the hydrogen ion concentration and the hydroxide ion concentration in a solution is called the ionic product for water, K_w. At 25 °C, the value of K_w is $10^{-14}\,\text{mol}^2\,\text{dm}^{-6}$.

$$[H_3O^+][OH^-] = K_w = 10^{-14}\,\text{mol}^2\,\text{dm}^{-6}$$

Therefore, in this solution,

$$[H_3O^+] = 10^{-14}/10^{-2} = 10^{-12}\,\text{mol dm}^{-2}$$
$$pH = -\lg[H_3O^+] = 12$$

EXAMPLE 1   Calculate the pH of a solution of $0.020\,\text{mol dm}^{-3}$ hydrochloric acid.

METHOD   If the concentration of hydrochloric acid is $0.020\,\text{mol dm}^{-3}$,

$$[H_3O^+] = 2 \times 10^{-2}\,\text{mol dm}^{-3}$$

ANSWER $pH = -\lg[H_3O^+] = -(0.301 + \bar{2}) = 1.70$

EXAMPLE 2   Calculate the pH of a $0.010\,\text{mol dm}^{-3}$ solution of calcium hydroxide.

METHOD   If the concentration of $Ca(OH)_2$ is $10^{-2}\,\text{mol dm}^{-3}$,

$$[OH^-] = 2 \times 10^{-2}\,\text{mol dm}^{-3}$$
$$pOH = -\lg(2 \times 10^{-2}) = 1.7$$

Since $\qquad [H_3O^+][OH^-] = K_w = 10^{-14} \, mol^2 \, dm^{-6}$

$$lg \, [H_3O^+] + lg \, [OH^-] = lg \, K_w = -14$$

$\therefore \qquad\qquad\qquad pH + pOH = 14$

Thus, if pOH = 1.7,

ANSWER   pH = $14 - 1.7 = 12.3$.

(For practice in using a calculator to find logarithms, see pp. 7–9 and 15–16.)

Weak acids and bases

For calculating the pH of a weak acid or base, the concentration of the solution is not sufficient information. The degree of dissociation must be taken into account. The following examples show how the pH can be calculated from the concentration and the degree of dissociation or the dissociation constant of a weak electrolyte.

The converse is also true. If the pH can be measured experimentally, the value of the pH can be used to calculate the dissociation constant of the weak electrolyte.

EXAMPLE 1   Calculate the pH of a solution of propanoic acid of concentration $0.0100 \, mol \, dm^{-3}$, given that the degree of dissociation is 0.116.

METHOD   Let the concentration of acid be c, and the degree of dissociation be α. Then in the equilibrium

$$HA + H_2O \rightleftharpoons H_3O^+ + A^-$$

we can put the concentrations

$$[HA] = c(1-\alpha); \quad [H_3O^+] = [A^-] = c\alpha.$$

Then $\qquad [H_3O^+] = 0.0100 \times 0.116 = 1.16 \times 10^{-3}$

$$pH = -lg \, [H_3O^+] = 2.9355$$

ANSWER   pH = 2.94.

EXAMPLE 2   Calculate the pH of a $0.0100 \, mol \, dm^{-3}$ solution of ethanoic acid, which has a dissociation constant of $1.76 \times 10^{-5} \, mol \, dm^{-3}$.

METHOD   When a weak acid ionises,

$$HA + H_2O \rightleftharpoons H_3O^+ + A^-$$

the dissociation constant K_a is given by the expression

$$K_a = \frac{[H_3O^+][A^-]}{[HA]}$$

One A^- is formed for each H_3O^+.

$\therefore \qquad [H_3O^+] = [A^-]$ and $[H_3O^+][A^-] = [H_3O^+]^2$

The degree of dissociation is so small that we can put [HA] equal to the total acid concentration, $0.0100 \, \text{mol dm}^{-3}$. In fact, it must be slightly less than this. Say the degree of dissociation is 0.001. Then $[HA] = 0.0100 - 0.00001 = 0.00999$. The error in assuming $[HA] = 0.0100$ is 1 part in 1000. This is smaller than the error in the practical measurements from which the values used in the calculation are obtained. For most solutions of weak acids and bases, we can make the approximation safely.

Thus,
$$K_a = \frac{[H_3O^+]^2}{10^{-2}}$$

$$[H_3O^+]^2 = 1.76 \times 10^{-7}$$

$$[H_3O^+] = 4.19 \times 10^{-4} \, \text{mol dm}^{-3}$$

$$pH = -\lg [H_3O^+]$$

ANSWER $pH = 3.34$.

EXAMPLE 3   Calculate the dissociation constant of phenol, given that a 1.00×10^{-2} mol dm^{-3} solution has a pH of 5.95.

METHOD   If $pH = 5.95$,

$$[H_3O^+] = \text{antilg} \, (-pH) = 1.13 \times 10^{-6}$$

If we write the dissociation of phenol

$$PhOH + H_2O \rightleftharpoons PhO^- + H_3O^+$$

$$K_a = \frac{[H_3O^+][PhO^-]}{[PhOH]} = \frac{[H_3O^+]^2}{[PhOH]}$$

$$K_a = \frac{(1.13 \times 10^{-6})^2}{10^{-2}} = 1.28 \times 10^{-10} \, \text{mol dm}^{-3}$$

ANSWER   The dissociation constant of phenol is $1.28 \times 10^{-10} \, \text{mol dm}^{-3}$.

EXAMPLE 4   A solution of pyridine of concentration $0.0100 \, \text{mol dm}^{-3}$ has a pH 8.63. Calculate the dissociation constant of pyridine.

METHOD   We can write the dissociation of pyridine as

$$P + H_2O \rightleftharpoons PH^+ + OH^-$$

Thus,
$$K_b = \frac{[PH^+][OH^-]}{[P]} = \frac{[OH^-]^2}{0.0100}$$

Since $\quad$ pH $=$ 8.63

$$\text{pOH} = 14.00 - 8.63 = 5.37$$

$$[\text{OH}^-] = 4.26 \times 10^{-6}$$

$$K_b = \frac{(4.26 \times 10^{-6})^2}{0.0100} = 1.82 \times 10^{-9}\,\text{mol dm}^{-3}$$

ANSWER $\quad$ The dissociation constant of pyridine is $1.82 \times 10^{-9}\,\text{mol dm}^{-3}$.

Conjugate pairs

Some tables list the base dissociation constants, K_b, of bases such as amines. Others list the acid dissociation constants, K_a.

The value of K_b refers to the equilibrium

$$\text{B} + \text{H}_2\text{O} \rightleftharpoons \text{BH}^+ + \text{OH}^-$$

where $\quad$ $$K_b = \frac{[\text{BH}^+]\,[\text{OH}^-]}{[\text{B}]}$$

The species BH^+ is referred to as the conjugate acid of the base, B. BH^+ dissociates according to the equilibrium

$$\text{BH}^+ + \text{H}_2\text{O} \rightleftharpoons \text{B} + \text{H}_3\text{O}^+$$

One can, therefore, write the acid dissociation constant of BH^+ as

$$K_a = \frac{[\text{B}]\,[\text{H}_3\text{O}^+]}{[\text{BH}^+]}$$

Multiplying K_a by K_b gives $[\text{H}_3\text{O}^+]\,[\text{OH}^-]$. That is, $K_a K_b = K_w$.

This is a useful relationship between the dissociation constants of conjugate pairs.

EXAMPLE 1 $\quad$ Calculate the pH of a solution of ammonia ($pK_a = 9.25$) of concentration $0.0100\,\text{mol dm}^{-3}$.

METHOD $\quad\quad\quad\quad$ $pK_a = 9.25 \quad\quad pK_b = 14.00 - 9.25 = 4.75$

$$K_b = 1.78 \times 10^{-5}$$

Since $\quad$ $$K_b = \frac{[\text{NH}_4^+]\,[\text{OH}^-]}{[\text{NH}_3]} = \frac{[\text{OH}^-]^2}{[\text{NH}_3]}$$

$$[OH^-]^2 = K_b \times c$$
$$= 1.78 \times 10^{-5} \times 0.0100$$
$$[OH^-] = 4.22 \times 10^{-4}$$
$$pOH = 3.37$$

ANSWER $$pH = 10.6$$

EXAMPLE 2 Calculate the degree of ionisation of phenylmethylamine (pK_a = 9.37) in a 0.100 mol dm^{-3} aqueous solution. Quote your answer to two significant figures.

METHOD   Since $$pK_a = 9.37$$
$$pK_b = 14.00 - 9.37 = 4.63$$
$$K_b = 2.34 \times 10^{-5}$$

Since $$K_b = \frac{[C_6H_5CH_2NH_3^+][OH^-]}{[C_6H_5CH_2NH_2]}$$
$$= \frac{\alpha^2 c}{(1-\alpha)}$$

making the approximation $1 - \alpha = 1$,

$$\alpha^2 = K_b/c$$
$$= 2.34 \times 10^{-5}/0.100 = 2.34 \times 10^{-4}$$
$$\alpha = 1.53 \times 10^{-2}$$

The approximation $(1 - \alpha) = 1$ is not entirely justified, but, since the answer is required to two significant figures only, this is close enough.

ANSWER   The degree of ionisation is 1.5×10^{-2}.

EXERCISE 42   Problems on pH

1. Calculate the pH of solutions with following H_3O^+ concentrations in mol dm^{-3}:

    a) 10^{-8} b) 10^{-4} c) 10^{-7}
    d) 6.8×10^{-3} e) 3.2×10^{-5} f) 0.035
    g) 0.25   h) 5.4×10^{-9} i) 7.1×10^{-7}
    j) 9.9×10^{-2}

    Now calculate the pOH of each of the solutions.

2. Calculate the pH of solutions with the following OH^- concentrations in $mol\,dm^{-3}$:

   a) 10^{-2} b) 10^{-3} c) 10^{-8}
   d) 0.055 e) 0.0010 f) 0.083
   g) 7.6×10^{-3} h) 4.9×10^{-5} i) 6.4×10^{-8}
   j) 3.7×10^{-10}

3. Calculate the H_3O^+ concentrations in solutions with the following pH values:

   a) 0.00 b) 4.30 c) 2.35
   d) 1.88 e) 4.15 f) 7.84
   g) 9.21 h) 13.7 i) 9.50
   j) 2.63

4. Calculate the pH of the solutions made by dissolving the following in distilled water and making up to $500\,cm^3$ of solution:

   a) $3.00\,g$ of hydrogen chloride
   b) $4.50\,g$ of chloric(VII) acid, $HClO_4$
   c) $4.00\,g$ of sodium hydroxide
   d) $1.00\,g$ of calcium hydroxide
   e) $6.30\,g$ of potassium hydroxide

5. An aqueous solution contains the acid, HX, at a concentration $0.100\,mol\,dm^{-3}$. The degree of ionisation of the acid is 0.0300. Calculate the pH of the solution.

6. If the acid HA is 1% ionised in a solution of concentration $0.0100\,mol\,dm^{-3}$, calculate: a) K_a, b) pK_a.

7. The degree of ionisation of phenol in water in a solution of concentration $1.10 \times 10^{-2}\,mol\,dm^{-3}$ is 1.1×10^{-4}. Calculate the value of pK_a.

8. Calculate the dissociation constants for each of the weak acids listed below:

	Electrolyte	*Concentration/$mol\,dm^{-3}$*	$[H_3O^+]$/$mol\,dm^{-3}$
a)	$HClO$	0.0100	1.92×10^{-5}
b)	CH_3CO_2H	0.100	1.32×10^{-3}
c)	HCN	1.00	1.99×10^{-5}
d)	$C_2H_5CO_2H$	0.200	1.61×10^{-3}

9. Calculate the dissociation constants of the weak acids listed below:

   a) a solution of $0.0100\,mol\,dm^{-3}$ CH_3CO_2H has a pH of 3.38
   b) a solution of $0.200\,mol\,dm^{-3}$ HCN has a pH of 5.05
   c) a solution of $0.0100\,mol\,dm^{-3}$ $CH_3CH_2CO_2H$ has a pH of 3.43
   d) a solution of $0.0100\,mol\,dm^{-3}$ $HOBr$ has a pH of 5.35

10. Calculate the values of K_b for the following weak bases from the data:

Base	Concentration/$mol\,dm^{-3}$	$[OH^-]/mol\,dm^{-3}$
a) CH_3NH_2	0.0100	4.78×10^{-7}
b) NH_3	1.00	2.37×10^{-5}
c) $C_2H_5NH_2$	0.0500	9.65×10^{-7}
d) $C_6H_5CH_2NH_2$	0.0100	2.06×10^{-6}

11. Calculate the degree of ionisation of each of the following in aqueous solution:
    a) $2.00 \times 10^{-3}\,mol\,dm^{-3}$ HCN   ($pK_a = 9.40$)
    b) $3.00 \times 10^{-2}\,mol\,dm^{-3}$ CH_3CO_2H ($pK_a = 4.76$)
    c) $5.00 \times 10^{-2}\,mol\,dm^{-3}$ NH_3 ($pK_a = 9.25$)
    d) $1.00 \times 10^{-2}\,mol\,dm^{-3}$ $(CH_3)_3N$ ($pK_a = 9.80$)
    (*Note.* You are given the values of pK_a for the conjugate acids of the bases.)

12. Calculate the pH of:
    a) $0.0100\,mol\,dm^{-3}$ hydrochloric acid,
    b) $0.0100\,mol\,dm^{-3}$ sodium hydroxide, and
    c) the solution obtained by diluting $5.00\,cm^3$ of hydrochloric acid of concentration $1.00\,mol\,dm^{-3}$ with conductivity water to $1.00\,dm^3$.

13. The dissociation constant of phenol in water at $20\,°C$ is 1.21×10^{-10} $mol\,dm^{-3}$. Calculate the percentage of phenol which is ionised in a solution of concentration $0.0100\,mol\,dm^{-3}$.

14. Given that the pK_a of the ammonium ion, NH_4^+, is 9.25 at $25\,°C$, find the pH of an aqueous solution of ammonia of concentration $0.100\,mol$ dm^{-3}. The ionic product for water at $25\,°C$ is $1.00 \times 10^{-14}\,mol^2\,dm^{-6}$.

15. If the acid HA is 1% ionised in a solution of concentration $0.0100\,mol$ dm^{-3}, calculate: a) K_a, b) pK_a.

16. Calculate the pH of hydrochloric acid of concentration 1.00×10^{-3} $mol\,dm^{-3}$. Calculate how many cm^3 of a) hydrochloric acid of concentration $1.00\,mol\,dm^{-3}$, b) sodium hydroxide of concentration $1.00\,mol\,dm^{-3}$ are needed to change the pH of $1.00\,dm^3$ of hydrochloric acid of concentration $1.00 \times 10^{-3}\,mol\,dm^{-3}$ by 1 unit. Ignore the small changes in volume.

BUFFER SOLUTIONS

A buffer solution is one which will resist changes in pH due to the addition of small amounts of acid and alkali. An effective buffer can be made by preparing a solution containing both a weak acid and also one of its salts with a strong base, e.g. ethanoic acid and sodium ethanoate. This will absorb hydrogen ions because they react with ethanoate ions to form molecules of ethanoic acid:

$$CH_3CO_2^- + H_3O^+ \rightleftharpoons CH_3CO_2H + H_2O$$

Hydroxide ions are absorbed by combining with ethanoic acid molecules to form ethanoate ions and water:

$$OH^- + CH_3CO_2H \rightleftharpoons CH_3CO_2^- + H_2O$$

A solution of a weak base and one of its salts formed with a strong acid, e.g. ammonia solution and ammonium chloride, will act as a buffer. If hydrogen ions are added, they combine with ammonia, and, if hydroxide ions are added, they combine with ammonium ions:

$$NH_3 + H_3O^+ \rightleftharpoons NH_4^+ + H_2O$$

$$OH^- + NH_4^+ \rightleftharpoons NH_3 + H_2O$$

The pH of a buffer solution consisting of a weak acid and its salt is calculated from the equation

$$K_a = \frac{[H_3O^+][A^-]}{[HA]}$$

$$[H_3O^+] = K_a \frac{[HA]}{[A^-]}$$

$$pH = pK_a + \lg \frac{[A^-]}{[HA]}$$

Since the salt is completely ionised and the acid only slightly ionised, one can assume that all the anions come from the salt, and put

$$[A^-] = [Salt]$$

$$[HA] = [Acid]$$

$$\therefore \qquad pH = pK_a + \lg \frac{[Salt]}{[Acid]}$$

For a buffer made from a base, B, and its salt with a strong acid, BH^+X^-,

$$K_b = \frac{[BH^+][OH^-]}{[B]}$$

$$pOH = pK_b + \lg \frac{[BH^+]}{[B]}$$

and
$$pH = pK_w - pK_b + \lg \frac{[B]}{[BH^+]}$$

Since the weak base is only slightly ionised, one can put

$$[B] = [Base\ added]$$
$$[BH^+] = [Salt\ added]$$

∴
$$pH = pK_w - pK_b + \lg \frac{[Base]}{[Salt]}$$

EXAMPLE 1 Three solutions contain propanoic acid ($K_a = 1.34 \times 10^{-5}\ mol\ dm^{-3}$) at a concentration of $0.10\ mol\ dm^{-3}$ and sodium propanoate at concentrations of a) $0.10\ mol\ dm^{-3}$, b) $0.20\ mol\ dm^{-3}$, c) $0.50\ mol\ dm^{-3}$ respectively. Calculate the pH values of the three solutions.

METHOD
$$pH = pK_a + \lg \frac{[Salt]}{[Acid]}$$

In solution a), $pH = 4.87 + \lg \dfrac{0.10}{0.10} = 4.87 + \lg 1.0$

ANSWER $pH(a) = 4.87$

In solution b), $pH = 4.87 + \lg \dfrac{0.20}{0.10} = 4.87 + \lg 2.0$

ANSWER $pH(b) = 5.17$

In solution c), $pH = 4.87 + \lg \dfrac{0.50}{0.10} = 4.87 + \lg 5.0$

ANSWER $pH(c) = 5.57$

EXAMPLE 2 Calculate the pH of solutions of $0.25\ mol\ dm^{-3}$ methylamine ($K_b = 4.54 \times 10^{-4}\ mol\ dm^{-3}$) containing a) $0.25\ mol\ dm^{-3}$ and b) $0.50\ mol\ dm^{-3}$ methylamine hydrochloride.

METHOD
$$pH = pK_w - pK_b + \lg \frac{[Base]}{[Salt]}$$

In solution a), $pH = 14.00 - 3.34 + \lg \dfrac{0.25}{0.25}$

ANSWER $pH(a) = 10.66$

In solution b), $pH = 14.00 - 3.34 + \lg \dfrac{0.25}{0.50}$

ANSWER $pH(b) = 10.36$

EXERCISE 43   Problems on Buffers

1. What fraction of a mole of sodium ethanoate must be added to $1 \, dm^3$ of ethanoic acid of concentration $0.10 \, mol \, dm^{-3}$ and $K_a = 2 \times 10^{-5}$ $mol \, dm^{-3}$ in order to produce a buffer solution of pH = 5?
    a 0.2   b 0.25   c 0.4   d 0.5   e 0.6

2. What amount of sodium ethanoate must be added to $1.00 \, dm^3$ of ethanoic acid of pK_a 4.73, concentration $0.100 \, mol \, dm^{-3}$, to produce a buffer of pH = 5.73?

3. Calculate the pH values of the following solutions.
    a) $20.0 \, cm^3$ of $1.00 \, mol \, dm^{-3}$ nitrous acid ($pK_a = 3.34$) added to $40.0 \, cm^3$ of $0.500 \, mol \, dm^{-3} \, mol \, dm^{-3}$ sodium nitrite solution.
    b) $10.0 \, cm^3$ of $1.00 \, mol \, dm^{-3}$ nitrous acid added to $20.0 \, cm^3$ of $2.00 \, mol \, dm^{-3}$ sodium nitrite solution.

4. a) What is the pH of a solution containing $0.100 \, mol \, dm^{-3}$ of ethanoic acid and $0.100 \, mol \, dm^{-3}$ of sodium ethanoate?
        ($K_a(CH_3CO_2H) = 1.86 \times 10^{-5} \, mol \, dm^{-3}$.)
    b) How many moles of sodium ethanoate must be added to $1.00 \, dm^3$ of $0.0100 \, mol \, dm^{-3}$ ethanoic acid to produce a buffer solution of pH 5.8?

SOLUBILITY AND SOLUBILITY PRODUCT

Many salts which we refer to as insoluble do in fact dissolve to a small extent. In a saturated solution, an equilibrium exists between the dissolved ions and the undissolved salt. For example, in a saturated solution of silver chloride in contact with undissolved silver chloride,

$$AgCl(s) \rightleftharpoons Ag^+(aq) + Cl^-(aq)$$

The product of the concentrations of silver ions and chloride ions is called the solubility product of silver chloride.

$$K_{sp} = [Ag^+][Cl^-]$$

The solubility product of a salt is the product of the concentrations of all the ions in a saturated solution of the salt.

It is different from solubility. *The solubility of a salt is expressed as either the amount of solute (in mol) or the mass of solute (in g) dissolved in $1 \, dm^3$ of solution at a stated temperature.*

Another example of a sparingly soluble salt is lead(II) chloride.

$$PbCl_2(s) \rightleftharpoons Pb^{2+}(aq) + 2Cl^-(aq)$$
$$K_{sp} = [Pb^{2+}][Cl^-]^2$$

If the solubility of $PbCl_2$ is a mol dm^{-3}, then $[Pb^{2+}] = a$ and $[Cl^-] = 2a$

$$K_{sp} = a \times (2a)^2 = 4a^3$$

Calculation of solubility product

EXAMPLE 1 The solubility of lead(II) hydroxide at 25 °C is 6.64×10^{-4} mol dm^{-3}. Calculate its solubility product.

METHOD $Pb(OH)_2(s) \rightleftharpoons Pb^{2+}(aq) + 2OH^-(aq)$

Since the solubility of $Pb(OH)_2$ is 6.64×10^{-4} mol dm^{-3}

$$[Pb^{2+}] = 6.64 \times 10^{-4} \text{ mol dm}^{-3}$$

$$[OH^-] = 2 \times 6.64 \times 10^{-4} \text{ mol dm}^{-3}$$

$$K_{sp} = [Pb^{2+}][OH^-]^2 = 6.64 \times 10^{-4} \times (1.33 \times 10^{-3})^2$$

ANSWER $K_{sp} = 1.17 \times 10^{-9} \text{ mol}^3 \text{ dm}^{-9}$

Note that the units in which the solubility product is expressed are the result of multiplying three concentrations together: $(\text{mol dm}^{-3})^3 = \text{mol}^3 \text{ dm}^{-9}$.

EXAMPLE 2 Given that the solubility product of lead(II) sulphate at 25 °C is 1.60×10^{-8} mol^2 dm^{-6}, calculate the solubility at this temperature.

METHOD Solubility of $PbSO_4$ = Concn of $PbSO_4$ in solution.

All the dissolved $PbSO_4$ is in the form of Pb^{2+} ions and $SO_4{}^{2-}$ ions.

$$\therefore \text{ Solubility of } PbSO_4 = [Pb^{2+}] = [SO_4{}^{2-}]$$

$$\text{Solubility product} = K_{sp} = [Pb^{2+}][SO_4{}^{2-}] = [PbSO_4]^2$$

$$[PbSO_4] = \sqrt{K_{sp}} = \sqrt{1.60 \times 10^{-8}} \text{ mol dm}^{-3}$$

ANSWER $[PbSO_4] = 1.26 \times 10^{-4} \text{ mol dm}^{-3}$

EXAMPLE 3 1.00 dm^3 of a solution of calcium chloride of concentration 0.100 mol dm^{-3} is added to 1.00 dm^3 of sodium hydroxide of concentration 0.100 mol dm^{-3}. If the solubility product of calcium hydroxide is 5.50×10^{-6} mol^3 dm^{-9}, calculate the mass of calcium hydroxide that will be precipitated.

METHOD The maximum concentration of Ca^{2+} which can remain in solution is given by

$$[Ca^{2+}][OH^-]^2 = 5.5 \times 10^{-6} \text{ mol}^3 \text{ dm}^{-9}$$

Let $[Ca^{2+}] = a$;   then $4a^3 = 5.5 \times 10^{-6} \text{ mol}^3 \text{ dm}^{-9}$

$$a = 1.11 \times 10^{-2} \text{ mol dm}^{-3}$$

Amount of Ca^{2+} in $2.00\,dm^3$ = $2.22 \times 10^{-2}\,mol$

Amount of Ca^{2+} added as $CaCl_2$ = $0.100\,mol$

Amount of Ca^{2+} precipitated as $Ca(OH)_2$ = $0.100 - 0.0222$
$$= 0.0778\,mol$$

ANSWER   Mass of $Ca(OH)_2$ precipitated = 0.0778×74.0 = $5.76\,g$

THE COMMON ION EFFECT

In a saturated solution of a salt, MA, in equilibrium with solid MA,

$$MA(s) \rightleftharpoons M^{2+}(aq) + A^{2-}(aq)$$

$$K_{sp} = [M^{2+}][A^{2-}]$$

If a solution containing M^{2+} ions is added, $[M^{2+}]$ is increased. The solubility product, K_{sp}, remains the same, even when the ions are not present in equimolar concentrations. So that the product $[M^{2+}][A^{2-}]$ shall not exceed K_{sp}, M^{2+} ions will be removed from solution as solid MA. Solute will be precipitated from solution. The addition of a solution containing A^{2-} ions will have the same effect. The separation of a solute from a solution on addition of an electrolyte solution which has an ion in common with the solute is an example of the *common ion effect.*

Another example of the common ion effect is the change in the concentration of ions produced by the dissociation of a weak acid in the presence of a solution of one of its ions.

EXAMPLE 1 Calculate the solubility at $25\,°C$ of silver chloride: a) in water, and b) in $0.10\,mol\,dm^{-3}$ hydrochloric acid. The solubility product of silver chloride at $25\,°C$ is $1.8 \times 10^{-10}\,mol^2\,dm^{-6}$.

METHOD   a) Since K_{sp} = $[Ag^+][Cl^-]$

$$[Ag^+]^2 = 1.8 \times 10^{-10}\,mol^2\,dm^{-6}$$

$$[Ag^+] = 1.3 \times 10^{-5}\,mol\,dm^{-3}$$

Concn of AgCl = $1.3 \times 10^{-5}\,mol\,dm^{-3}$

Molar mass of AgCl = $143.5\,g\,mol^{-1}$

ANSWER   Solubility of AgCl = $1.3 \times 10^{-5} \times 143.5$ = $1.9 \times 10^{-3}\,g\,dm^{-3}$.

b) The value of $[Cl^-]$ is the sum of that from the $0.10\,mol\,dm^{-3}$ HCl, and that from the dissolved AgCl. The latter is of the order of $10^{-5}\,mol\,dm^{-3}$, and can be neglected in comparison with $0.10\,mol$ dm^{-3} from the acid

$$K_{sp} = [Ag^+][Cl^-] = [Ag^+](0.10) \, mol^2 \, dm^{-6}$$
$$[Ag^+] = 1.8 \times 10^{-10}/0.10 = 1.8 \times 10^{-9} \, mol \, dm^{-3}$$
$$\text{Concn of AgCl} = 1.8 \times 10^{-9} \, mol \, dm^{-3}$$

ANSWER   Solubility of AgCl $= 1.8 \times 10^{-9} \times 143.5 = 2.6 \times 10^{-7} g \, dm^{-3}$.

The qualitative analysis scheme for the identification of metal cations is an application of solubility products. Many metals are precipitated as sulphides when a solution of the metal cations is treated with hydrogen sulphide. This large group of sparingly soluble sulphides can be divided into two groups. Hydrogen sulphide in acid solution brings down the least soluble sulphides in Group II of the qualitative analysis scheme. The rest are precipitated in Group IV from an alkaline solution of hydrogen sulphide. The effect of hydrogen ion concentration on the ionisation of hydrogen sulphide can be calculated.

EXAMPLE 2   In a saturated solution of hydrogen sulphide,
$$[H_3O^+]^2[S^{2-}] = 1.1 \times 10^{-23} \, mol^3 \, dm^{-9}$$

Calculate the sulphide ion concentration: a) at pH 7, b) at pH 8, and c) at pH 2.

METHOD   a) At pH 7,
$$[H_3O^+] = 10^{-7}$$

ANSWER $[S^{2-}] = 1.1 \times 10^{-23}/(10^{-7})^2 = 1.1 \times 10^{-9} \, mol \, dm^{-3}$

b) At pH 8,

ANSWER $[S^{2-}] = 1.1 \times 10^{-23}/(10^{-8})^2 = 1.1 \times 10^{-7} \, mol \, dm^{-3}$

c) At pH 2,

ANSWER $[S^{2-}] = 1.1 \times 10^{-23}/(10^{-2})^2 = 1.1 \times 10^{-19} \, mol \, dm^{-3}$

The very low sulphide ion concentration at pH 2 will precipitate only the most insoluble metal sulphides.

EXERCISE 44   Problems on Solubility Products

1. Given the following solubilities in $mol \, dm^{-3}$ of solution, calculate the solubility products of the solids listed:

   a) CaS 1.3×10^{-14} b) CoS 6.3×10^{-10}
   c) Ag_2S 1.1×10^{-17} d) $Pb(OH)_2$ 5.0×10^{-6}

2. Given the following solubilities in g per dm^3 of solution, calculate the solubility products of the solids listed:

   a) PbS 1.20×10^{-11}

   b) AgI 2.14×10^{-6}

   c) $BaSO_4$ 2.41×10^{-3}

   d) CaF_2 1.47×10^{-2}

   e) AgCN 1.50×10^{-6}

The following questions require a knowledge of the solubility products listed below:

CuS	$6.3 \times 10^{-36}\, mol^2\, dm^{-6}$	Bi_2S_3	$1.0 \times 10^{-97}\, mol^5\, dm^{-15}$
Ag_2S	$6.3 \times 10^{-51}\, mol^3\, dm^{-9}$	HgS	$1.6 \times 10^{-52}\, mol^2\, dm^{-6}$
NiS	$3.2 \times 10^{-19}\, mol^2\, dm^{-6}$	FeS	$6.3 \times 10^{-18}\, mol^2\, dm^{-6}$
$BaSO_4$	$1.1 \times 10^{-10}\, mol^2\, dm^{-6}$	$Al(OH)_3$	$6.3 \times 10^{-32}\, mol^4\, dm^{-12}$
$CaSO_4$	$2.4 \times 10^{-5}\, mol^2\, dm^{-6}$	SrF_2	$2.4 \times 10^{-9}\, mol^3\, dm^{-9}$
Ag_2SO_4	$1.7 \times 10^{-5}\, mol^3\, dm^{-9}$	$PbCl_2$	$1.6 \times 10^{-5}\, mol^3\, dm^{-9}$

3. What concentration of sulphide ion is needed to precipitate the metal as its sulphide from each of the following solutions?

   a) $CuSO_4(aq)$ $1.0 \times 10^{-2}\, mol\, dm^{-3}$

   b) $AgNO_3(aq)$ $1.0 \times 10^{-4}\, mol\, dm^{-3}$

   c) $NiSO_4(aq)$ $1.0 \times 10^{-5}\, mol\, dm^{-3}$

   d) $Bi(NO_3)_3(aq)$ $1.0 \times 10^{-4}\, mol\, dm^{-3}$

   e) $Hg(NO_3)_2(aq)$ $1.0 \times 10^{-6}\, mol\, dm^{-3}$

   f) $FeSO_4(aq)$ $1.0 \times 10^{-6}\, mol\, dm^{-3}$

4. Will a precipitate appear when the following solutions are added?

   a) $10\, cm^3\, BaCl_2$ $(0.01\, mol\, dm^{-3})$ and
      $10\, cm^3\, Na_2SO_4$ $(0.1\, mol\, dm^{-3})$

   b) $25\, cm^3\, Ca(OH)_2$ $(8 \times 10^{-3}\, mol\, dm^{-3})$ and
      $25\, cm^3\, Na_2SO_4$ $(0.01\, mol\, dm^{-3})$

   c) $50\, cm^3\, AlCl_3$ $(10^{-3}\, mol\, dm^{-3})$ and
      $50\, cm^3\, NaOH$ $(10^{-2}\, mol\, dm^{-3})$

   d) $10\, cm^3\, AgNO_3$ $(10^{-3}\, mol\, dm^{-3})$ and
      $40\, cm^3\, Na_2SO_4$ $(0.1\, mol\, dm^{-3})$

   e) $100\, cm^3\, Sr(NO_3)_2$ $(10^{-2}\, mol\, dm^{-3})$ and
      $100\, cm^3\, KF$ $(2 \times 10^{-2}\, mol\, dm^{-3})$

   f) $250\, cm^3\, Pb(NO_3)_2$ $(2 \times 10^{-2}\, mol\, dm^{-3})$ and
      $150\, cm^3\, NaCl$ $(0.01\, mol\, dm^{-3})$

Show how you arrive at your conclusions.

5. In a saturated aqueous solution of hydrogen sulphide, the product
$$[H_3O^+]^2[S^{2-}] = 1.1 \times 10^{-23} \, mol^3 \, dm^{-9}$$
   The solubility products of four sulphides are:

   CdS, $3.6 \times 10^{-29} \, mol^2 \, dm^{-6}$
   FeS, $3.7 \times 10^{-19} \, mol^2 \, dm^{-6}$
   MnS, $1.4 \times 10^{-15} \, mol^2 \, dm^{-6}$
   NiS, $1.4 \times 10^{-24} \, mol^2 \, dm^{-6}$

   A solution contains each of the metal ions at a concentration of $0.10 \, mol \, dm^{-3}$ and $0.25 \, mol \, dm^{-3}$ hydrochloric acid. The solution is saturated with hydrogen sulphide. Calculate which of the sulphides will be precipitated.

6. In the estimation of chlorides by titration with a standard silver nitrate solution, using a chromate indicator, the precipitation of silver chloride is complete before the precipitation of silver chromate begins. Explain why this is so, using the solubility of silver chloride ($2.009 \times 10^{-3} \, g \, dm^{-3}$) and silver chromate ($3.207 \times 10^{-2} \, g \, dm^{-3}$) at 25 °C to calculate the solubility products and then: a) the concentration of silver ions needed to precipitate silver chloride from a neutral solution of chloride ions of concentration $0.100 \, mol \, dm^{-3}$, and b) the concentration of silver ions required to precipitate silver chromate, Ag_2CrO_4, from a neutral solution of chromate ions containing $5.00 \times 10^{-3} \, mol \, dm^{-3}$.

7. A solution contains $0.10 \, mol \, dm^{-3}$ of sodium carbonate and $0.10 \, mol \, dm^{-3}$ of sodium sulphate. To $1 \, dm^3$ of the solution is added $0.10 \, mol$ calcium chloride. The solubility products are: $CaCO_3$, 1.7×10^{-8}; $CaSO_4$, $2.3 \times 10^{-4} \, mol^2 \, dm^{-6}$. Find out which salt will be precipitated and calculate the mass of the precipitate.

8. The solubility product of mercury(II) sulphide is quoted in one reference book as $2 \times 10^{-49} \, mol^2 \, dm^{-6}$. If this value is correct, how many mercury(II) ions will be present in $1 \, dm^3$ of a saturated solution of this salt? (The Avogadro constant is $6 \times 10^{23} \, mol^{-1}$.)

9. The solubility product of $PbBr_2$ is 7.9×10^{-5}; that of PbI_2 is $1.0 \times 10^{-9} \, mol^3 \, dm^{-9}$; $1.0 \, dm^3$ of a solution containing $0.20 \, mol \, dm^{-3}$ of sodium bromide and $0.20 \, mol \, dm^{-3}$ of sodium iodide is added to $1.0 \, dm^3$ of lead(II) nitrate solution of concentration $0.10 \, mol \, dm^{-3}$. Which salt is precipitated? What is the mass of the precipitate?

10. Calculate the solubility of magnesium hydroxide: a) in water, b) in $0.10 \, mol \, dm^{-3}$ sodium hydroxide solution, c) in $0.010 \, mol \, dm^{-3}$ magnesium chloride solution, all at 25 °C. The solubility product of magnesium hydroxide is $1.1 \times 10^{-11} \, mol^3 \, dm^{-9}$ at 25 °C.

11. $K_{sp}(SrSO_4) = 4.0 \times 10^{-7} mol^2 dm^{-6}$. Calculate the solubility in mol dm^{-3} of $SrSO_4$ a) in water, b) in $0.10 \, mol \, dm^{-3}$ aqueous sodium sulphate.

12. $K_{sp}(MgF_2) = 7.2 \times 10^{-9} mol^3 dm^{-9}$. Calculate the solubility in mol dm^{-3} of MgF_2 a) in water, b) in a $0.20 \, mol \, dm^{-3}$ solution of NaF.

13. Calculate the solubility of calcium fluoride in $mol \, dm^{-3}$ a) in water, b) in a $0.010 \, mol \, dm^{-3}$ solution of sodium fluoride, c) in a $1.0 \, mol \, dm^{-3}$ solution of hydrogen fluoride. ($K_{sp}(CaF_2) = 4.0 \times 10^{-11} mol^3 dm^{-9}$, $K_a(HF) = 5.6 \times 10^{-4} mol \, dm^{-3}$.)

ELECTRODE POTENTIALS

If a strip of metal is placed in a solution of its ions, atoms of the metal may dissolve as positive ions, leaving a build-up of electrons on the metal:

$$M(s) \longrightarrow M^{2+}(aq) + 2e^-$$

The metal will become negatively charged. Alternatively, metal ions may take electrons from the strip of metal and be discharged as metal atoms:

$$M^{2+}(aq) + 2e^- \longrightarrow M(s)$$

In this case, the metal will become positively charged. The potential difference between the strip of metal and the solution depends on the nature of the metal and on the concentration of the ions involved in the equilibrium at the metal surface. Zinc acquires a more negative potential than copper, since it has a greater tendency to dissolve as ions and a smaller tendency to be deposited as metal. In order to compare electrode potentials for different metals, *standard electrode potentials* are quoted at 25 °C with an ionic concentration of 1 mol dm^{-3}. The zero on the standard electrode potential scale is the potential of a strip of platinum in contact with hydrogen gas at 1 atm pressure and hydrogen ions at a concentration of 1 mol dm^{-3}.

Metals are reducing agents. Other oxidation–reduction systems also have electrode potentials, the value of which depend on the standard electrode potential for the system and the concentrations of the ions in the equilibrium. The standard electrode potential of a redox system is the potential acquired by a piece of platinum immersed in a solution of the redox system in which the concentration of each dissolved component is 1 mol dm^{-3}. A powerful oxidising agent removes electrons and gives the platinum a high positive potential. When all the redox systems are arranged in order of their standard electrode potentials, the *electrochemical serie* is obtained. Table 12.1 shows some of the redox systems in the series.

Table 12.1   Values of standard electrode potential $E^{\ominus}$ at 298 K

Reaction	$E^{\ominus}/V$
$K^+(aq) + e^- \longrightarrow K(s)$	-2.92
$Ca^{2+}(aq) + 2e^- \longrightarrow Ca(s)$	-2.87
$Na^+(aq) + e^- \longrightarrow Na(s)$	-2.71
$Mg^{2+}(aq) + 2e^- \longrightarrow Mg(s)$	-2.36
$Al^{3+}(aq) + 3e^- \longrightarrow Al(s)$	-1.66
$Zn^{2+}(aq) + 2e^- \longrightarrow Zn(s)$	-0.76
$Fe^{2+}(aq) + 2e^- \longrightarrow Fe(s)$	-0.44
$Cr^{3+}(aq) + e^- \longrightarrow Cr^{2+}(aq)$	-0.41
$Ni^{2+}(aq) + 2e^- \longrightarrow Ni(s)$	-0.25
$Sn^{2+}(aq) + 2e^- \longrightarrow Sn(s)$	-0.14
$Pb^{2+}(aq) + 2e^- \longrightarrow Pb(s)$	-0.13
$2H_3O^+(aq) + 2e^- \longrightarrow H_2(g) + 2H_2O(l)$	0.00
$Sn^{4+}(aq) + 2e^- \longrightarrow Sn^{2+}(aq)$	0.15
$Cu^{2+}(aq) + 2e^- \longrightarrow Cu(s)$	0.34
$I_2(s) + 2e^- \longrightarrow 2I^-(aq)$	0.54
$Fe^{3+}(aq) + e^- \longrightarrow Fe^{2+}(aq)$	0.77
$Ag^+(aq) + e^- \longrightarrow Ag(s)$	0.80
$Br_2(l) + 2e^- \longrightarrow 2Br^-(aq)$	1.09
$MnO_2(s) + 4H^+(aq) + 2e^- \longrightarrow Mn^{2+}(aq) + 2H_2O(l)$	1.23
$Cr_2O_7^{2-}(aq) + 14H^+(aq) + 6e^- \longrightarrow 2Cr^{3+}(aq) + 7H_2O(l)$	1.33
$Cl_2(g) + 2e^- \longrightarrow 2Cl^-(aq)$	1.36
$Ce^{4+}(aq) + e^- \longrightarrow Ce^{3+}(aq)$ (in $H_2SO_4(aq)$)	1.44
$PbO_2(s) + 4H^+(aq) + 2e^- \longrightarrow Pb^{2+}(aq) + 2H_2O(l)$	1.46
$MnO_4^-(aq) + 8H^+(aq) + 5e^- \longrightarrow Mn^{2+}(aq) + 4H_2O(l)$	1.51
$Ce^{4+}(aq) + e^- \longrightarrow Ce^{3+}(aq)$ (in $HNO_3(aq)$)	1.61
$H_2O_2(aq) + 2H^+(aq) + 2e^- \longrightarrow 2H_2O(l)$	1.78
$F_2(g) + 2e^- \longrightarrow 2F^-(aq)$	2.85

GALVANIC CELLS

When two electrodes are combined to form a cell, their standard electrode potentials will tell you which will be the positive and which the negative electrode. An easy way to work out which of the possible reactions will happen is to use the *anticlockwise rule*. Write down the two redox systems, with the more negative standard electrode potential at the top. Then draw a circle anticlockwise. For example, when copper and silver are in contact with solutions of their ions,

$$Cu^{2+}(aq) \; + \; 2e^- \; \rightleftharpoons \; Cu(s) \quad E^{\ominus} = +0.34\,V$$
$$Ag^+(aq) \; + \; e^- \; \rightleftharpoons \; Ag(s) \quad E^{\ominus} = +0.80\,V$$

The circle tells you that the reaction which takes place is

$$Cu(s) \; + \; 2Ag^+(aq) \; \longrightarrow \; Cu^{2+}(aq) \; + \; 2Ag(s)$$

The silver electrode is positive; the copper electrode is negative.

Reaction will take place between two redox systems which differ by 0.3 V or more.

Fig. 12.2 shows two metals inserted into solutions of their ions. The two solutions are joined by a salt bridge, and the two metal electrodes are connected by an external circuit. The cell has an e.m.f. which is equal to the difference between the standard electrode potentials of the two metals, and a current flows through the external circuit.

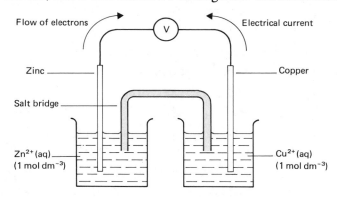

Fig. 12.2   A galvanic cell

The cell shown in Fig. 12.2 can be represented by

$$Zn(s) \mid Zn^{2+}(aq)\,(1\,mol\,dm^{-3}) \; \vdots \; Cu^{2+}(aq)\,(1\,mol\,dm^{-3}) \mid Cu(s)$$

By convention, the e.m.f. of the cell is taken as

$$E = E(RHS\ electrode) - E(LHS\ electrode)$$
$$= E^{\ominus}_{Cu} - E^{\ominus}_{Zn} \quad \text{where } E^{\ominus} \text{ is the standard electrode potential}$$
$$= 0.34 - (-0.76) = +1.10\,V$$

The flow of electrons is clockwise through the external circuit (from zinc to copper). Conventional electricity flows anticlockwise through the external circuit (from copper to zinc).

If the cell is written as

$$Cu(s) \mid Cu^{2+}(aq)\ (1\ mol\ dm^{-3}) \mathrel{\vdots} Zn^{2+}(aq)\ (1\ mol\ dm^{-3}) \mid Zn(s)$$

then the e.m.f. is given by

$$E = E^{\ominus}_{Zn} - E^{\ominus}_{Cu} = -0.76 - 0.34 = -1.10\ V$$

In general, in a cell A | B | C | D if the reactions which occur are A $\longrightarrow$ B and C $\longrightarrow$ D, then the e.m.f. is positive; if B $\longrightarrow$ A and D $\longrightarrow$ C, then the e.m.f. is negative.

EXERCISE 45   Problems on Standard Electrode Potentials

Refer to the table of values on p. 176.

1. Which of the following species are oxidised by manganese(IV) oxide?
   Br⁻,   Ag,   I⁻,   Cl⁻

2. Which of the following species are reduced by Sn^{2+}?
   I_2, Ni^{2+}, Cu^{2+}, Fe^{3+}

3. Calculate the standard e.m.f.'s of the following cells at 298 K:
   a) $Ni(s) \mid Ni^{2+}(aq) \mathrel{\vdots} Sn^{2+}(aq), Sn^{4+}(aq) \mid Pt$
   b) $Pt \mid I_2(s), I^-(aq) \mathrel{\vdots} Ag^+(aq) \mid Ag(s)$
   c) $Pt \mid Cl_2(g), Cl^-(aq) \mathrel{\vdots} Br_2(l), Br^-(aq) \mid Pt$

4. Calculate the standard e.m.f. of each of the cells:
   a) $Sn(s) \mid Sn^{2+}(aq) \mathrel{\vdots} Ag^+(aq) \mid Ag(s)$
   b) $Ag(s) \mid Ag^+(aq) \mathrel{\vdots} Cu^{2+}(aq) \mid Cu(s)$
   c) $Ce^{3+}(aq) \mid Ce^{4+}(aq) \mathrel{\vdots} Fe^{3+}(aq) \mid Fe^{2+}(aq)$
   d) $Fe(s) \mid Fe^{2+}(aq) \mathrel{\vdots} Cu^{2+}(aq) \mid Cu(s)$
   e) $Zn(s) \mid Zn^{2+}(aq) \mathrel{\vdots} Pb^{2+}(aq) \mid Pb(s)$

5. Iron filings are added to a solution containg the ions Cu^{2+}, Fe^{2+}, Fe^{3+}, H_3O^+ and Zn^{2+}, all at a concentration of $1\ mol\ dm^{-3}$. From the standard electrode potentials of the redox systems, deduce what reaction occurs, and write the equation.

6. A solution contains Fe^{2+}, Fe^{3+}. Cr^{3+} and $Cr_2O_7{}^{2-}$ in their standard states and dilute sulphuric acid. Deduce what happens, and write the equation for the reaction.

7. Predict the reactions between:
   a) $Fe^{3+}(aq)$ and $I^-(aq)$ b) $Ag^+(aq)$ and $Cu(s)$
   c) $Fe^{3+}(aq)$ and $Br^-(aq)$ d) $Ag(s)$ and $Fe^{3+}(aq)$
   e) $Br_2(aq)$ and $Fe^{2+}(aq)$
   From the standard electrode potentials, predict which of the halogens, Cl_2, Br_2, I_2, will oxidise i) Fe^{2+} to Fe^{3+} ii) Sn^{2+} to Sn^{4+}.

EXERCISE 46   Questions from A-level Papers

1. The electrochemical cell

$$Pb(s) \mid Pb(NO_3)_2(aq, 1.0\,M) \parallel SnCl_2(aq, 1.0\,M) \mid Sn(s)$$

was set up in the laboratory. A salt bridge was used to connect the two electrode compartments. The standard electrode potentials for lead and tin are

$$Pb^{2+}(aq) + 2e^- \longrightarrow Pb(s) \qquad E^\ominus = -0.126\,V$$
$$Sn^{2+}(aq) + 2e^- \longrightarrow Sn(s) \qquad E^\ominus = -0.136\,V$$

a) Name the reference electrode against which other standard electrode potentials are measured.

b) State the function of the salt bridge.

c) What is the e.m.f. of the cell given above?

d) If the cell is part of an electrical circuit, from which electrode will electrons move in the external circuit? Explain your answer.

(JMB92)

2. a) Define pH.

b) i) Define the ionic product of water, K_w, and state appropriate units for it.

ii) If the pH of pure water at 5 °C is 7.6, derive the value for K_w at this temperature.

c) Calculate the pH of 0.05 M aqueous sodium hydroxide at 5 °C.

(JMB91)

3. The apparatus shown below was used to measure the standard electrode potential of cobalt $Co^{2+}(aq) + 2e^- = Co(s)$.

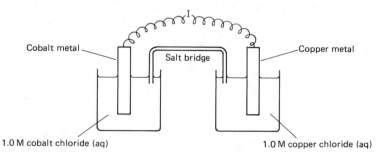

Cobalt metal — Salt bridge — Copper metal

1.0 M cobalt chloride (aq)   1.0 M copper chloride (aq)

a) i) Which instrument would you use at position I to measure the e.m.f. of the cell?

ii) Which salt would you choose for the salt bridge? Give a reason for your choice.

iii) The e.m.f. of the cell was measured as 0.62 volts. Taking the

standard electrode potential of Cu/Cu^{2+} as +0.34 volts calculate the standard electrode potential of the Co/Co^{2+} system.
Carefully indicate whether your answer is positive or negative.

iv) State the direction in which the electrons flow.

b) i) What would you *see* if pieces of copper metal were added to an
aqueous solution of cobalt(II) chloride?

ii) What would you *see* if pieces of cobalt metal were added to an
aqueous solution of copper(II) chloride?

c) What would be the e.m.f. of the cell if the cobalt electrode system
were removed and replaced by a standard hydrogen electrode?

(O&C90,AS)

*4. a) Read the passage below and answer the questions that follow.

When an ionic salt dissolves in water, the resulting solution conducts
electricity. The *conductivity* (conducting power) of the solution
depends both on the nature of the ionic species present and on
their concentration. The table below shows the relative conductivities in water of a number of ions at equal concentrations.

	H$^+$(aq)	OH$^-$(aq)	Cl$^-$(aq)	NH$_4^+$(aq)	Na$^+$(aq)	CH$_3$COO$^-$(aq)
Relative conductivity	1.00	0.56	0.22	0.21	0.14	0.12

In a *conductimetric titration,* the conductivity of an acid or base
(often at a concentration of 0.10 mol l^{-1}) is monitored as portions
of base or acid (usually at a concentration of 1.0 mol l^{-1}) are
added. The concentration difference is chosen to minimise the
diluting effect that might otherwise mask the underlying conductivity changes as ions of one kind are replaced by those of another.

The graph below shows the results of four conductimetric titrations
(*I* to *IV*) in which 50 cm^3 portions of dilute acid or base (0.10 mol l^{-1})
were titrated with more concentrated solutions (1.0 mol l^{-1}) of
base or acid.

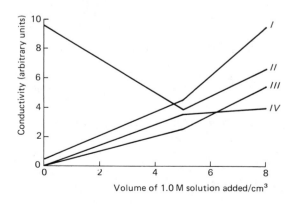

The solutions were taken from among those shown below.

	HCl(aq)	CH₃COOH(aq)	NaOH(aq)	NH₃(aq)
0.10 M solution	A	B	C	D
1.0 M solution	Q	R	S	T

Estimate the order of relative conductivity of the four possible salt solutions which could be obtained at equivalence in order to determine which acid–base pairs are involved in I to IV. Then examine the graph and, on the basis of the conductivity changes shown both before and after equivalence (no further calculation required), determine which solution (Q, R, S or T) is being added to which acid or base (A, B, C or D). Justify your reasoning.

b) The table below shows degrees of dissociation, α, of aqueous solutions of methanoic (*formic*) acid at different concentrations c.

$1000c/\text{mol l}^{-1}$	6.80	11.0	16.2	25.9	68.4	194
10α	1.5	1.2	1.0	0.80	0.50	0.30

Draw up a table containing values of $1000\alpha^2$ and $1/c$ and then plot a graph of these two quantities. Use the graph to determine the acid dissociation constant, K_c, for methanoic acid. Justify the method you use.   (JMB90,S)

5. The water of Lake Nakuru in the Kenyan rift valley contains dissolved sodium carbonate and sodium hydrogencarbonate. The following equilibrium exists:

$$HCO_3^-(aq) \rightleftharpoons H^+(aq) + CO_3^{2-}(aq)$$

a) Explain how this solution acts as a buffer on the addition of either acid or alkali.

b) The pH of Lake Nakuru is 10.3 and the ratio $\dfrac{[CO_3^{2-}(aq)]}{[HCO_3^-(aq)]}$ is 0.958. Calculate the equilibrium constant for the above reaction.

c) When 10.0 cm³ of lake water were titrated with 0.20 mol dm⁻³ HCl, 22.0 cm³ of acid were required to neutralise all the carbonate and hydrogencarbonate ions according to the following equations:

$$H^+(aq) + HCO_3^-(aq) \longrightarrow H_2O(l) + CO_2(g)$$
$$2H^+(aq) + CO_3^{2-}(aq) \longrightarrow H_2O(l) + CO_2(g)$$

Calculate the total number of moles of acid used, and thus by using the ratio quoted in part b), calculate $[HCO_3^-(aq)]$ and $[CO_3^{2-}(aq)]$ in the lake.   (C91)

6. Apple juice has a pH of 3.5.

a) i) Define pH.

ii) Calculate the molar concentration of hydrogen ions in apple juice.

Apple juice can be titrated with standard alkali.

A $25.0 \, cm^3$ sample of apple juice was exactly neutralised by $27.5 \, cm^3$ of $0.10 \, mol \, dm^{-3}$ sodium hydroxide using phenolphthalein as indicator.

b) Assuming that apple juice contains a single acid which is monobasic, calculate the molar concentration of the acid in the juice.

c) i) How can you explain the difference between the two results you have obtained in a) ii) and b)?

ii) What constant can be determined from these two results?

iii) Calculate a numerical value of this constant.

d) Suggest *two* reasons why phenolphthalein is a suitable indicator for this titration.   (C90)

7. The first part of this question concerns bonding, the second is about pH and pH calculations.

a) Give an account of the nature of the ionic bond *and* of the covalent bond. Consider in your answer, for *both* types of bond,

i) the processes which are thought to occur when these bonds form

ii) the nature of the accompanying energy changes

iii) the ways in which the strengths of the two types of bond are described

iv) the electron density distributions in the bonds.

Wherever possible, discuss the similarities and differences between the two types of bond.

b) i) Calculate the pH of an aqueous solution containing 0.1 mol dm^{-3} of HCl.

ii) Calculate the pH *change* of a given volume of pure water when an equal volume of $0.1 \, mol \, dm^{-3}$ HCl solution is added to it. (N.B. The final volume is twice the original volume of the HCl solution.)

iii) Calculate the pH of a solution which is $1.0 \, mol \, dm^{-3}$ with respect to sodium ethanoate and $1.0 \, mol \, dm^{-3}$ with respect to ethanoic acid, given that the acid dissociation constant (K_a) for ethanoic acid has a value of $1.8 \times 10^{-5} \, mol \, dm^{-3}$ at 298 K.

iv) Thus calculate the pH *change* in the solution in b) iii) above when an equal volume of $0.1 \, mol \, dm^{-3}$ HCl solution is added to it, given that the final pH of this mixture is 4.70.

v) Compare the pH changes calculated in b) ii) and b) iv) above and use your answers to discuss the mode of action of buffer solutions. State specifically what happens to most of the hydrogen ions present in the $0.1 \, mol \, dm^{-3}$ HCl solution when it is mixed with the solution described in b) iii).   (WJEC91)

*8. a) The pH of acid rain is 5. What is the hydrogen ion concentration of such rain?

b) The hydrogen ion concentration of a solution can be measured using an electrochemical cell, and this is the basis of the pH meter. Describe the electrode sensitive to hydrogen ions which is used, and state the other essential parts of the meter.

c) Such a pH meter is usually calibrated by using a buffer solution.

1.00 dm^3 of a buffer solution, pH 5.00, was prepared using ethanoic acid and sodium ethanoate. The solution contained 50.0 g of anhydrous sodium ethanoate. What mass of ethanoic acid did it contain? The K_a for ethanoic acid is 1.8×10^{-5} mol dm^{-3}.

d) i) If 1.00 cm^3 of 10.0 mol dm^{-3} hydrochloric acid is added to 1.00 dm^3 of the buffer solution, what will be the new pH of the solution?

ii) What would have been the pH of a solution obtained by adding 1.00 cm^3 of 10.0 mol dm^{-3} hydrochloric acid to 1.00 dm^3 of aqueous acid of pH 5.00.

iii) Explain the difference between these two results.   (O91,S)

*9. a) The standard electrode potentials at 298 K for two half-reactions are shown below.

$$Zn^{2+}(aq) \ + \ 2e^- \longrightarrow \ Zn(s) \qquad -0.76 \, V$$
$$Cu^{2+}(aq) \ + \ 2e^- \longrightarrow \ Cu(s) \qquad +0.34 \, V$$

Describe what is observed when a piece of zinc is placed in an aqueous solution of copper(II) sulphate, and account for what happens in terms of the given data.

b) Draw a labelled diagram of a simple electrochemical cell which has two electrodes, one of copper and one of zinc immersed in electrolytes, separated by a porous barrier, and explain carefully how the cell operates to generate an electromotive force.

What will eventually cause such a cell to fail as a power source?

c) As an alternative to the porous barrier in b), a cell could be formed from two half-cells connected by a *salt bridge*. Of what could such a salt bridge consist, what is its function, and how does it achieve it?

d) An electric current is passed in series through i) an aqueous solution of silver nitrate and ii) an aqueous solution of copper(II) sulphate, using platinum electrodes. After 60 s, 35 mg of silver has been deposited in the first cell. How much copper will have been deposited in the second cell?   (O90,S)

10. a) Describe how you could determine a value for the Avogadro constant by electrolysis. You should describe the apparatus and chemicals that you would use, the measurements you would make, and any other data you need to know, and the calculations you would carry out.

b) Calculate the ratio of the mass of silver to the mass of nickel deposited on the cathodes when the same current is passed through electrolytic cells containing aqueous silver(I) nitrate and aqueous nickel(II) sulphate, connected in series.   (C92)

11. One method of rust prevention involves electroplating of the iron with chromium. The electrolyte used for this is sodium chromate (Na_2CrO_4). The mechanism of the electroplating process is complex, but the principal two stages can be summarised as follows:

$$Na_2CrO_4 \longrightarrow Cr^{n+} \longrightarrow Cr \text{ metal}$$

(where Cr^{n+} is a positive ion formed by chromium).

a) What are the oxidation states of chromium in:
   i) sodium chromate
   ii) chromium metal?

b) Give the formulas of both ions in sodium chromate.

c) Apart from rust protection, what other advantages does chromium plating have?

d) Draw a labelled diagram to show the circuit used for electroplating an iron object with chromium.

e) When a current of 2.50 A is passed through a solution containing cation Cr^{n+} for 50.0 minutes, it is found that 1.35 g of chromium is deposited. Use this information to calculate n. (The Faraday constant = 96 500 C mol^{-1}.)

f) Hence write the ionic equation for the reaction which produces the chromium metal.

g) Construct the ion-electron equation for the process in which the chromate ion is converted to Cr^{n+}. Justify the statement that this process is reduction.

h) State *two* other methods of rust prevention.   (O91,AS)

12. Hydrochloric acid, HCl, of concentration 0.10 mol dm^{-3} has a pH of 1.0 whereas ethanoic acid, CH_3CO_2H, of the same concentration has a pH of 2.9.

a) Define pH.

b) Calculate the concentration, in mol dm^{-3}, of a solution of hydrochloric acid which has a pH of 2.9.

c) Why do you need different concentrations of these two acids to give solutions of the same pH value?

d) 25 cm^3 of hydrochloric acid, of concentration 0.10 mol dm^{-3}, and 25 cm^3 of ethanoic acid of the same concentration are separately reacted with excess zinc powder. What will be the difference in the total volume of hydrogen produced in the two experiments?

e) A solution of chloroethanoic acid, $ClCH_2-CO_2H$, of concentration 0.1 mol dm^{-3} has a pH of 1.95.

   i) What does this tell you about the relative strengths of chloroethanoic acid and ethanoic acid?

   ii) Suggest an explanation for this difference in strengths.

                                               (C91,AS)

13. The standard electrode potentials for copper and zinc are:

$$Cu^{2+}(aq) + 2e^- = Cu(s) \qquad E^\ominus \ +0.34 \text{ volts}$$

$$Zn^{2+}(aq) + 2e^- = Zn(s) \qquad E^\ominus \ -0.76 \text{ volts}$$

a) Two sheets of metal, one of copper and one of zinc, were suspended in a dilute solution of copper sulphate as shown in the diagram:

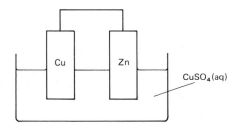

   i) Write the equation for the reaction that will take place.

   ii) Describe what you would expect to see 1) happening to the zinc, 2) happening to the copper.

   iii) What would be the e.m.f. of a cell made from a zinc half-cell and a copper half-cell if it were operating under standard conditions?

   iv) The system was left until no further change took place. A sample of the solution was removed and treated with dilute sodium hydroxide solution. A mixture of two precipitates was produced — one a white precipitate and the other a blue one. Name the two precipitates.

b) The experiment was repeated but this time the arrangement was as in the following diagram:

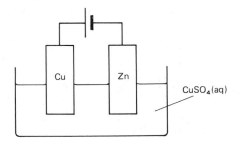

An electrical current, at a potential of 2.5 volts, was passed through the system such that the zinc was the negative electrode (cathode) and the copper the positive electrode (anode).

i) Write the ionic half-equation for what will happen at the copper electrode.

ii) Write the ionic half-equation for what will happen at the zinc electrode.

iii) The system was left until no further change took place. A sample of the solution was removed and treated with a dilute solution of sodium hydroxide. Only one precipitate was produced. Give the name of that precipitate and its colour. Give a reason for your answer. (C91,AS)

14. a) i) Explain what is meant by the term *acid dissociation constant*, K_a, and indicate how pK_a values can be used to compare the relative strengths of weak acids.

ii) The pH of solutions of ethanoic acid and trichloroethanoic acid, both of concentration $1.00 \, mol \, dm^{-3}$, are 2.38 and 0.43 respectively.
Calculate the K_a values for both acids and explain their relative values.

b) Explain the words *acid* and *conjugate base* according to the Bronsted–Lowry theory. Illustrate your answer with a suitable example.

Explain how the equation given below can be said to represent a reaction between an acid and a base.

$$NH_4^+Cl^- \; + \; Na^+NH_2^- \longrightarrow Na^+Cl^- \; + \; 2NH_3$$

(AEB90)

*15. This question is about the equilibrium between dichromate(VI) ions and chromate(VI) ions in aqueous solution.

a) i) Write an expression for the equilibrium constant K_c.

ii) Calculate the value of K_c in a 1.5 M solution of potassium dichromate(VI) which has a pH of 3.13 at 25 °C.

iii) Barium dichromate(VI) is freely soluble in water; barium chromate(VI) is not. When a certain solution of barium chloride is added to the potassium dichromate(VI) solution in a) above, the pH falls from 3.13 to less than 3.00. When an equally concentrated solution of sodium chloride is added to the potassium dichromate(VI) solution, the pH rises. Explain these observations.

b) If an ionic solid AB is in contact with its saturated aqueous solution then, at constant temperature, the *ionic product* K_s equals $[A^{n+}(aq)] \, [B^{n-}(aq)]$ and is the equilibrium constant for the reaction

$$AB(s) \rightleftharpoons A^{n+}(aq) \; + \; B^{n-}(aq)$$

The solubility s of AB at the given temperature can be calculated from this equilibrium constant. Explain why this is so and show how K_s is related to s.

The table below shows equilibrium concentrations c of dichromate(VI) ions and the ratio $r(= [H^+]/[Ba^{2+}])$ of the concentrations, observed simultaneously, of hydrogen ions and barium ions which co-exist in the presence of solid barium chromate(VI) at a fixed temperature.

$c/\text{mol}\,1^{-1}$	0.049	0.13	0.20	0.24	0.30
r	1.10	1.80	2.22	2.44	2.73

Plot a graph of c against the square of r and from it determine the solubility in $\text{mol}\,1^{-1}$ of barium chromate(VI) at this temperature.

(JMB91,S)

*16. This question involves some calculations and inferences on the 'greenhouse effect', i.e. the effect of burning fossil fuels on the Earth's environment.

a) At present the atmosphere contains about 350 ppm (parts per million) of carbon dioxide, giving a total mass of 2×10^{18} g. Man is currently burning fuels equivalent to 5×10^{15} g of carbon per annum.

Calculate the annual increase in ppm, of CO_2 assuming that all the gas formed remains in the atmosphere. ($A_r(C) = 12.01$, $A_r(O) = 16.00$.)

b) Measurements of atmospheric CO_2 content show a 'zig-zag' increase with annual minima in mid-summer.

Suggest a reason for this fact.

c) In reality about half of the CO_2 formed dissolves in the oceans to form initially the weak dibasic carbonic acid (H_2CO_3) for which the acid dissociation constants are $K_1 = 5 \times 10^{-7}\,\text{mol}\,\text{dm}^{-3}$, and $K_2 = 5 \times 10^{-11}\,\text{mol}\,\text{dm}^{-3}$ at 20°C.

Given that the pH of sea water is 8.0 and that the concentration of H_2CO_3 in it is $1.0 \times 10^{-5}\,\text{mol}\,\text{dm}^{-3}$ at 20°C and 1 atmosphere pressure, calculate the concentration of carbonate ion, CO_3^{2-}, in sea water.

d) The calcium ion, Ca^{2+}, concentration in sea water is $1.0 \times 10^{-2}\,\text{mol}\,\text{dm}^{-3}$ at 20°C and the solubility product K_{sp} of $CaCO_3$ at 20°C is $5 \times 10^{-9}\,\text{mol}^2\,\text{dm}^{-6}$. Calculate whether, or not, solid $CaCO_3$ should be precipitating from the sea on the basis of these figures and your value in c).

State, giving a reason, (based on your own visual experience) whether such precipitation is occurring and suggest an explanation for any discrepancy.

e) Almost all the CO_2 on the Earth is fixed as carbonate rock (chalk, limestone etc.). Discuss whether an increase in atmospheric CO_2 could cause these rocks to begin to dissolve.

(*Hints*   1. It will be helpful first to evaluate K for the equilibrium

$$2HCO_3^- \rightleftharpoons H_2CO_3 + CO_3^{2-}$$

using the information given, before answering this section.

2. Remember that $CO_2(g) + H_2O(l) \rightleftharpoons H_2CO_3(aq)$ is an ordinary equilibrium.)

f) Draw a diagram linking together all the equilibria involved in this question.

g) Both CO_2 and $CaCO_3$ dissolve exothermically in water. If the 'greenhouse effect' increases the temperature of the oceans, discuss as far as you can the effects that this will have on the linked equilibria in section f).   (WJEC90,S)

13 Thermochemistry

INTERNAL ENERGY AND ENTHALPY

Matter contains energy. You are familiar in physics with kinetic energy — the energy which enables an object to move — and potential energy — the energy which an object possesses due to its position. Matter possesses energy of both kinds. Matter possesses kinetic energy because the atoms and molecules of which it is composed are in motion. It possesses potential energy due to the positions which atoms and molecules occupy relative to one another; that is to the chemical bonds. During the course of a chemical reaction, one type of matter changes into another, and energy is either given out or taken in from the surroundings (see Fig. 13.1). A reaction in which energy is given out is termed an *exothermic reaction*; a reaction in which energy is taken in is termed an *endothermic reaction*.

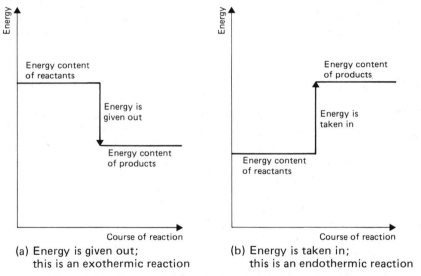

(a) Energy is given out;
this is an exothermic reaction

(b) Energy is taken in;
this is an endothermic reaction

Fig. 13.1   Energy changes during chemical reactions

The heat absorbed during a reaction is equal to the internal energy of the products minus the internal energy of the reactants plus any work done by the system on the surroundings. Since most laboratory work is done at constant pressure, any gases formed are allowed to escape into the atmosphere and work is done in expansion:

$$\begin{pmatrix} \text{Heat absorbed at} \\ \text{constant pressure} \end{pmatrix} = \begin{pmatrix} \text{Change in} \\ \text{internal energy} \end{pmatrix} + \begin{pmatrix} \text{Work done on} \\ \text{surroundings} \end{pmatrix}$$

The heat absorbed at constant pressure is given the name *change in enthalpy* and the symbol ΔH. Enthalpy is defined by the equation

$$H = U + PV$$

where H = Enthalpy, U = Internal energy, P = Pressure, and V = Volume.

Then,

$$\Delta H = \Delta U + P\Delta V$$

When expansion occurs, ΔV is positive and $\Delta H > \Delta U$.

When contraction occurs, ΔV is negative and $\Delta H < \Delta U$.

If reaction takes place at constant volume, $\Delta V = 0$, and $\Delta H = \Delta U$.

Reactions of solids and liquids do not involve large changes in volume, and ΔH is close to ΔU. Reactions in which ΔV is large are those involving gases, and the value of ΔV can be calculated from the ideal gas equation. Since

$$PV = nRT$$

$$P\Delta V = \Delta nRT$$

Δn, the increase in the number of molecules of gas, is indicated by the equation for the reaction. For example, in the reaction

$$CaCO_3(s) \longrightarrow CaO(s) + CO_2(g)$$

$\Delta n = 1$.

The enthalpy of a substance is quoted for the substance in its standard state. The *standard state* of a substance is 1 mole of the substance in a specified state (solid, liquid or gas) at 1 atmosphere pressure. The value of an enthalpy change is quoted for standard conditions: gases at 1 atmosphere, solutions at unit concentration, and substances in their normal states at a specified temperature. $\Delta H_T^\ominus$ means the standard enthalpy change at a temperature T. $\Delta H_{298}^\ominus$ is sometimes written as $\Delta H^\ominus$.

Definitions of some standard enthalpy changes follow:

Standard enthalpy of formation is the heat absorbed when 1 mole of a substance is formed from its elements *in their standard states* at constant pressure. (If the reaction is exothermic, the heat absorbed is negative, and $\Delta H_F^\ominus$ has a negative value. All elements in their standard states are assigned a value of zero for their standard enthalpies of formation.)

Standard enthalpy of combustion is the heat absorbed when 1 mole of a substance is completely burned in oxygen at constant pressure.

Standard enthalpy of hydrogenation is the heat absorbed when 1 mole of an unsaturated compound is converted into a saturated compound by reaction with gaseous hydrogen at constant pressure.

Standard enthalpy of neutralisation is the heat absorbed when an acid and a base react at constant pressure to form 1 mole of water.

Standard enthalpy of reaction is the heat absorbed in a reaction at constant pressure between the number of moles of reactants shown in the equation for the reaction. In the reaction

$$4H_2O(g) \ + \ 3Fe(s) \longrightarrow Fe_3O_4(s) \ + \ 4H_2(g)$$

the standard enthalpy change refers to the reaction between 4 moles of steam and 3 moles of iron.

Standard enthalpy of solution is the heat absorbed when 1 mole of a substance is dissolved at constant pressure in a stated amount of solvent. This may be 100 g or 1 000 g of solvent or it may be an 'infinite' amount of solvent, i.e. a volume so large that on further dilution there is no further heat change.

STANDARD ENTHALPY CHANGE FOR A CHEMICAL REACTION

The standard enthalpy change for a chemical reaction can be calculated from the standard enthalpies of formation of all the products and reactants involved. For example, in the addition of hydrogen chloride to ethene,

$$CH_2 = CH_2(g) \ + \ HCl(g) \longrightarrow C_2H_5Cl(g)$$
$$(+52.3) \qquad\quad (-92.3) \qquad\quad (-105)$$

The standard enthalpies of formation in $kJ \, mol^{-1}$ are shown under each species.

The standard enthalpy of reaction $\Delta H^{\ominus}$ is given by

$$\Delta H^{\ominus} \ = \ (-105) - (52.3 + (-92.3)) \ = \ -65 \, kJ \, mol^{-1}$$

The negative sign means that the products contain less energy than the reactants and the difference is the heat energy given out in the reaction: the reaction is exothermic. A positive value for $\Delta H^{\ominus}$ indicates an endothermic reaction.

The standard enthalpy of reaction depends only on the difference between the standard enthalpy of the reactants and the standard enthalpy of the products and not on the route by which the reaction occurs.

This idea is embodied in Hess's law, which states that, if a reaction can take place by more than one route, the overall change in enthalpy is the same, whichever route is followed.

STANDARD ENTHALPY OF NEUTRALISATION

EXAMPLE   250 cm^3 of sodium hydroxide of concentration 0.400 mol dm^{-3} were added to 250 cm^3 of hydrochloric acid of concentration 0.400 mol dm^{-3} in a calorimeter. The temperature of the two solutions and the calorimeter was 17.05 °C. The mass of the calorimeter was 500 g, and its specific heat capacity was 400 J kg^{-1} K^{-1}. The temperature rose to 19.55 °C. Assuming that the specific heat capacity[†] of all the solutions is 4200 J kg^{-1} K^{-1} calculate the standard enthalpy of neutralisation.

METHOD   Mass of solutions = 500 g

Heat capacity of solutions = 0.500 × 4200 = 2100 J

Mass of calorimeter = 500 g

Heat capacity of calorimeter = 0.500 × 400 = 200 J

Rise in temperature = 2.50 °C

Heat evolved = (2100 + 200) × 2.50 = 5750 J

Amount of water formed = 250 × 10^{-3} × 0.400 = 0.100 mol

Heat evolved per mole = 5750/0.100 = 57 500 J

ANSWER   The standard enthalpy of neutralisation = 57.5 kJ mol^{-1}.

[†]The *heat capacity* of a mass of substance is the quantity of heat needed to raise its temperature by 1 K or 1 °C.

The *specific heat capacity* of a substance is the quantity of heat required to raise the temperature of 1 kg of the substance by 1 K or 1 °C.

Heat capacity = Mass × Specific heat capacity.

EXERCISE 47   Problems on Standard Enthalpy of Neutralisation

1. 50.0 cm^3 of sodium hydroxide solution of concentration 0.400 mol dm^{-3} required 20.0 cm^3 of sulphuric acid of concentration 0.500 mol dm^{-3} for neutralisation. A temperature rise of 3.4 °C was observed if both solutions and the calorimeter were initially at the same temperature. Calculate the standard enthalpy of neutralisation of sodium hydroxide with sulphuric acid. The heat capacity of the calorimeter is 39.0 J K^{-1}. (The specific heat capacity of all the solutions is 4.2 J K^{-1} g^{-1}.)

2. 100 cm^3 of potassium hydroxide solution of concentration 1.00 mol dm^{-3} and 100 cm^3 of hydrochloric acid of concentration 1.00 mol dm^{-3} were mixed in a calorimeter. All three were at the same temperature. The heat capacity of the calorimeter was 95 J K^{-1}, and the rise in temperature was 6.25 K. Calculate the standard enthalpy of neutralisation. (Specific heat capacity of water = 4.2 J K^{-1} g^{-1}.)

3. 100 cm³ of 1.00 mol dm⁻³ sodium hydroxide solution and 100 cm³ of 1.00 mol dm⁻³ ethanoic acid were mixed in a calorimeter. All three were at the same temperature. The heat capacity of the calorimeter was 90 J K⁻¹, and the rise in temperature was 5.3 K. Calculate the standard enthalpy of neutralisation.

4. A calorimeter has a mass of 200 g and a specific heat capacity of 0.42 J g⁻¹. Into it are put 50 cm³ of 1.25 mol dm⁻³ hydrochloric acid and 50 cm³ of 1.25 mol dm⁻³ potassium hydroxide solution at the same temperature. The temperature of the calorimeter and contents rises by 7.0 °C. Calculate the standard enthalpy of neutralisation.

5. Fig. 13.2 shows the results of a thermometric titration to find a value of the standard enthalpy of neutralisation. 50.0 cm³ of a solution of

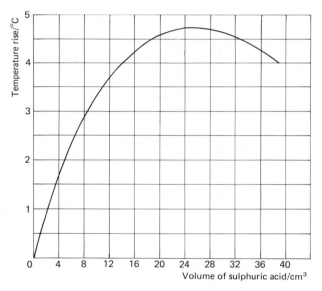

Fig. 13.2

sodium hydroxide of concentration 0.500 mol dm⁻³ were titrated against a 0.568 mol dm⁻³ solution of sulphuric acid. Calculate the standard enthalpy of neutralisation. Assume that the specific heat capacity of the solutions is 4.2 J K⁻¹g⁻¹, and assume that no heat passes to the container.

FINDING THE STANDARD ENTHALPY OF A COMPOUND INDIRECTLY

Sometimes, the standard enthalpy of formation of a compound can be measured directly by allowing known amounts of elements to combine and measuring the amount of heat evolved. Other reactions are difficult to study, and the standard enthalpy of reaction must be found indirectly.

To find the standard enthalpy of formation of ethyne from practical measurements is impossible as attempts to make ethyne from carbon and hydrogen,

$$2C(s) + H_2(g) \longrightarrow C_2H_2(g)$$

will result in the formation of a mixture of hydrocarbons. The standard enthalpy of combustion of ethyne can, however, be measured experimentally, and from it can be calculated the standard enthalpy of formation. The standard enthalpies of combustion of carbon and hydrogen are also required.

EXAMPLE 1 Find the standard enthalpy of formation of ethyne, given the standard enthalpies of combustion (in $kJ\,mol^{-1}$): $C_2H_2(g) = -1300$; $C(s) = -394$; $H_2(g) = -286$.

METHOD 1 The method of calculation is based on the three equations for the combustion of ethyne, carbon and hydrogen:

$$C(s) + O_2(g) \longrightarrow CO_2(g); \qquad \Delta H_1^{\ominus} = -394\,kJ\,mol^{-1} \quad [1]$$

$$H_2(g) + \tfrac{1}{2}O_2(g) \longrightarrow H_2O(l) \qquad \Delta H_2^{\ominus} = -286\,kJ\,mol^{-1} \quad [2]$$

$$C_2H_2(g) + 2\tfrac{1}{2}O_2(g) \longrightarrow 2CO_2(g) + H_2O(l) \quad \Delta H_3^{\ominus} = -1300\,kJ\,mol^{-1} \quad [3]$$

Looking at equation [1] one can see that the standard enthalpy of combustion of carbon is the same as the standard enthalpy of formation of carbon dioxide.

Likewise, equation [2] shows that the standard enthalpy of combustion of hydrogen is the same as the standard enthalpy of formation of water.

The standard enthalpy content of a substance is equal to the standard enthalpy of formation of the substance from its elements in their standard states.

Putting the standard enthalpy content of each substance into equation [3] gives

$$C_2H_2(g) + 2\tfrac{1}{2}O_2(g) \longrightarrow 2CO_2(g) + H_2O(l); \Delta H_3^{\ominus} = -1300\,kJ\,mol^{-1}$$
$$\Delta H_F^{\ominus}(C_2H_2) \quad 0 \qquad\qquad\qquad 2(-394) \ (-286)$$

Since

$$\begin{pmatrix}\text{Standard}\\ \text{enthalpy change}\end{pmatrix} = \begin{pmatrix}\text{Standard}\\ \text{enthalpy content}\\ \text{of products}\end{pmatrix} - \begin{pmatrix}\text{Standard}\\ \text{enthalpy content}\\ \text{of reactants}\end{pmatrix}$$

$$\Delta H_3^{\ominus} = -1300 = 2(-394) + (-286) - \Delta H_F^{\ominus}(C_2H_2)$$

$$\Delta H_F^{\ominus}(C_2H_2) = 226\,kJ\,mol^{-1}$$

ANSWER The standard enthalpy of formation of ethyne is $226\,kJ\,mol^{-1}$. Since $\Delta H_F^{\ominus}$ is positive, ethyne is referred to as an endothermic compound.

METHOD 2 Another method of tackling the problem is to construct an enthalpy diagram:

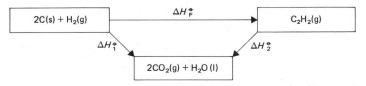

According to Hess's law, the change in standard enthalpy when carbon and hydrogen burn to form carbon dioxide and water is the same as the sum of the standard enthalpy changes when carbon and hydrogen combine to form ethyne and then ethyne burns to form carbon dioxide and water. Thus, in the above diagram,

$$\Delta H_1^{\ominus} = \Delta H_F^{\ominus} + \Delta H_2^{\ominus}$$

Putting

$$\Delta H_1^{\ominus} = 2(\Delta H^{\ominus} \text{ for combustion of C}) + (\Delta H^{\ominus} \text{ for combustion of } H_2)$$

gives

$$\Delta H_1^{\ominus} = 2(-394) + (-286) = -1074$$

$$\Delta H_F^{\ominus} = \Delta H_1^{\ominus} - \Delta H_2^{\ominus} = -1074 - (-1300)$$

ANSWER $\Delta H_F^{\ominus} = +226 \text{ kJ mol}^{-1}$ (as before)

EXAMPLE 2 Calculate the standard enthalpy of formation of propan-1-ol, given the standard enthalpies of combustion, in kJ mol^{-1}: $C_3H_7OH(l)$, -2010; $C(s)$, -394; $H_2(g)$, -286.

METHOD 1 Again, as the equation for combustion is the basis for the calculation, it must be carefully balanced:

$$C_3H_7OH(l) + 4\tfrac{1}{2}O_2(g) \longrightarrow 3CO_2(g) + 4H_2O(l); \quad \Delta H^{\ominus} = -2010 \text{ kJ mol}^{-1}$$

Putting the standard enthalpies of formation of $CO_2(g)$ and $H_2O(l)$ into the equation, as in Example 1, gives

$$\begin{array}{cc} C_3H_7OH(l) + 4\tfrac{1}{2}O_2(g) \longrightarrow & 3CO_2(g) + 4H_2O(l); \quad \Delta H^{\ominus} = -2010 \text{ kJ mol}^{-1} \\ \Delta H_F^{\ominus}(C_3H_7OH) \quad 0 & 3(-394) \quad 4(-286) \end{array}$$

Since

$$\begin{pmatrix} \text{Standard} \\ \text{enthalpy change} \\ \text{for reaction} \end{pmatrix} = \begin{pmatrix} \text{Standard} \\ \text{enthalpy content} \\ \text{of products} \end{pmatrix} - \begin{pmatrix} \text{Standard} \\ \text{enthalpy content} \\ \text{of reactants} \end{pmatrix}$$

$$-2010 = 3(-394) + 4(-286) - \Delta H_F^{\ominus}(C_3H_7OH(l))$$

$$\Delta H_F^{\ominus}(C_3H_7OH(l)) = -316 \text{ kJ mol}^{-1}$$

ANSWER The standard enthalpy of formation of liquid propan-1-ol is -316 kJ mol^{-1}.

METHOD 2   The enthalpy diagram for the formation of propanol is

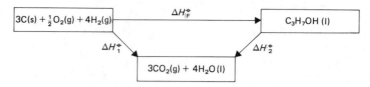

$$\Delta H_1^{\ominus} = 3(\Delta H^{\ominus} \text{ for combustion of } C) + 4(\Delta H^{\ominus} \text{ for combustion of } H_2)$$

$$= 3(-394) + 4(-286) = -2326$$

According to Hess's law,

$$\Delta H_1^{\ominus} = \Delta H_F^{\ominus} + \Delta H_2^{\ominus}$$

$$\Delta H_F^{\ominus} = \Delta H_1^{\ominus} - \Delta H_2^{\ominus}$$

$$\Delta H_F^{\ominus} = -2326 - (-2010)$$

ANSWER

$$\Delta H_F^{\ominus} = -316 \text{ kJ mol}^{-1} \quad \text{(as before)}$$

You will have noticed in both Examples 1 and 2 that

$$
\begin{pmatrix} \text{Standard enthalpy} \\ \text{of reaction} \end{pmatrix} = \begin{pmatrix} \text{Sum of standard} \\ \text{enthalpies of} \\ \text{combustion of} \\ \text{reactants} \end{pmatrix} - \begin{pmatrix} \text{Sum of standard} \\ \text{enthalpies of} \\ \text{combustion of} \\ \text{products} \end{pmatrix}
$$

STANDARD ENTHALPY OF REACTION FROM STANDARD ENTHALPIES OF FORMATION

The standard enthalpies of formation of the reactants and products can be used to give the standard enthalpy of a reaction.

EXAMPLE 1   Calculate the standard enthalpy of the reaction

$$CH_2 = CH_2(g) + H_2(g) \longrightarrow CH_3CH_3(g)$$

given that the standard enthalpies of formation are: ethene, $+52$, ethane, -85 kJ mol^{-1}.

METHOD   Put the standard enthalpy content of each species into the equation (units kJ mol^{-1}):

$$CH_2 = CH_2(g) + H_2(g) \longrightarrow CH_3CH_3(g)$$
$$+52 \qquad\qquad 0 \qquad\qquad\qquad -85$$

$$
\begin{pmatrix} \text{Standard enthalpy} \\ \text{of reaction} \end{pmatrix} = \begin{pmatrix} \text{Standard enthalpy} \\ \text{of product} \end{pmatrix} - \begin{pmatrix} \text{Standard enthalpy} \\ \text{of reactants} \end{pmatrix}
$$

$$= -85 - (52 + 0) = -137$$

ANSWER   Standard enthalpy $= -137\,\text{kJ mol}^{-1}$

The method of calculation is simply:

$$
\begin{pmatrix}\text{Standard enthalpy} \\ \text{of reaction}\end{pmatrix} = \begin{pmatrix}\text{Sum of standard} \\ \text{enthalpies of} \\ \text{formation of} \\ \text{products}\end{pmatrix} - \begin{pmatrix}\text{Sum of standard} \\ \text{enthalpies of} \\ \text{formation of} \\ \text{reactants}\end{pmatrix}
$$

EXAMPLE 2   Calculate the standard enthalpy change in the reaction

$$SO_2(g) + 2H_2S(g) \longrightarrow 3S(s) + 2H_2O(l)$$

The standard enthalpy of combustion of sulphur is $-297\,\text{kJ mol}^{-1}$, and the standard enthalpies of formation of hydrogen sulphide and water are $-20.2\,\text{kJ mol}^{-1}$ and $-286\,\text{kJ mol}^{-1}$.

METHOD   This problem is tackled by putting the standard enthalpies of formation of each species into the equation (units kJ mol^{-1}):

$$
\begin{array}{ccccccc}
SO_2(g) & + & 2H_2S(g) & \longrightarrow & 3S(s) & + & 2H_2O(l) \\
(-297) & + & 2(-20.2) & & 3 \times 0 & & 2(-286)
\end{array}
$$

$$
\begin{pmatrix}\text{Standard enthalpy} \\ \text{of reaction}\end{pmatrix} = \begin{pmatrix}\text{Standard enthalpy} \\ \text{of products}\end{pmatrix} - \begin{pmatrix}\text{Standard enthalpy} \\ \text{of reactants}\end{pmatrix}
$$

$$= -572 + 297 + 40.4$$

ANSWER   Standard enthalpy change $= -235\,\text{kJ (mol of the equation)}^{-1}$

STANDARD BOND DISSOCIATION ENTHALPIES

The standard bond dissociation enthalpy is the energy that must be absorbed to separate the two atoms in a bond. When hydrogen chloride dissociates,

$$HCl(g) \longrightarrow H(g) + Cl(g); \qquad \Delta H^{\ominus} = 429.7\,\text{kJ mol}^{-1}$$

The standard bond dissociation enthalpy of the H—Cl bond in HCl is $429.7\,\text{kJ mol}^{-1}$.

AVERAGE STANDARD BOND ENTHALPIES

When you want to assign a value to the standard enthalpy of dissociation of the C—H bond in methane, the problem is different. The energy required to break the first C—H bond in methane is not the same as that required to remove a hydrogen atom from a methyl radical. In the dissociation,

$$CH_4(g) \longrightarrow C(g) + 4H(g); \quad \Delta H^{\ominus} = +1662 \, kJ \, mol^{-1}$$

Dividing the standard enthalpy change between the four bonds gives an average value for the C—H bond of $416 \, kJ \, mol^{-1}$. This value is called the average standard bond enthalpy for the C—H bond.

Tables of average standard bond enthalpies make the assumption that the standard enthalpy of a bond is independent of the molecule in which it exists. This is only roughly true. Since standard bond enthalpies vary from one compound to another, the use of average standard bond enthalpies gives only approximate values for standard enthalpies of reaction calculated from them. Experimental methods are used to obtain standard enthalpies of reaction whenever possible. Calculations based on average standard bond enthalpies are used only for reactions which cannot be studied experimentally — for example, the reactions of a substance which has not been isolated in a pure state.

Average standard bond enthalpy is often called the *bond energy term*. One can say that the bond energy term for the C—H bond is $416 \, kJ \, mol^{-1}$. The sum of all the bond energy terms for a compound is the standard enthalpy change absorbed in atomising that compound *in the gaseous state*. The standard enthalpy of formation of a compound includes the sum of the bond energy terms and also the standard enthalpy of atomisation of the carbon atoms and the standard enthalpy of atomisation of the hydrogen atoms.

EXAMPLE   Calculate the standard enthalpy of formation of methane. C—H bond energy term = $416 \, kJ \, mol^{-1}$; standard enthalpies of atomisation are $C(s) = 716 \, kJ \, mol^{-1}$; $\frac{1}{2}H_2(g) = 217.5 \, kJ \, (mol \, H \, atoms)^{-1}$.

METHOD 1   The sum of the bond energy terms in methane = $1662 \, kJ \, mol^{-1}$. Putting this information into the form of an equation, and writing the standard enthalpy content of each species underneath its formula, we get

$$C(g) + 4H(g) \longrightarrow CH_4(g); \quad \Delta H^{\ominus} = -1662 \, kJ \, mol^{-1}$$
$$(716) \quad 4(217.5) \qquad\qquad \Delta H_F^{\ominus}$$

The values 716 and 217.5 are the standard enthalpies of formation of gaseous carbon and hydrogen atoms from the elements in their standard states.

Since

$$\begin{pmatrix}\text{Standard enthalpy}\\\text{change}\end{pmatrix} = \begin{pmatrix}\text{Sum of standard}\\\text{enthalpies of}\\\text{products}\end{pmatrix} - \begin{pmatrix}\text{Sum of standard}\\\text{enthalpies of}\\\text{reactants}\end{pmatrix}$$

$$-1662 = \Delta H_F^\ominus - 716 - 4(217.5)$$

ANSWER $\Delta H_F^\ominus = -76\,\text{kJ mol}^{-1}$

METHOD 2 The information can also be represented in the form of an enthalpy diagram:

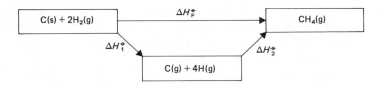

$$\Delta H_1^\ominus = \Delta H^\ominus \text{ of atomisation of C} + 4\Delta H^\ominus \text{ of atomisation of H}$$

$$\Delta H_2^\ominus = -(\text{Sum of bond energy terms for CH}_4)$$

According to Hess's law,

$$\Delta H_F^\ominus = \Delta H_1^\ominus + \Delta H_2^\ominus$$

$$= 716 + 4(217.5) - 1662$$

ANSWER $\Delta H_F^\ominus = -76\,\text{kJ mol}^{-1}$ (as before)

STANDARD ENTHALPY OF REACTION FROM AVERAGE STANDARD BOND ENTHALPIES

Mean standard bond enthalpies can be used to give an approximate estimate of the standard enthalpy change which occurs in a reaction. During a reaction, energy is supplied to break the bonds in the reactants, and energy is given out when the bonds in the products form. The difference between the sum of the standard bond enthalpies of the products and the standard bond enthalpies of the reactants is the standard enthalpy of the reaction. The value obtained is less reliable than an experimental measurement.

EXAMPLE 1 Calculate the standard enthalpy of the reaction

$$CH_2 = CH_2(g) + H_2(g) \longrightarrow CH_3CH_3(g)$$

Mean standard bond enthalpies are (in kJ mol^{-1}): C—H, 416; C=C, 612; C—C, 348; H—H, 436.

METHOD   Bonds broken are:

one $C=C$ bond, of standard enthalpy $= 612\,kJ\,mol^{-1}$

one $H—H$ bond, of standard enthalpy $= 436$

Total enthalpy absorbed $= 1048\,kJ\,mol^{-1}$

Bonds created are:

one $C—C$ bond, of standard enthalpy $= 348\,kJ\,mol^{-1}$

two $C—H$ bonds, of standard enthalpy $= 832$

Total enthalpy released $= -1180\,kJ\,mol^{-1}$

ANSWER   Standard enthalpy of reaction $= -1180 + 1048 = -132\,kJ\,mol^{-1}$

EXAMPLE 2   Benzene has a standard enthalpy of formation of $83\,kJ\,mol^{-1}$. Calculate the standard enthalpy of formation from the following data:

Mean standard bond enthalpies are: $(C—C) = 348; (C=C) = 615; (C—H) = 412\,kJ\,mol^{-1}$.

$\Delta H^{\ominus}$ for vaporisation of carbon $= 715\,kJ\,mol^{-1}$

$\Delta H^{\ominus}$ for atomisation of hydrogen (per mole of H atoms) $= 217.5\,kJ\,mol^{-1}$

Compare the experimental value and the theoretical value for $\Delta H_F^{\ominus}$.

METHOD   Enthalpy is absorbed in atomising carbon and hydrogen.

Standard enthalpy absorbed $= (6 \times 715) + (6 \times 217.5)$
$= 5595\,kJ\,mol^{-1}$

Enthalpy is released when bonds are formed.

Standard enthalpy released $= 6(C—H) = -2472\,kJ\,mol^{-1}$
$+ 3(C—C) = -1044\,kJ\,mol^{-1}$
$+ 3(C=C) = -1845\,kJ\,mol^{-1}$
Total $= -5361\,kJ\,mol^{-1}$

ANSWER $\Delta H_F^{\ominus} = +5595 - 5361 = 234\,kJ\,mol^{-1}$.

The calculated value for the standard enthalpy of formation is higher than the experimental value: the benzene molecule is more stable than it is calculated to be. The difference is the value of the energy of electron delocalisation or 'resonance', $151\,kJ\,mol^{-1}$.

THE BORN–HABER CYCLE

The Born–Haber cycle is a technique for applying Hess's law to the standard enthalpy changes which occur when an ionic compound is formed. Consider the reaction between sodium and chlorine to form sodium chloride. The steps which are involved in this reaction are:

a) Vaporisation of sodium

$Na(s) \longrightarrow Na(g); \quad \Delta H_s^{\ominus} =$ standard enthalpy of sublimation

b) Ionisation of sodium

$$Na(g) \longrightarrow Na^+(g) + e^-; \quad \Delta H_I^{\ominus} = \text{ionisation energy of sodium}$$

c) Dissociation of chlorine molecules

$$\tfrac{1}{2}Cl_2(g) \longrightarrow Cl(g); \quad \Delta H_D^{\ominus} = \tfrac{1}{2} \text{ standard bond dissociation enthalpy of chlorine}$$

d) Ionisation of chlorine atoms

$$Cl(g) + e^- \longrightarrow Cl^-(g); \quad \Delta H_E^{\ominus} = \text{electron affinity of chlorine}$$

e) Reaction between ions

$$Na^+(g) + Cl^-(g) \longrightarrow NaCl(s); \quad \Delta H_L^{\ominus} = \text{standard lattice enthalpy}$$

Definitions of the standard enthalpies used above are:

The *standard enthalpy of sublimation* is the heat absorbed when one mole of sodium atoms are vaporised.

The *ionisation energy* of sodium is the energy required to remove a mole of electrons from a mole of sodium atoms in the gas phase.

The *standard enthalpy of bond dissociation* of chlorine is the enthalpy required to dissociate one mole of chlorine molecules into atoms.

The *electron affinity* of chlorine is the energy absorbed when a mole of chlorine atoms form chloride ions. It has a negative value, showing that this reaction is exothermic.

The *standard lattice enthalpy* is the energy absorbed when one mole of gaseous sodium ions and one mole of gaseous chloride ions form one mole of crystalline sodium chloride. It has a negative value.

The steps in the Born–Haber cycle are represented as going upwards if they absorb energy and downwards if they give out energy (see Fig. 13.3).

According to Hess's law, the standard enthalpy of formation of sodium chloride is equal to the sum of the enthalpy changes in the various steps:

$$\Delta H_F^{\ominus} = \Delta H_S^{\ominus} + \tfrac{1}{2}\Delta H_D^{\ominus} + \Delta H_I^{\ominus} + \Delta H_E^{\ominus} + \Delta H_L^{\ominus}$$
$$= +109 + 121 + 494 - 380 - 755 = -411 \text{ kJ mol}^{-1}$$

In practice, it is easier to measure standard enthalpies of formation than to measure some of the other steps. The electron affinity is the hardest term to measure experimentally, and the Born–Haber cycle is often used to calculate electron affinities.

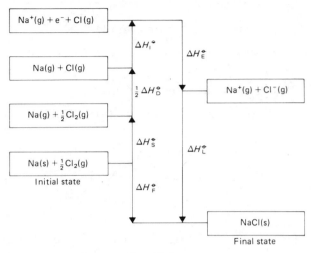

Fig. 13.3

EXERCISE 48   Problems on Standard Enthalpy of Reaction and Average Standard Bond Enthalpies

1. The following are standard enthalpies of combustion at 298 K, in kJ mol^{-1}:

C(graphite)	-394	$C_2H_6(g)$	-1561	$C_4H_{10}(l)$	-3510
$H_2(g)$	-286	$CH_2{=}CH_2(g)$	-1393	$CH{\equiv}CH(g)$	-1299
$CH_3CO_2H(l)$	-876	$C_2H_5OH(l)$	-1400	$CH_3OH(l)$	-715
$C_4H_6(g)$	-2542	$CH_3OCH_3(g)$	-1455	$C_2H_5OH(g)$	-1444
$CH_4(g)$	-891	$C_3H_8(g)$	-2220		
$CH_3CO_2C_2H_5(l)$	-2246	$C_6H_{12}(l)$	-3924		

a) Calculate the standard enthalpy change for the reaction:

$$2C(graphite) \ + \ 2H_2(g) \ + \ O_2(g) \longrightarrow CH_3CO_2H(l)$$

b) Calculate the standard enthalpy change of formation of buta-1, 3-diene, $C_4H_6(g)$.

c) Calculate the standard enthalpy of formation of methane, $CH_4(g)$ and of ethene, $CH_2{=}CH_2(g)$.

d) Calculate the standard enthalpy change in the hydrogenation of ethene(g) to ethane(g).

e) Calculate the standard enthalpy change for the theoretical reaction:

$$CH_3OCH_3(g) \longrightarrow C_2H_5OH(g)$$

    f) Calculate the standard enthalpy of formation of propane(g) and of butane(l).

    g) Calculate the standard enthalpy of formation of methanol(l), ethanol(l), ethylethanoate(l) and cyclohexane(l).

2. Calculate the standard enthalpy change of the reaction

    Anhydrous copper(II) sulphate $+$ Water $\longrightarrow$ Copper(II) sulphate-5-water

    Use the values for the standard enthalpy of solution:

    a) anhydrous copper(II) sulphate, $-66.5 \text{ kJ mol}^{-1}$

    b) copper(II) sulphate-5-water, 11.7 kJ mol^{-1}.

3. Calculate the standard enthalpies of formation of: a) sulphur dioxide, b) carbon dioxide, and c) steam. On burning in excess oxygen under standard conditions (1 atm, 298 K): 1.00 g of sulphur evolves 9.28 kJ; 1.00 g of carbon evolves 32.8 kJ; and 1.00 dm^3 (at 1 atm, 298 K) of hydrogen evolves 12.76 kJ of heat.

4. Calculate the standard enthalpy change in the reaction

$$PbO(s) \; + \; CO(g) \longrightarrow Pb(s) \; + \; CO_2(g)$$

    The standard enthalpies of formation of lead(II) oxide, carbon monoxide and carbon dioxide are -219, -111, and -394 kJ mol^{-1}, respectively.

5. Calculate the standard enthalpy change for the reaction

$$Fe_2O_3(s) \; + \; 2Al(s) \longrightarrow Al_2O_3(s) \; + \; 2Fe(s)$$

    The standard enthalpies of formation of iron(III) oxide and aluminium oxide are -822 and $-1669 \text{ kJ mol}^{-1}$. State whether the reaction is exothermic or endothermic.

6. The standard enthalpy of combustion of rhombic sulphur is $-296.9 \text{ kJ mol}^{-1}$ and the standard enthalpy of combustion of monoclinic sulphur is $-297.2 \text{ kJ mol}^{-1}$. Calculate the standard enthalpy of conversion of monoclinic sulphur to rhombic sulphur.

7. The standard enthalpies of formation of $CO_2(g)$ and $H_2O(g)$ are -394 and -242 kJ mol^{-1}. The standard enthalpy of combustion of ethane is $-1560 \text{ kJ mol}^{-1}$. The standard enthalpy of reduction of ethene to ethane by gaseous hydrogen is -138 kJ mol^{-1}. Calculate the standard enthalpy of formation of ethene.

8. Given the standard enthalpy change of formation of $MgO = -602$ kJ mol^{-1} and of $Al_2O_3 = -1700 \text{ kJ mol}^{-1}$, calculate the standard enthalpy change for the reaction

$$Al_2O_3 \; + \; 3Mg \longrightarrow 2Al \; + \; 3MgO$$

    Does your answer tell you whether magnesium will reduce aluminium oxide?

9. The following are standard enthalpies of formation, $\Delta H_F^{\ominus}$, in kJ mol^{-1} at 298 K:

$CH_4(g)$; -76; $CO_2(g)$, -394; $H_2O(l)$, -286; $H_2O(g)$, -242; $NH_3(g)$, -46.2; $HNO_3(l)$, -176; $C_2H_5OH(l)$, -278; $C_8H_{18}(l)$, -210.

a) Calculate the standard enthalpy change at 298 K for the reaction

$$CH_4(g) + 2O_2(g) \longrightarrow CO_2(g) + 2H_2O(l)$$

b) Calculate the standard enthalpy change for the reaction

$$\tfrac{1}{2}N_2(g) + \tfrac{3}{2}H_2O(g) \longrightarrow NH_3(g) + \tfrac{3}{4}O_2(g)$$

c) Calculate the standard enthalpy change for the reaction

$$\tfrac{1}{2}N_2(g) + \tfrac{1}{2}H_2O(g) + \tfrac{5}{4}O_2(g) \longrightarrow HNO_3(l)$$

d) Calculate the enthalpy change which occurs when each of the following is burned completely under standard conditions: i) 1.00 kg hydrogen, ii) 1.00 kg ethanol(l), iii) 1.00 kg octane(l).

10. What is meant by the terms *standard bond dissociation enthalpy* and *bond energy term*?

The standard bond dissociation enthalpies for the first, second, third and fourth C—H bonds in methane are 423, 480, 425 and 335 kJ mol^{-1} respectively. Calculate the C—H bond energy term for methane.

11. Consult the average standard bond enthalpies and standard enthalpies of atomisation (in kJ mol^{-1}) listed below:

C—C	348	C=O	743	C(graphite)	718
C=C	612	H—Cl	432	$\tfrac{1}{2}H_2(g)$	218
C≡C	837	C—Cl	338	$\tfrac{1}{2}O_2(g)$	248
C—H	412	C—Br	276	$\tfrac{1}{2}Br_2(g)$	96.5
C—O	360	H—Br	366	$\tfrac{1}{2}Cl_2(g)$	121
H—O	463				

a) Calculate the standard enthalpy of formation of ethane and of ethene.

b) Find the standard enthalpy change for the reaction,

$$CH_2{=}CH{-}CH_3(g) + Br_2(g) \longrightarrow CH_2BrCHBrCH_3(g)$$

c) Find the standard enthalpy of formation of methoxymethane, $CH_3OCH_3(g)$.

d) Calculate the standard enthalpy of formation of gaseous ethyl ethanoate, $CH_3CO_2C_2H_5(g)$.

e) Calculate the standard enthalpy of formation of benzene, assuming its structure is

Explain the difference between the value you have calculated and the value of $83 \, kJ \, mol^{-1}$ obtained from measurements of the standard enthalpy of combustion.

f) Find the standard enthalpy of formation of gaseous buta-1,3-diene, $CH_2{=}CH{-}CH{=}CH_2(g)$. How does this value compare with the value you obtained in Question 1(b) from the standard enthalpy of combustion? How do you explain the difference?

g) Estimate the standard enthalpy changes for the reactions:

i) $Cl\cdot \; + \; CH_4 \longrightarrow CH_3Cl \; + \; H\cdot$

ii) $Cl\cdot \; + \; CH_4 \longrightarrow CH_3\cdot \; + \; HCl$

Which of the two reactions will occur more readily?

12. Use the data below to draw an energy diagram for the formation of potassium chloride. Calculate the electron affinity of chlorine.

Standard enthalpy of sublimation of potassium	$= \quad 90 \, kJ \, mol^{-1}$
Standard enthalpy of ionisation of potassium	$= \quad 420 \, kJ \, mol^{-1}$
Standard enthalpy of dissociation of chlorine	$= \quad 244 \, kJ \, mol^{-1}$
Standard lattice enthalpy of potassium chloride	$= -706 \, kJ \, mol^{-1}$
Standard enthalpy of formation of potassium chloride	$= -436 \, kJ \, mol^{-1}$

13. Using the following data, which is a set of standard enthalpy changes, calculate the standard enthalpy of formation of potassium chloride, $KCl(s)$:

$$\Delta H^{\ominus}/kJ \, mol^{-1}$$

$KOH(aq) \; + \; HCl(aq) \longrightarrow KCl(aq) \; + \; H_2O(l)$	-57.3
$H_2(g) \; + \; \tfrac{1}{2}O_2(g) \longrightarrow H_2O(l)$	-286
$\tfrac{1}{2}H_2(g) \; + \; \tfrac{1}{2}Cl_2(g) \; + \; aq \longrightarrow HCl(aq)$	-164
$K(s) \; + \; \tfrac{1}{2}O_2(g) \; + \; \tfrac{1}{2}H_2(g) \; + \; aq \longrightarrow KOH(aq)$	-487
$KCl(s) \; + \; aq \longrightarrow KCl(aq)$	$+18$

14. Use the data below to calculate the electron affinity of chlorine:

Standard enthalpy of formation of rubidium chloride	$-431\,kJ\,mol^{-1}$
Lattice energy of rubidium chloride	$-675\,kJ\,mol^{-1}$
First ionisation energy of rubidium	$+408\,kJ\,mol^{-1}$
Standard enthalpy of atomisation of rubidium	$+86\,kJ\,mol^{-1}$
Bond dissociation enthalpy of molecular chlorine	$+242\,kJ\,mol^{-1}$

15. From a Born–Haber cycle calculation, it can be estimated that the standard enthalpy of formation of magnesium(I) chloride, MgCl, would be $-130\,kJ\,mol^{-1}$. The standard enthalpy of formation of magnesium(II) chloride $MgCl_2$, is $-640\,kJ\,mol^{-1}$.

a) Why do you think that $MgCl_2$ is formed, and not MgCl, when magnesium reacts with chlorine?

b) Calculate the standard enthalpy change in the theoretical reaction

$$2MgCl(s) \longrightarrow Mg(s) + MgCl_2(s)$$

16. Calculate the lattice energy of sodium chloride from the following data:

			$\Delta H^{\ominus}/kJ\,mol^{-1}$
$Na(s)$	$\longrightarrow$	$Na(g)$	$+109$
$Na(g)$	$\longrightarrow$	$Na^+(g) + e^-$	$+494$
$Cl_2(g)$	$\longrightarrow$	$2Cl(g)$	$+242$
$Cl(g) + e^-$	$\longrightarrow$	$Cl^-(g)$	-360
$Na(s) + \frac{1}{2}Cl_2(s)$	$\longrightarrow$	$NaCl(s)$	-411

17. a) Data for the Born–Haber cycle for the formation of calcium chloride are

$Ca(s)$	$\longrightarrow$	$Ca(g)$	$\Delta H^{\ominus} = +190\,kJ\,mol^{-1}$
$Ca(g)$	$\longrightarrow$	$Ca^{2+}(g) + 2e^-$	$\Delta H^{\ominus} = +1730\,kJ\,mol^{-1}$
$\frac{1}{2}Cl_2(g)$	$\longrightarrow$	$Cl(g)$	$\Delta H^{\ominus} = +121\,kJ\,mol^{-1}$
$Ca^{2+}(g) + 2Cl^-(g)$	$\longrightarrow$	$CaCl_2(s)$	$\Delta H^{\ominus} = -2184\,kJ\,mol^{-1}$
$Ca(s) + Cl_2(g)$	$\longrightarrow$	$CaCl_2(s)$	$\Delta H^{\ominus} = -795\,kJ\,mol^{-1}$

Calculate the electron affinity of chlorine.

b) For the reactions

$Ca(g)$	$\longrightarrow$	$Ca^+(g) + e^-$	$\Delta H^{\ominus} = +590\,kJ\,mol^{-1}$
$Ca^+(g) + Cl^-(g)$	$\longrightarrow$	$CaCl(s)$	$\Delta H^{\ominus} = -760\,kJ\,mol^{-1}$

use these standard enthalpy changes and those given in a) to calculate the standard enthalpy of formation of CaCl(s). Why do you think CaCl$_2$ is formed in preference to CaCl?

18. When an ionic compound dissolves, an amount of energy equal to the lattice energy must be supplied to separate the ions. When the ions dissolve they are hydrated by water molecules, and energy is released. If the enthalpy of hydration is greater than the lattice enthalpy, there is a net release of energy and a decrease in the enthalpy content of the system, and this favours solution.

The values below (in $kJ\,mol^{-1}$) relate to the solubility of lithium chloride, sodium chloride and sodium fluoride:

	LiCl	NaCl	NaF
Standard lattice enthalpy	−843	−775	−968
Sum of standard hydration enthalpies of separate ions	−883	−778	−965

What can you predict from these values for the standard enthalpy changes about the relative solubilities of: a) LiCl and NaCl, b) NaF and NaCl? Explain your answer.

19. Given the standard enthalpy changes for the reactions

$$H_2(g) \longrightarrow 2H(g); \qquad \Delta H^{\ominus} = 436\,kJ\,mol^{-1}$$
$$Br_2(g) \longrightarrow 2Br(g); \qquad \Delta H^{\ominus} = 193\,kJ\,mol^{-1}$$
$$H_2(g) + Br_2(g) \longrightarrow 2HBr(g); \qquad \Delta H^{\ominus} = -104\,kJ\,mol^{-1}$$

calculate the standard enthalpy change for the reaction

$$H(g) + Br(g) \longrightarrow HBr(g)$$

20. The following values for standard enthalpy change relate to the hydrogenation of cyclohexene and benzene. Comment on the values of $\Delta H^{\ominus}$.

$$C_6H_{10}(l) + H_2(g) \longrightarrow C_6H_{12}(l) \qquad \Delta H^{\ominus} = -120\,kJ\,mol^{-1}$$
$$C_6H_6(l) + H_2(g) \longrightarrow C_6H_8(l) \qquad \Delta H^{\ominus} = +31\,kJ\,mol^{-1}$$
$$C_6H_6(l) + 3H_2(g) \longrightarrow C_6H_{12}(l) \qquad \Delta H_c^{\ominus} = -208\,kJ\,mol^{-1}$$

FREE ENERGY AND ENTROPY

Some reactions which happen spontaneously are endothermic. The difference in enthalpy between the products and the reactants cannot be the only factor which decides whether a chemical reaction takes place. There must be an additional factor involved. It is often observed that reactions which occur spontaneously increase the randomness or disorder of the system. For example, when an ionic solid dissolves, it passes from the regular arrangement of a crystalline lattice to a random solution of ions. This is termed an increase in *entropy* of the system. The two factors combine to give the change in the *free energy* of the system:

Free energy G = Enthalpy H − Temperature/K × Entropy S

$$G = H - TS$$

It follows that $\Delta G = \Delta H - T\Delta S$

For a physical or a chemical change to occur, ΔG for that change must be negative. The change is therefore assisted by a decrease in enthalpy (ΔH negative) and by an increase in entropy (ΔS positive).

If the change takes place under standard conditions, i.e. with each reactant and product at unit concentration (or pressure), then the free energy change is equal to the standard free energy change, $\Delta G^{\ominus}$. When reaction takes place under non-standard conditions, ΔG, the free energy change differs from $\Delta G^{\ominus}$ as ΔG depends on the concentrations (or pressures) of the reactants and products. It is easy to obtain $\Delta G^{\ominus}$ from tables of standard enthalpies and standard entropies, but one really wants to know the value of ΔG for the real conditions, and this is not easy to compute. However, if $\Delta G^{\ominus}$ has a sufficiently large positive or negative value, $\Delta G^{\ominus}$ may determine the feasibility of reaction over a large range of concentrations (or pressures).

CALCULATION OF CHANGE IN STANDARD ENTROPY

The standard entropy change of a process is given by:

$$\begin{pmatrix} \text{Standard} \\ \text{entropy change} \end{pmatrix} = \begin{pmatrix} \text{Sum of standard} \\ \text{entropies of products} \end{pmatrix} - \begin{pmatrix} \text{Sum of standard} \\ \text{entropies of reactants} \end{pmatrix}$$

EXAMPLE 1 Calculate the standard entropy change for the reaction of chlorine and ethene, given the values (in $J\,K^{-1}\,mol^{-1}$):

$S^{\ominus}(Cl_2(g)) = 223$; $S^{\ominus}(CH_2{=}CH_2(g)) = 219$; $S^{\ominus}(CH_2ClCH_2Cl(l)) = 208$.

METHOD The equation for the reaction is

$$CH_2{=}CH_2(g) \ + \ Cl_2(g) \ \longrightarrow \ CH_2ClCH_2Cl(l)$$

$$S^{\ominus}(\text{product}) \ = \ 208\,J\,K^{-1}\,mol^{-1}$$

$$S^{\ominus}(\text{reactants}) \ = \ 219 + 223 \ = \ 442\,J\,K^{-1}\,mol^{-1}$$

$$\Delta S^{\ominus} \ = \ 208 - 442 \ = \ -234\,J\,K^{-1}\,mol^{-1}$$

ANSWER The standard entropy change for the reaction is $-234\,J\,K^{-1}\,mol^{-1}$. The negative sign means a decrease in disorder. Since two moles of gas have formed one mole of liquid, this is what one would expect.

CALCULATION OF CHANGE IN STANDARD FREE ENERGY

The change in standard enthalpy, the change in standard entropy and the temperature must be known and inserted into the equation

$$\Delta G^{\ominus} \ = \ \Delta H^{\ominus} - T\Delta S^{\ominus}$$

EXAMPLE Calculate the change in standard free energy and determine whether the reaction

$$Fe_2O_3(s) + 3H_2(g) \longrightarrow 2Fe(s) + 3H_2O(g)$$

will take place at a) 20 °C, b) 500 °C. Use the values (in kJ mol^{-1}):

	Fe_2O_3	H_2	Fe	H_2O
Standard enthalpy:	-822	0	0	-242
Standard entropy:	0.090	0.131	0.027	0.189

METHOD $\Delta G^\ominus = \Delta H^\ominus - T\Delta S^\ominus$

$\Delta H^\ominus = (0 + 3(-242)) - (-822 + 0) = +96\,\text{kJ mol}^{-1}$

$\Delta S^\ominus = (2 \times 0.027) + (3 \times 0.189) - 0.090 - (3 \times 0.131)$

$\qquad = 0.054 + 0.567 - 0.090 - 0.393 = +0.138\,\text{kJ mol}^{-1}$

a) At 20 °C,

$\Delta G^\ominus = \Delta H^\ominus - T\Delta S^\ominus$

$\qquad = +96 - (293 \times 0.138) = 96 - 40.43 = +55.57\,\text{kJ mol}^{-1}$

ANSWER $\Delta G^\ominus$ is 42.3 kJ mol^{-1} which is positive, and the reaction will therefore not occur at 20 °C.

b) At 500 °C,

$\Delta G^\ominus = +96 - (773 \times 0.138) = -10.7\,\text{kJ mol}^{-1}$

ANSWER $\Delta G^\ominus$ is -10.7 kJ mol^{-1} which is negative, the reaction will occur at 500 °C.

(*Note* the assumption that $\Delta H^\ominus$ does not vary with temperature.)

EXERCISE 49 Problems on Standard Entropy Change and Standard Free Energy Change

1. Refer to the following values of standard entropy (J mol^{-1} K^{-1}) at 298 K:

$H_2(g)$	131	$H_2O(l)$	70	$NH_4Cl(s)$	94.6
$Cl_2(g)$	223	$H_2O(g)$	189	$N_2O_4(g)$	304
$N_2(g)$	192	$HCl(g)$	187	$C_2H_4(g)$	220
$O_2(g)$	205	$NH_3(g)$	193	$C_2H_6(g)$	230
$Na(s)$	51	$NO_2(g)$	240	$HNO_3(l)$	156
		$NaCl(s)$	72.4		

Calculate the standard entropy changes for the following reactions:

a) $H_2(g) + Cl_2(g) \longrightarrow 2HCl(g)$

b) $N_2(g) + 3H_2(g) \longrightarrow 2NH_3(g)$

c) $H_2(g) + \frac{1}{2}O_2(g) \longrightarrow H_2O(l)$

d) $H_2(g) + C_2H_4(g) \longrightarrow C_2H_6(g)$

e) $N_2O_4(g) \longrightarrow 2NO_2(g)$

f) $Na(s) + \frac{1}{2}Cl_2(g) \longrightarrow NaCl(s)$

g) $NH_4Cl(s) \longrightarrow NH_3(g) + HCl(g)$

h) $4HNO_3(l) \longrightarrow 4NO_2(g) + O_2(g) + 2H_2O(l)$

2. Predict whether the following reactions will have a positive or negative value of $\Delta S^\ominus$:

a) $NH_4NO_3(s) \longrightarrow N_2O(g) + 2H_2O(g)$

b) $2H_2O_2(aq) \longrightarrow 2H_2O(l) + O_2(g)$

c) $PH_3(g) + HI(g) \longrightarrow PH_4I(s)$

d) $3O_2(g) \longrightarrow 2O_3(g)$

e) $CO_2(g) + C(s) \longrightarrow 2CO(g)$

f) $Ni(s) + 4CO(g) \longrightarrow Ni(CO)_4(g)$

3. Use the following values of standard entropy content and standard enthalpy of formation to calculate standard free energy changes:

Substance	$\Delta H_F^\ominus$/kJ mol^{-1}	$S^\ominus$/J K^{-1} mol^{-1}
HgO(s) (red)	−90.7	72.0
HgO(s) (yellow)	−90.2	73.0
HgS(s) (red)	−58.2	77.8
HgS(s) (black)	−54.0	83.3

a) Calculate the value of $\Delta G^\ominus$ for the change

$$HgO(s) \ (red) \longrightarrow HgO(s) \ (yellow)$$

at 25 °C and at 100 °C. At what temperature will the change take place?

b) Calculate the value of $\Delta G^\ominus$ for the change

$$HgS(s) \ (red) \longrightarrow HgS(s) \ (black)$$

at 25 °C. At what temperature will the change occur?

4. *Cis*-but-2-ene has $\Delta H_F^\ominus = -5.7$ kJ mol^{-1} and $S^\ominus = 301$ J K^{-1} mol^{-1}; *trans*-but-2-ene has $\Delta H_F^\ominus = -10.1$ kJ mol^{-1} and $S^\ominus = 296$ J K^{-1} mol^{-1}. Calculate

a) $\Delta G^\ominus$ for the transition *cis*-but-2-ene $\longrightarrow$ *trans*-but-2-ene and

b) for the transition *trans*-but-2-ene $\longrightarrow$ *cis*-but-2-ene

Which is the more stable isomer?

EXERCISE 50   Questions from A-level Papers

1. a) State Hess's law.
   b) i) Define bond dissociation enthalpy for a diatomic molecule.
       ii) Write the equation for the reaction for which the enthalpy change is equal to that of the bond dissociation enthalpy of iodine ($+151.1 \text{ kJ mol}^{-1}$).
       iii) Write the equation for the sublimation of iodine and calculate the value of ΔH for this process, given that

   $$\tfrac{1}{2}I_2(s) \longrightarrow I(g) \qquad \Delta H = +106.8 \text{ kJ mol}^{-1}. \quad \text{(JMB91)}$$

*2. a) The carbon–carbon bond lengths in ethane, ethene and benzene are 154 pm, 133 pm and 140 pm respectively. Discuss the bonding in these three compounds, and show how it accounts for the observed bond lengths.
   b) The enthalpy change for the combustion of hydrocarbons in excess oxygen in the vapour phase at 298 K can be estimated by assuming the following contributions for each type of bond.

Type of bond	Contribution to $\Delta H_{\text{combustion}}$/kJ mol^{-1}
C—H	−226
C—C	−205
C=C	−489

   The measured value of $\Delta H_{\text{combustion}}$ for benzene in the vapour phase at 298 K is $-3298 \text{ kJ mol}^{-1}$.

   Estimate the enthalpy change for the combustion of benzene, and comment on any difference in relation to the structure of benzene.

   c) i) How may benzene be converted into phenylethanone, $C_6H_5COCH_3$, and what is the mechanism of the reaction?
          Why is a substitution rather than an addition product formed?
       ii) Excess phenylethanone reacts with hydrazine, NH_2—NH_2, to give a compound $C_{16}H_{16}N_2$. Give a structural formula for this compound. (O90,S)

3. a) State and explain the similarities and differences between the crystal structures of sodium chloride and caesium chloride, using diagrams where appropriate.
   b) Some energy data are tabulated below.

Process	$\Delta H^{\ominus}(298 \text{ K})$/kJ mol^{-1}
$Na(s) \longrightarrow Na(g)$	+108
$\tfrac{1}{2}Cl_2(g) \longrightarrow Cl(g)$	+121
$Na(g) \longrightarrow Na^+(g) + e^-$	+496
$Cl(g) + e^- \longrightarrow Cl^-(g)$	−349
$Ca(g) \longrightarrow Ca^{2+}(g) + 2e^-$	+1736
$Ca^{2+}(g) \longrightarrow Ca^{3+}(g) + e^-$	+4941
$Ca^{2+}(g) + 2Cl^-(g) \longrightarrow CaCl_2(s)$	−2220

$$Ca^{3+}(g) + 3Cl^-(g) \longrightarrow CaCl_3(s) \qquad -4800 \text{ (est)}$$
$$NaCl(s) \longrightarrow Na^+(g) + Cl^-(g) \qquad +787$$
$$NaCl(s) + water \longrightarrow Na^+(aq) + Cl^-(aq) \qquad +4$$

Using this information,
  i) calculate the standard molar enthalpy change for the process

$$Na(s) + \tfrac{1}{2}Cl_2(g) \longrightarrow Na^+(g) + Cl^-(g)$$

  ii) explain why $CaCl_3(s)$ does not exist but $CaCl_2(s)$ does
  iii) comment on the difference between the values of the enthalpy change of lattice breaking of $NaCl(s)$ and the enthalpy of solution of $NaCl(s)$ in water and define a term which is useful in this context
  iv) discuss the processes occurring at the molecular level when solid sodium chloride dissolves in water.

  c) State and discuss the general principles which govern the extent to which compounds are soluble in water.   (WJEC90)

4. a) i) Define the term *lattice enthalpy.*
       ii) State and explain the effect of ionic charge and ionic radius on the magnitude of the lattice enthalpy of a salt.

   b) Explain briefly why the entropy change (ΔS) is positive for the dissolution of an ionic solid in water.

   c) Calculate the temperature at which a reaction for which the enthalpy change (ΔH) is $+100 \text{ kJ mol}^{-1}$ and the entropy change (ΔS) is $+0.04 \text{ kJ K}^{-1} \text{mol}^{-1}$ would become energetically feasible. Explain the reasoning behind your calculation.   (AEB90)

5. The industrial preparation of the polymer, poly(tetrafluoroethene) or PTFE, is based on the synthesis of the monomer tetrafluoroethene, $CF_2{=}CF_2$, which is produced by thermal cracking of chlorodifluoromethane, $CHClF_2$, according to reaction (1) below.

$$2CHClF_2(g) \rightleftharpoons CF_2{=}CF_2(g) + 2HCl(g) \qquad (1)$$

Here the $CHClF_2$ is diluted by superheated steam, which also acts as the heat source.

The monomer $CF_2{=}CF_2$ is also obtained via reaction (2).

$$2CHF_3(g) \rightleftharpoons CF_2{=}CF_2(g) + 2HF(g)$$
$$\Delta H^\ominus = +198.1 \text{ kJ mol}^{-1} \qquad (2)$$

Consider this information, together with the data in the table below, and answer the following questions.

Compound	$\Delta H_F^\ominus$ /kJ mol^{-1}	Compound	$\Delta H_F^\ominus$ /kJ mol^{-1}	Molecule $X-X$	$D(X-X)$ /kJ mol^{-1}
HCl(g)	-92.3	$CF_4(g)$	-679.6	F$-$F(g)	154.7
$CHClF_2(g)$	-485.2	$CCl_4(g)$	-106.6	Cl$-$Cl(g)	246.7
$CF_2{=}CF_2(g)$	-658.3				

a) i) Calculate the value of the enthalpy change, $\Delta H^{\ominus}$, for reaction (1). State, giving your reasons, how you would expect the yield of the tetrafluoroethene monomer to be affected by: 1. increase of temperature and 2. increase of pressure. In the latter case explain how your conclusion is compatible with the experimental conditions described.

ii) Indicate and explain whether there are any drawbacks to the use of reaction (2) which would make reaction (1) preferable.

b) i) Use the expressions

$$CX_4(g) \longrightarrow C(s) + 2X_2(g) \qquad \Delta H^{\ominus} = -\Delta H^{\ominus}_F$$

$$C(s) \longrightarrow C(g) \qquad \Delta H^{\ominus} = +718.0 \, \text{kJ mol}^{-1}$$

and

$$2X_2(g) \longrightarrow 4X(g) \qquad \Delta H^{\ominus} = 2D(X-X)$$

where $X = $ F, Cl, to calculate $\Delta H^{\ominus}$ for the *two* processes

$$CX_4(g) \longrightarrow C(g) + 4X(g).$$

Hence find the average $C-X$ bond energies for the species $CX_4(g)$ (where $X = $ F and $X = $ Cl). Given that the average $C-H$ bond energy is $416.1 \, \text{kJ mol}^{-1}$, explain the implications of your results for the relative chemical reactivities of $C-H$, $C-F$ and $C-Cl$ bonds.

ii) Chlorofluorocarbons (CFCs) are widely used as propellent gases for aerosols. In the upper atmosphere, photochemically induced homolytic fission of one of the carbon–halogen bonds of CFCs produces halogen radicals which then attack the ozone layer. Use your results from b) i) above to suggest *which* halogen is likely to be the dominant cause of such damage. (WJEC92)

6. The Born–Haber cycle below represents the energy changes occurring at 298 K when potassium hydride, KH, is formed from its elements.

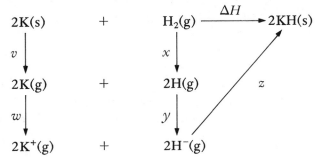

Enthalpy of atomization of potassium	=	$90 \, \text{kJ mol}^{-1}$
First ionization energy of potassium	=	$418 \, \text{kJ mol}^{-1}$
Bond enthalpy of hydrogen	=	$436 \, \text{kJ mol}^{-1}$
First electron affinity of hydrogen	=	$-78 \, \text{kJ mol}^{-1}$
Lattice enthalpy of potassium hydride	=	$-710 \, \text{kJ mol}^{-1}$

a) In terms of the letters v to z write down expressions for:
   i) ΔH for the reaction $2K(s) + H_2(g) \qquad 2KH(s)$
   ii) the first ionization energy of potassium
   iii) the first electron affinity of hydrogen
   iv) the lattice enthalpy of KH.
   (*The lattice enthalpy is the enthalpy change which accompanies the formation of 1 mol of KH(s) from its gaseous ions.*)

b) Which of v to y is
   i) the most exothermic
   ii) the most endothermic?

c) i) Calculate the value of ΔH, showing all your working.
    ii) Calculate the standard enthalpy of formation of KH.

d) i) Write a balanced equation for the reaction of potassium hydride with water.
    ii) On complete reaction with water, 0.10 g of potassium hydride yielded a solution requiring 25 cm³ of 0.10 mol dm⁻³ hydrochloric acid for neutralisation. Calculate the relative atomic mass of potassium from this information.   (L92)

7. a) Distinguish clearly, by means of examples, between the terms 'bond enthalpy' and 'mean bond enthalpy'.

   You will need the following data for the calculations which follow:

   Mean bond enthalpies/kJ mol⁻¹: C=C 612, C—C 348, C—H 412, H—H 436.

   Enthalpy of atomisation of carbon/kJ mol⁻¹: 715.

   b) Using these mean bond enthalpies, calculate the enthalpy change for the dimerisation of ethene to give cyclobutane:

$$2CH_2{=}CH_2(g) \longrightarrow \begin{array}{l} CH_2{-}CH_2 \\ \;|\qquad\;| \\ CH_2{-}CH_2 \end{array} (g)$$

   c) i) Given that the mean enthalpy of the C—C bond in cyclobutane is, in fact, 320 kJ mol⁻¹, calculate the enthalpy change for the reaction below.

$$4H_2(g) \;+\; 4C(s) \longrightarrow \begin{array}{l} CH_2{-}CH_2 \\ \;|\qquad\;| \\ CH_2{-}CH_2 \end{array} (g)$$

   ii) What does your answer suggest about the above reaction as a way of making cyclobutane? Explain your reasoning.
   iii) The C—C bond enthalpy in cyclobutane is 28 kJ mol⁻¹ smaller than in butane. Comment on this difference.   (L92)

8. This question is concerned with the element titanium, Ti, and with some of its compounds.

   a) i) State the electronic configuration of a titanium atom in its ground state.

ii) Is titanium likely to display more than one oxidation number in its compounds? Justify your answer.

b) Titanium occurs naturally as the mineral rutile, TiO_2. One possible method suggested for the extraction of the metal is to reduce the rutile by heating it with carbon:

$$TiO_2(s) + 2C(s) \longrightarrow Ti(s) + 2CO(g)$$

i) Calculate ΔH for this reaction given that

$$\Delta H^{\ominus}_{f,298}[TiO_2] (s) = -940 \text{ kJ mol}^{-1}$$

and $\quad \Delta H^{\ominus}_{f,298}[CO] (g) = -110 \text{ kJ mol}^{-1}$

ii) Calculate ΔG at 2200 K for this reaction using your value for ΔH, $\Delta S = +365 \text{ J K}^{-1} \text{ mol}^{-1}$, and the relationship: $\Delta G = \Delta H - T\Delta S$.

iii) Is this reaction feasible at 2200 K? Justify your answer.

iv) Explain the pollution problem which might be caused by this process.

c) This method is unsatisfactory because of the formation of titanium carbide. In this compound the carbon atoms occupy the octahedral sites in the close-packed structure of the metal. Explain the term *octahedral site*, illustrating your answer with a simple diagram.   (L92,N)

9. The enthalpy of hydration of anhydrous copper(II) sulphate is defined as the heat absorbed or evolved, at constant pressure, when one mole of anhydrous solid is converted into one mole of the crystalline hydrated solid.

$$CuSO_4(s) + 5H_2O(l) = CuSO_4 \cdot 5H_2O(s)$$

It cannot be measured directly.

In an experiment to determine the enthalpy of hydration indirectly, 4.0 g of anhydrous solid was added to 50.0 g of water and the rise in temperature noted as 8°C whereas when 4.0 g of the hydrated solid was added to 50.0 g of water the temperature fell by 1.3°C. In each case the known mass of water was measured into a polystyrene cup, the solid was added to the water and the mixture was stirred continuously with the thermometer until a steady temperature was noted.

a) Why was it better to use a polystyrene cup than a copper calorimeter?

b) Give *three* reasons why the temperature rise of 8°C might be inaccurate.

c)   i) Calculate the heat produced by dissolving 4.0 g of anhydrous solid in 50.0 g of water.

ii) Calculate the enthalpy of solution, in kJ mol^{-1}, of anhydrous copper(II) sulphate.

d) Given that the enthalpy of solution of the hydrated copper(II) sulphate is $+11.3 \text{ kJ mol}^{-1}$, calculate the enthalpy of hydration of the anhydrous solid.

e) Comment on the following statements, which may be either true or false:

   i) 'If the enthalpy change for a reaction is negative then that reaction will take place very quickly.'

   ii) 'The C—Cl bond energy is very high, making that bond very difficult to break and so compounds containing the C—Cl bond are generally unreactive.'

   iii) 'A catalyst speeds up a chemical reaction by making the enthalpy change for the reaction, ΔH, more negative.'

   (Specific heat capacity of water $= 4.18 \text{ J g}^{-1}\text{K}^{-1}$.)   (O&C90,AS)

10. Chemical companies manufacture containers filled with liquid butane for use by campers. The enthalpy change of combustion of butane is $-3000 \text{ kJ mol}^{-1}$.

   a) Write an equation for the complete combustion of butane.

   A camper estimates that the liquid butane left in a container would give 1.2 dm^3 of butane gas (measured at ordinary temperature and pressure).

   b) Calculate the mass of water at $20°C$ that could be brought to the boiling point by burning this butane: use the following information.

   Assume that
   80% of the heat from the butane is absorbed by the water,
   the specific heat capacity of water is $4.2 \text{ J g}^{-1}\text{K}^{-1}$,
   1 mol of a gas occupies 24 dm^3 at ordinary temperatures and pressures.

   c) Suggest how the camper might have estimated how much butane was left in the container.

   d) When burnt in a limited supply of air, butane forms carbon and steam.

   i) Construct a balanced equation for this reaction.

   The enthalpy change of this reaction is $-1400 \text{ kJ mol}^{-1}$.

   ii) Explain why the enthalpy changes of these two combustion reactions are different.

   iii) What additional quantitative information can be calculated from this difference?   (C91)

11. a) State the first law of thermodynamics and discuss the relationship between this law and Hess's law.

   b) Describe how you could measure the molar enthalpy of combustion (ΔH_c) of ethanol by a simple laboratory experiment. Discuss the practical precautions which would be necessary to minimise experimental error. Explain how a value for ΔH_c could be calculated from the experimental results.

c) Methanol can be produced from methane by a two-step process.

Step 1 $CH_4(g) + H_2O(g) \rightleftharpoons CO(g) + 3H_2(g)$

Step 2 $CO(g) + 2H_2(g) \rightleftharpoons CH_3OH(g)$

i) Use the following enthalpies of combustion to calculate the enthalpy change, ΔH, for each of the two steps.

	$CH_4(g)$	$CO(g)$	$H_2(g)$	$CH_3OH(g)$
ΔH_c/kJ mol^{-1}	-808	-283	-245	-671

(*Note* Where water is a product of combustion the figures refer to the formation of $H_2O(g)$.)

ii) Discuss how changes in temperature and pressure will affect the yield of products in each step.

iii) Discuss *two* economic advantages of operating these two steps in reaction vessels close to each other in an industrial plant.

(JMB92)

12. a) The ionisation energy of hydrogen atoms is $+1310$ kJ mol^{-1}.

i) Write the equation which defines the ionisation process.

ii) Given that the electron affinity of chlorine is -364 kJ mol^{-1} and the following additional information:

$H(g) + Cl(g) \longrightarrow HCl(g)$ $\Delta H_{298}^{\ominus} = -432$ kJ mol^{-1}

$HCl(g) \longrightarrow H^+(aq) + Cl^-(aq)$ $\Delta H_{298}^{\ominus} = -75$ kJ mol^{-1}

calculate the standard enthalpy change for the process

$H^+(g) + Cl^-(g) \longrightarrow H^+(aq) + Cl^-(aq)$

b) The table below gives the standard enthalpy changes of hydration of some gaseous ions.

Ion	Cl^-	Br^-	I^-	Li^+	Na^+	K^+
Enthalpy of hydration/kJ mol^{-1}	-380	-350	-310	-520	-400	-320

i) Using the result from a) ii) calculate the enthalpy of hydration of the proton.

ii) Suggest a reason why your answer is quite different from any of the values in the table.

c) Ethyne, C_2H_2, can be converted into ethene, C_2H_4, by the following reaction:

$$H-C \equiv C-H(g) \ + \ H_2(g) \longrightarrow \underset{H}{\overset{H}{\diagdown}} C = C \underset{H}{\overset{H}{\diagup}} \ (g)$$

    i) Why is it difficult to determine experimentally an accurate value for the standard enthalpy of this reaction?

    ii) How can mean bond enthalpies be used to estimate a value for this standard enthalpy change? (L91)

13. a) The diagram below shows an outline, *not to scale*, of the Born–Haber cycle used for the calculation of the lattice energy of strontium chloride from experimental data.

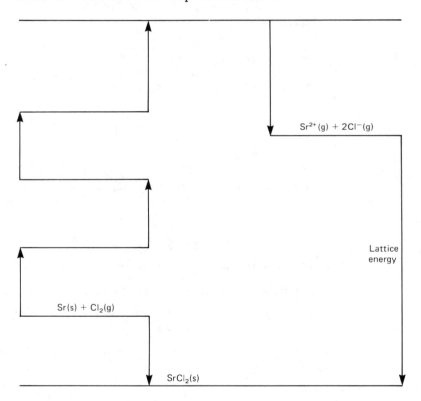

    i) On each of the four empty lines in a copy of the Born–Haber cycle diagram above, write in the formulae for the species present at that stage in the cycle. The diagram is based on the ionisation of strontium being a two-stage process,

    ii) From the table below, select the data required for the calculation of the lattice energy of strontium chloride, and write these in the correct spaces on the Born–Haber cycle diagram.

$$\Delta H^{\ominus}_{\text{at, 298}}[\text{Sr(s)}] \qquad\qquad\qquad = +164.4\,\text{kJ mol}^{-1}$$
$$\Delta H^{\ominus}_{\text{at, 298}}[\tfrac{1}{2}\text{Cl}_2\,(\text{g})] \qquad\qquad\quad = +121.7\,\text{kJ mol}^{-1}$$

First ionisation energy of strontium $\quad = +550.0\,\text{kJ mol}^{-1}$

Second ionisation energy of strontium $= +1064\,\text{kJ mol}^{-1}$

Electron affinity of chlorine $\qquad\quad\; = -348.8\,\text{kJ mol}^{-1}$

$$\Delta H^{\ominus}_{\text{f, 298}}[\text{SrCl}_2(\text{s})] \qquad\qquad\quad\;\; = -828.9\,\text{kJ mol}^{-1}$$

iii) Using your completed Born–Haber cycle, calculate a value for the lattice energy of strontium chloride.

b) Theoretical values have been obtained for the standard enthalpy changes of formation of the two hypothetical compounds SrCl(s) and $\text{SrCl}_3(\text{s})$:

$$\Delta H^{\ominus}_{\text{f, 298}}[\text{SrCl(s)}] \;\; = -198\,\text{kJ mol}^{-1}$$
$$\Delta H^{\ominus}_{\text{f, 298}}[\text{SrCl}_3(\text{s})] = +571\,\text{kJ mol}^{-1}$$

i) Comment on the likely energetic stability of these compounds in relation to:
1) the elements strontium and chlorine
2) $\text{SrCl}_2(\text{s})$.

ii) Theoretical values for the lattice energies for these two compounds have been calculated:

$$\text{Lattice energy for SrCl(s)} = -632\,\text{kJ mol}^{-1}$$
$$\text{Lattice energy for SrCl}_3(\text{s}) = -4560\,\text{kJ mol}^{-1}$$

Suggest reasons for:
1) the large difference in the values of the lattice energies between $\text{SrCl}_2(\text{s})$ and SrCl(s)
2) the large difference in the values of the standard enthalpy changes of formation between $\text{SrCl}_2(\text{s})$ and $\text{SrCl}_3(\text{s})$.

c) When 1 mole of rubidium chloride is dissolved in water at 298 K to form a solution of concentration $1\,\text{mol dm}^{-3}$, the enthalpy change is $+19\,\text{kJ mol}^{-1}$:

$$\text{RbCl(s)} \;+\; \text{aq} \longrightarrow \text{Rb}^+(\text{aq}) \;+\; \text{Cl}^-(\text{aq})$$
$$\Delta H^{\ominus}_{298} = +19\,\text{kJ mol}^{-1}$$

i) Calculate the entropy change in the surroundings when this process takes place.
ii) Calculate the entropy change in the system for this process from the data:

$$S^{\ominus}[\text{RbCl(s)}] \;\; = +95.9\,\text{J mol}^{-1}\,\text{K}^{-1}$$
$$S^{\ominus}[\text{Rb}^+(\text{aq})] = +121.5\,\text{J mol}^{-1}\,\text{K}^{-1}$$
$$S^{\ominus}[\text{Cl}^-(\text{aq})] \; = +56.5\,\text{J mol}^{-1}\,\text{K}^{-1}$$

iii) Use the results of your calculations to explain why rubidium chloride dissolves readily in water in spite of this being an endothermic process.   (L91,N)

14. a) The cyclic unsaturated hydrocarbon, naphthalene ($C_{10}H_8$), which

may be written as 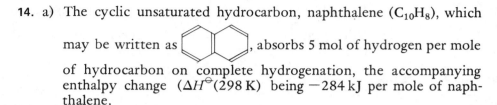, absorbs 5 mol of hydrogen per mole

of hydrocarbon on complete hydrogenation, the accompanying enthalpy change ($\Delta H^\ominus$(298 K) being -284 kJ per mole of naphthalene.

The average enthalpy of hydrogenation of a C=C double bond in a ring is -120 kJ mol^{-1}.

Use this information to calculate the delocalisation or resonance energy in naphthalene and explain the basis of your calculation.

b) Bearing in mind your result in d) i) above and your answers in a) and b) above, state what you would expect to be the characteristic chemical behaviour of naphthalene. Give a reason.

c) A simple substituted naphthalene is used in the manufacture of a range of useful chemical compounds. Name this naphthalene derivative and state the way in which it is used.   (WJEC91,p)

15. Ellingham diagrams, showing the variation of standard free energy change, $\Delta G^\ominus$, with temperature, have proved useful in deciding the best conditions for the extraction of metals from their ores. An Ellingham diagram for the oxides of aluminium, carbon, hydrogen and zinc is shown opposite.

a) Discuss the advantages and disadvantages of using aluminium, hydrogen and carbon as reducing agents in the extraction of metals.

b) Write the equation for the reaction between zinc oxide and carbon to form zinc and carbon monoxide.

Use the Ellingham diagram above to obtain a value for $\Delta G^\ominus$ for this reaction at 1100 K. Would aluminium or hydrogen reduce zinc oxide at this temperature?

c) By considering both $\Delta H^\ominus$ and $\Delta S^\ominus$ explain why $\Delta G^\ominus$ varies with temperature for the reactions between
   i) zinc and oxygen
   ii) carbon and oxygen forming carbon monoxide.   (L91,N)

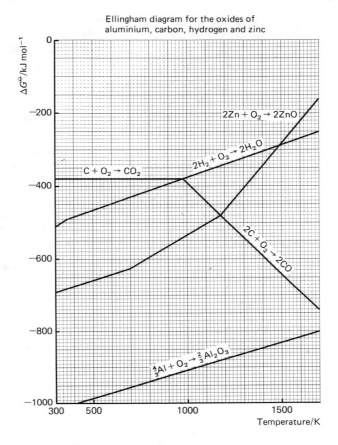

Ellingham diagram for the oxides of aluminium, carbon, hydrogen and zinc

16. Iodine monochloride (ICl) may be prepared using the apparatus shown below. It is a solid at room temperature with a melting point of 27 °C and a boiling point of 101 °C.

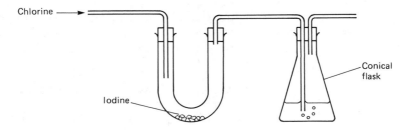

Chlorine gas is passed over iodine.

$$I_2(s) \; + \; Cl_2(g) \longrightarrow 2ICl(s)$$

Excess chlorine is absorbed in the conical flask.

a)   i) Name a substance that could be used in the flask to absorb the chlorine.

   ii) Name the product(s) formed when the substance used in a) i) reacts with chlorine.

   iii) Describe a suitable test that could be used to determine whether chlorine was escaping from the apparatus.

b) Iodine monochloride is a polar molecule and adds to alkenes in a similar way to hydrogen chloride.

   i) Show the polarisation within a molecule of iodine monochloride.

   ii) Write a balanced equation, showing the structures of the reactants and product, for the reaction of but-2-ene with iodine monochloride.

c) The boiling point of iodine monochloride is 101 °C which is almost equal to that of water. Explain, using intermolecular forces, why iodine monochloride has such a similar boiling point although a molecule of ICl has a mass six times greater than that of H_2O.

d) The reaction between iodine vapour and chlorine gas is weakly exothermic.

$$I_2(g) \ + \ Cl_2(g) \ \longrightarrow \ 2ICl(g) \qquad \Delta H = -30\,kJ$$

   i) Use the following bond energies to calculate E_{I-Cl} (the I—Cl bond energy).

Bond	Energy/kJ mol^{-1}
I—I	151
Cl—Cl	243

   ii) Use the additional thermochemical data below to calculate $\Delta H_F(ICl)$ (the standard enthalpy change of formation for iodine monochloride).

$$I_2(s) \ \longrightarrow \ I_2(g) \qquad\qquad \Delta H = +62\,kJ$$
$$2ICl(g) \ \longrightarrow \ 2ICl(s) \qquad\quad \Delta H = -102\,kJ$$

   iii) The enthalpy change for the reaction of $I_2(g)$ with $Cl_2(g)$ has been indirectly measured using the following reactions. ΔH_1 and ΔH_2 represent the enthalpy changes of the reactions.

$$2NOCl(g) \ + \ I_2(g) \ \longrightarrow \ 2NO(g) \ + \ 2ICl(g) \qquad \Delta H_1$$
$$2NOCl(g) \ \longrightarrow \ 2NO(g) \ + \ Cl_2(g) \qquad \Delta H_2$$

Explain how this is possible. An experimental method is not required.
                                                     (NI90)

14 Reaction Kinetics

Reaction Kinetics is the study of the factors which affect the rates of chemical reactions.

REACTION RATE

The rate of a chemical reaction is the rate of change of concentration. Consider a reaction of the type $A \longrightarrow B$, where one molecule of the reactant forms one molecule of the product, Fig. 14.1 shows how the concentration of product, x, increases as the time, t, which has passed since the start of the reaction increases. The initial concentration of reactant (the concentration at the start of the reaction) is a, and at any time after the start of the reaction, the concentration of reactant is $(a - x)$.

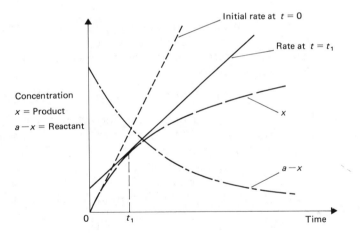

Fig. 14.1   Variation of concentrations of reactant and product with time

You can see that the rate of reaction is decreasing as the reaction proceeds and the reactant is being used up. One can only state the rate of reaction between certain times.

One can calculate the average rate of reaction over a certain interval of time in this way.

To $1 \, dm^3$ of solution containing $0.300 \, mol$ methyl ethanoate is added a small amount of mineral acid. This catalyses the hydrolysis reaction

$$CH_3CO_2CH_3(aq) + H_2O(l) \rightleftharpoons CH_3CO_2H(aq) + CH_3OH(aq)$$

After 100 seconds, the concentration has decreased to $0.292 \text{ mol dm}^{-3}$. This means that $0.008 \text{ mol dm}^{-3}$ of methyl ethanoate has reacted, and $0.008 \text{ mol dm}^{-3}$ of methanol and ethanoic acid have been formed.

$$\left(\begin{array}{l} \text{Average rate of reaction} \\ \text{over this time interval} \end{array} \right) = \frac{\text{Change in concentration}}{\text{Time}}$$

$$= \frac{0.008}{100} = 8 \times 10^{-5} \text{ mol dm}^{-3} \text{ s}^{-1}$$

Since the rate of reaction varies with time, it is usual to quote the initial rate of the reaction. This is the rate at the start of the reaction when an infinitesimally small amount of the reactant has been used up. In Fig. 10.1, the gradient of the tangent to the curve at $t = 0$ gives the initial rate of the reaction.

THE EFFECT OF CONCENTRATION ON RATE OF REACTION

Consider a reaction between A and B to form C:

$$A + B \longrightarrow C$$

The rate of formation of C depends on the concentrations of A and B, but one cannot simply say that the rate of formation of C is proportional to the concentration of A and proportional to the concentration of B. The relationship is

$$\text{Rate of formation of C} \propto [A]^m [B]^n$$

where m and n are usually integers, often 0, 1 or 2, and are characteristic of the reaction. One says that the reaction is of order m with respect to A and of order n with respect to B. The order of reaction is $(m + n)$. One cannot tell the order simply by looking at the chemical equation for the reaction. For example, the reaction between bromate(V) ions and bromide ions and acid to give bromine

$$BrO_3^-(aq) + 5Br^-(aq) + 6H^+(aq) \longrightarrow 3Br_2(aq) + 3H_2O(l)$$

has a rate of disappearance of BrO_3^-

$$\frac{-d[BrO_3^-]}{dt} \propto [BrO_3^-] [Br^-] [H^+]^2$$

It is first order with respect to bromate(V), first order with respect to bromide, second order with respect to hydrogen ion and fourth order overall. The negative sign means that $[BrO_3^-]$ decreases with time.

If   Reaction rate $\propto [A]^m [B]^n$ it follows that

Reaction rate $= k[A]^m [B]^n$

The proportionality constant k is called the *rate constant* for the reaction or the *rate coefficient* for the reaction.

ORDER OF REACTION

As a reaction proceeds, the concentrations of the reactants decrease, and the rate of reaction decreases, as shown in Fig. 14.1. The shape of the curve depends on the order of the reaction (see Fig. 14.2).

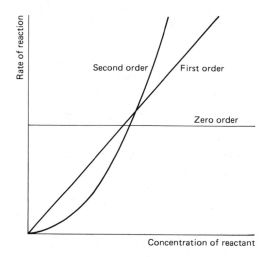

Fig. 14.2   Graphs of rate against concentration

FIRST-ORDER REACTIONS

If the reaction

$$A \longrightarrow Products$$

is a first-order reaction, the rate equation will be

$$\text{Rate} = k[A] \quad \text{i.e.} \quad \frac{-d[A]}{dt} = k[A]$$

If $[A]_0 =$ initial concentration of A, the integrated form of this equation is

$$kt = \ln \frac{[A]_0}{[A]} = 2.303 \lg \frac{[A]_0}{[A]}$$

The units of k, the first-order rate constant, are s^{-1}.

HALF-LIFE

Let $t_{1/2}$ be the time taken for half the amount of A to react. $t_{1/2}$ is called the *half-life* of the reaction.

After $t_{1/2}$ seconds, $[A] = [A]_0/2$

$\therefore$ $kt_{1/2} = \ln 2 = 2.303 \lg 2$

$t_{1/2} = 0.693/k$

The half-life of a first-order reaction is independent of the initial concentration of the reactant. Radioactive decay is an example of first-order kinetics.

PSEUDO-FIRST-ORDER REACTIONS

The acid-catalysed hydrolysis of an ester, e.g. ethyl ethanoate,

$$CH_3CO_2C_2H_5(aq) + H_2O(l) \longrightarrow CH_3CO_2H(aq) + C_2H_5OH(aq)$$

is first order with respect to ester and first order with respect to water. If water is present in excess, so that the fraction of the water which is used up in the reaction is small, the concentration of water is practically constant, and, since the acid catalyst is not used up, the rate depends only on the concentration of ester:

$$\frac{-d[CH_3CO_2C_2H_5]}{dt} = k'[CH_3CO_2C_2H_5]$$

k' is constant for a certain concentration of acid, and the reaction obeys a first-order rate equation.

EXAMPLE 1   The rate constant of a first-order reaction is $2.0 \times 10^{-6}\,s^{-1}$. The initial concentration of the reactant is $0.10\,mol\,dm^{-3}$. What is the value of the initial rate in $mol\,dm^{-3}\,s^{-1}$?

METHOD   The rate equation has the form

$$Rate = k[A]$$

Putting the values of $[A]$ and k into this equation gives

ANSWER $Rate = 2.0 \times 10^{-6} \times 0.10 = 2.0 \times 10^{-7}\,mol\,dm^{-3}\,s^{-1}$.

EXAMPLE 2   The half-life for the radioactive decay of thorium-234 is 24 days. a) Calculate the rate constant for the decay. b) What time will elapse before 3/4 of the thorium has decayed?

METHOD   a) Radioactive decay follows the first-order law:

$$kt_{1/2} = 2.303 \lg 2$$

ANSWER $$k = \frac{2.303 \lg 2}{24 \times 24 \times 60 \times 60} = 3.34 \times 10^{-7}\,s^{-1}$$

b) When 3/4 of the thorium has decayed, the initial amount of thorium, A_0, has become $A_0/4$.

$$kt = 2.303 \lg \frac{A_0}{A}$$

$$3.34 \times 10^{-7} \times t = 2.303 \lg \frac{A_0}{A_0/4}$$

ANSWER $$t = 4.15 \times 10^6 \, s = 48 \, days$$

EXAMPLE 3 Carbon-14 is radioactive. The half-life is 5600 years. Calculate the age of a piece of wood which gives 10 counts per minute per gram of carbon, compared with 15 c.p.m. per gram of carbon from a sample of new wood.

METHOD a) First, find the first-order rate constant for the decay.
b) Use the rate constant to find the age of the sample.

a) $$2.303 \lg \frac{A_0}{A} = kt$$

$\therefore$ $$2.303 \lg \frac{N_0}{N} = kt$$

where N_0 = Count rate for new wood and

N = Count rate after time t

At the half-life, $$2.303 \lg 2 = kt_{1/2}$$

Since $t_{1/2} = 5600$ years, $k = 1.24 \times 10^{-4} \, year^{-1}$

b) Since $N_0/N = 15/10$,

$$2.303 \lg 1.5 = 1.24 \times 10^{-4} t$$

ANSWER Age of wood, $$t = 3270 \, years.$$

SECOND-ORDER REACTIONS, WITH REACTANTS OF EQUAL CONCENTRATION

In the simplest case, the reaction

$$A + B \longrightarrow Products$$

has a rate equation $$\frac{-d[A]}{dt} = \frac{-d[B]}{dt} = k[A][B]$$

EXAMPLE 1 The initial rate of a second-order reaction is $8.0 \times 10^{-3} \, mol \, dm^{-3} \, s^{-1}$. The initial concentrations of the two reactants, A and B, are 0.20 mol dm^{-3}. What is the rate constant in $dm^3 \, mol^{-1} \, s^{-1}$?

METHOD   Since $$-\frac{d[A]}{dt} = k\,[A]\,[B]$$

putting

$$\frac{d[A]}{dt} = 8.0 \times 10^{-3}\,\text{mol dm}^{-3}\,\text{s}^{-1} \text{ and } [A] = [B] = 0.20\,\text{mol dm}^{-3}$$

gives $$8.0 \times 10^{-3} = k \times 0.20 \times 0.20$$

ANSWER   Rate constant, $$k = 0.20\,\text{dm}^3\,\text{mol}^{-1}\,\text{s}^{-1}$$

EXAMPLE 2   In the alkaline hydrolysis of an ester, both ester and alkali had the initial concentration of $0.0500\,\text{mol dm}^{-3}$. The following results were obtained.

Time/s	100	200	300	400	600
% of ester remaining	70.5	55.8	44.5	37.8	29.7

The order of reaction is 2. Plot a graph of the percentage of ester remaining against time. By drawing a tangent to the graph, find the initial rate of reaction. Calculate the rate constant of this second order reaction.

METHOD   The graph of percentage of ester remaining against time is shown in Fig. 14.3. The tangent to the graph at time $t = 0$ has a gradient

$$42\% \times 0.05\,\text{mol dm}^{-3}/100\,\text{s} = 2.1 \times 10^{-4}\,\text{mol dm}^{-3}\,\text{s}^{-1}$$

Since the reaction is second order, the rate constant, k, is given by

$$k \times 0.05 \times 0.05 = 2.1 \times 10^{-4}$$
$$k = 8.4 \times 10^{-2}\,\text{dm}^3\,\text{mol}^{-1}\,\text{s}^{-1}$$

(*Note*   In more advanced work, you will learn a more accurate graphical method of finding the rate constant of a second order reaction.)

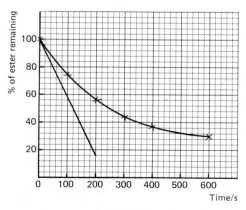

Fig. 14.3

ZERO-ORDER REACTIONS

In a zero-order reaction, the rate is independent of the concentration of the reactant. In the reaction between propanone and iodine

$$CH_3COCH_3(aq) \; + \; I_2(aq) \longrightarrow CH_3COCH_2I(aq) \; + \; HI(aq)$$

the reaction rate does not change if the concentration of iodine is changed. The rate of reaction is independent of the iodine concentration, and the reaction is said to be zero order with respect to iodine.

*THE EFFECT OF TEMPERATURE ON REACTION RATES

An increase in temperature increases the rate of a reaction by increasing the rate constant. A plot of the logarithm of the rate constant, k, against $1/T$ is a straight line, with a negative gradient (see Fig. 14.4).

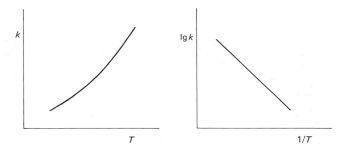

Fig. 14.4   Dependence of rate constant on temperature

The variation of rate constant with temperature obeys the *Arrhenius equation*

$$k = A e^{-E/RT}$$

A and E are constant for a given reaction; R is the gas constant. In order to react, two molecules must collide with a minimum amount of energy E, which is called the *activation energy*. The fraction of molecules possessing energy E is given by $e^{-E/RT}$. The constant A, called the *pre-exponential factor*, represents the maximum rate which the reaction can reach when all the molecules have energy equal to or greater than E.

The Arrhenius equation can be written as

$$\lg k = \lg A - \frac{E}{2.303RT}$$

A plot of $\lg k$ against $1/T$ is linear with a gradient of $-E/2.303R$ and intercept $\lg A$. The values of A and E can thus be found from a plot of $\lg k$ against $1/T$.

EXAMPLE   The reaction

$$2N_2O_5(g) \longrightarrow 2N_2O_4(g) + O_2(g)$$

was studied at a number of temperatures, and the following values for the rate constant were obtained:

Temperature/K	293	308	318	338
Rate constant/s^{-1}	1.76×10^{-5}	1.35×10^{-4}	4.98×10^{-4}	4.87×10^{-3}

Calculate the activation energy and the pre-exponential factor for the reaction.

METHOD   Since

$$\lg k = \lg A - \frac{E}{2.303RT}$$

a plot of $\lg k$ against $1/T$ gives a straight line with a gradient of $-E/2.303\,R$ and an intercept on the $\lg k$ axis of $\lg A$.

The table below gives the values which must be plotted:

T/K	293	308	318	338
k/s^{-1}	1.76×10^{-5}	1.35×10^{-4}	4.98×10^{-4}	4.87×10^{-3}
$1/T\big/1/K$	3.41×10^{-3}	3.25×10^{-3}	3.14×10^{-3}	2.96×10^{-3}
$\lg k$	-4.755	-3.870	-3.303	-2.312

Fig 14.5 shows the plot of $\lg k$ against $1/T$. The gradient is -5550 K.

$$\therefore \qquad -5500 = -\frac{E}{2.303 \times 8.31}$$

$$E = 106\,000 \text{ J mol}^{-1} = 106 \text{ kJ mol}^{-1}$$

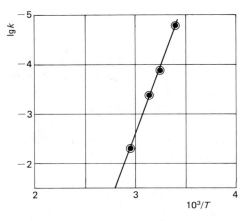

Fig. 14.5   Plot of $\lg k$ against $10^3/T$ for this example

A can be found from the intercept or by substituting values of $\lg k$ and $1/T$ in the equation. At $1/T = 3.28 \times 10^{-3}$, $\lg k = -4.00$.

$$\therefore \qquad \lg A = -4.00 + \frac{106 \times 10^3 \times 3.28 \times 10^{-3}}{2.303 \times 8.314} = 14.16$$

$$A = 1.45 \times 10^{14}\,\text{s}^{-1}$$

ANSWER The activation energy is $106\,\text{kJ mol}^{-1}$; the pre-exponential factor is $1.45 \times 10^{14}\,\text{s}^{-1}$.

EXERCISE 51   Problems on Finding the Order of Reaction

1. X and Y react together. For a three-fold increase in the concentration of X, there is a nine-fold increase in the rate of reaction. What is the order of reaction with respect to X?

2. A and B react to form C. In one run, the concentration of A is doubled, while B is kept constant, and the initial rate is doubled. In a second run, the concentration of B is doubled while that of A is kept constant, and the initial rate is quadrupled. What can you deduce about the order of the reaction?

3. Fig. 14.6 shows that the rate of reaction is:

   a proportional to $[I_2]$
   b proportional to $[I_2]^2$
   c proportional to $1/[I_2]$
   d independent of $[I_2]$

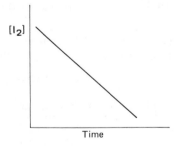

Fig. 14.6

4. X decomposes to form $Y + Z$. The following results were obtained in a study of the reaction:

Initial [X] /mol dm^{-3}	2.0×10^{-3}	4.0×10^{-3}	5.0×10^{-3}
Initial rate/mol dm^{-3} s^{-1}	1.3×10^{-6}	1.3×10^{-6}	1.3×10^{-6}

What is the rate expression? What is the order of the reaction?

5. The reaction $A + B \longrightarrow C$ is first order with respect to A and to B. When the initial concentrations are $[A] = 1.5 \times 10^{-2}\,\text{mol dm}^{-3}$ and $[B] = 2.5 \times 10^{-3}\,\text{mol dm}^{-3}$, the initial rate of reaction is found to be $3.75 \times 10^{-4}\,\text{mol dm}^{-3}\,\text{s}^{-1}$. Calculate the rate constant for the reaction.

6. In the reaction

$$A + B \longrightarrow P + Q$$

the following results were obtained for the initial rates of reaction for different initial concentrations:

[A] /mol dm^{-3}	[B] /mol dm^{-3}	Initial rate/mol dm^{-3} s^{-1}
1.0	1.0	2.0×10^{-3}
2.0	1.0	4.0×10^{-3}
4.0	2.0	16×10^{-3}

Deduce the rate equation and calculate the rate constant.

7. The rate of a reaction depends on the concentrations of the reactants. In the reaction between X and Y, the following results were obtained for runs at the same temperature.

Initial concentration of X/mol dm^{-3}	Initial concentration of Y/mol dm^{-3}	Initial rate/ mol dm^{-3} h^{-1}
2×10^{-3}	3×10^{-3}	3.0×10^{-3}
2×10^{-3}	6×10^{-3}	1.2×10^{-2}
4×10^{-3}	6×10^{-3}	2.4×10^{-2}

Deduce the order of the reaction with respect to: a) X, b) Y. Calculate the rate constant for the reaction.

8. The following results were obtained for the decomposition of nitrogen(V) oxide

$$2N_2O_5(g) \longrightarrow 4NO_2(g) + O_2(g)$$

Concentration of N$_2$O$_5$/mol dm^{-3}	Initial rate/mol dm^{-3} s^{-1}
1.6×10^{-3}	0.12
2.4×10^{-3}	0.18
3.2×10^{-3}	0.24

What is the rate expression for the reaction? What is the order of reaction? What is the initial rate of reaction when the concentration of N_2O_5 is:

a) 2.0×10^{-3} mol dm^{-3} b) 2.4×10^{-2} mol dm^{-3}?

EXERCISE 52   Problems on First-order Reactions

1. A isomerises to form B. The reaction is first order. If 75% of A is converted to B in 2.5 hours, what is the value of the rate constant for the isomerisation?

2. The reaction

$$2N_2O_5 \longrightarrow 4NO_2 + O_2$$

is first order. If the initial concentration of N_2O_5 is 1.25 mol dm^{-3} and the initial rate is $1.38 \times 10^{-5} \text{ mol dm}^{-3} \text{s}^{-1}$, what is the rate constant for the decomposition?

3.

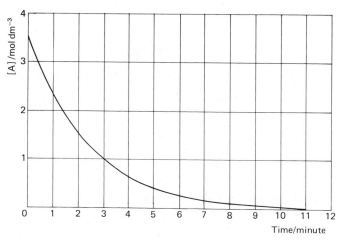

Fig. 14.7

The curve shown in Fig. 14.7 represents the decomposition of A at a certain temperature.

Calculate the gradients of the curve at: a) 1 minute, b) 3 minutes, c) 5 minutes, and d) 11 minutes. Plot a graph of the gradient against the concentration of A at 1, 3, 5 and 11 minutes. Calculate the rate constant for the reaction from the slope of the graph. What is the order of the reaction?

4. Fig. 14.8 shows the results of a study of the reaction

$$A + B \longrightarrow C$$

The experimental conditions were:

	Initial concentrations/mol dm^{-3}	
	[A]	[B]
Curve 1	0.10	0.10
Curve 2	0.10	0.20
Curve 3	0.10	0.30

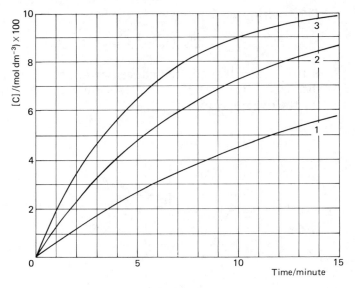

Fig. 14.8

a) Find the initial rates of curves 1, 2 and 3.

b) What is the order of reaction with respect to B?

c) Inspect curve 3. What is the time required for completion of $\frac{1}{2}$ reaction, and of $\frac{3}{4}$ reaction? What is the order with respect to A?

d) Write an overall rate equation for the reaction.

e) Find the rate constant for the reaction.

EXERCISE 53   Problems on Second-order Reactions

1. The following results were obtained from a study of the reaction between P and Q.

Concentrations/mol dm^{-3}		Initial rate/mol dm^{-3} s^{-1}
[P]	[Q]	
2.00×10^{-3}	2.00×10^{-3}	2.00×10^{-4}
1.80×10^{-3}	1.80×10^{-3}	1.62×10^{-4}
1.40×10^{-3}	1.40×10^{-3}	9.80×10^{-5}
1.10×10^{-3}	1.10×10^{-3}	6.05×10^{-5}
0.80×10^{-3}	0.80×10^{-3}	3.20×10^{-5}

Prove that the reaction is second order. Calculate the rate constant.

2. The following results were obtained for a reaction between A and B.

Concentrations/mol dm^{-3}		Initial rate/mol dm^{-3} s^{-1}
[A]	[B]	
0.5	1.0	2
0.5	2.0	8
0.5	3.0	18
1.0	3.0	36
2.0	3.0	72

What is the order of reaction with respect to A and with respect to B? What is the rate equation for the reaction? Calculate the rate constant. State the units in which it is expressed.

EXERCISE 54   Problems on Radioactive Decay

1. Plot a graph, using the following figures, to show the radioactive decay of krypton. From the graph, find the half-life.

Time/minute	0	20	40	60	80	100	120
Activity/count per second	100	92	85	78	72	66	61

2. A sample of gold was irradiated in a nuclear reactor. It gave the following results when its radioactivity was measured at various intervals. Plot the results, and deduce the half-life of the radioactive isotope of gold formed.

Time/hour	0	1	5	10	25	50	75	100
Radioactivity/count per minute	300	296	285	270	228	175	133	103

3. A sample of bromine was irradiated in a nuclear reactor. The following results were obtained when the radioactivity was measured after various time intervals. Plot the results, and deduce what you can about the decay of radioactive bromine.

Time/hour	0	0.1	0.2	0.5	1	2	5	10	25	50	75	100
Radio-activity/count per minute	500	442	399	320	268	242	225	204	154	95	55	35

4. A radioactive source, after storing for 42 days, is found to have 1/8th of its original activity. What is the half-life of the radioactive isotope present in the source?

5. Actinium B has a half-life of 36.0 min. What fraction of the original quantity of actinium remains after: a) 180.0 min, b) 1080 min?

6. The half-life of carbon-14 is 5580 years. A 10 g sample of carbon prepared from newly cut timber gave a count rate of 2.04 s^{-1}. A 10 g sample of carbon from an ancient relic gave a count rate of 1.84 s^{-1}. Calculate the age of the relic.

7. A dose of 1.00×10^{-4} g of astatine-211 is given to a patient for treatment of cancer of the thyroid gland. How much of this radioactive isotope ($t_{1/2} = 7.21$ h) will remain in the body 24 hours later?

*8. Tritium has a half-life of 12.3 years. When tritiated water is used in tracer experiments, what percentage of the original activity will remain after: a) 5 years, b) 50 years?

*9. The half-life of carbon-14 is 5600 years. A piece of wood from an ancient ship gives a count of 10 counts per minute, while carbon obtained from new wood gives 15 counts per minute. What is the age of the ship?

EXERCISE 55   Problems on Rates of Reaction

1. A first-order reaction is 50% complete at the end of 30 minutes. What is the value of the rate constant? In how many minutes is reaction 80% complete?

2. The half-life for the disintegration of bismuth-214 is 19.7 minutes. Calculate the rate constant for the decay in s^{-1}.

3. The half-life for the radioactive disintegration of bismuth-210 is 5.0 days. Calculate: a) the rate constant in s^{-1}, b) the time needed for 0.016 mg of bismuth-210 to decay to 0.001 mg.

4. Hydrogen and iodine combine to form hydrogen iodide. The reaction is first order with respect to hydrogen and first order with respect to iodine. The rate constant is 2.78×10^{-4} mol dm^{-3} s^{-1}. If the concentrations are $[H_2] = 0.85 \times 10^{-2}$ mol dm^{-3}, and $[I_2] = 1.25 \times 10^{-2}$ mol dm^{-3}, what is the initial rate of reaction?

5. The reaction $2NO(g) + Cl_2(g) \longrightarrow 2NOCl(g)$ is third order. The rate constant is 1.7×10^{-5} dm^6 mol^{-2} s^{-1}. If the concentrations of the reactants are each 0.20 mol dm^{-3}, what is the initial rate of reaction?

6. A and B react in the gas phase. In experiment 1, a glass vessel was used. In experiment 2, the glass was coated with another material. The results of the two experiments are shown below. Deduce the rate equations for the two experiments. Can you explain how they come to differ?

	[A]/mol dm^{-3}	[B]/mol dm^{-3}	Initial rate/mol dm^{-3} s^{-1}
Experiment 1	0.20	0.12	2×10^{-3}
	0.40	0.12	8×10^{-3}
	0.20	0.24	4×10^{-3}
Experiment 2	0.20	0.12	2×10^{-3}
	0.40	0.24	8×10^{-3}
	0.80	0.24	32×10^{-3}

7. Hydrogen peroxide decomposes in aqueous solution:

$$2H_2O_2(aq) \longrightarrow 2H_2O(l) + O_2(g)$$

The following results show how the rate of decomposition varies with the initial hydrogen peroxide concentration:

Rate/mol dm^{-3} s^{-1}	[H$_2$O$_2$]/mol dm^{-3}
3.64×10^{-5}	0.05
7.41×10^{-5}	0.10
1.51×10^{-4}	0.20
2.21×10^{-4}	0.30

Plot the rate of decomposition against the concentration. Deduce: a) the order of the reaction, and b) the rate constant under the conditions of the experiment.

*EXERCISE 56   Problems on Activation Energy

1. The reaction

$$2A(g) + B(g) \longrightarrow C(g)$$

was studied at a number of temperatures, and the following results were obtained:

Temperature/°C	12	60	112	203	292
Rate constant/dm^6 mol^{-2} s^{-1}	2.34	13.2	52.5	316	1000

Calculate the activation energy.

2. The following results were obtained in the study of the effect of temperature on the rate of a chemical reaction.

Temperature/K	2000	1136	909	676	540
Rate constant/dm^3 mol^{-1} s^{-1}	1.00	0.316	0.158	0.050	0.0158

From a plot of lg$k$ against $1/T$, find the activation energy for the reaction.

3. Use the following results to calculate the energy of activation of the reaction:

Temperature/K	800	625	472	377	333
Rate constant/dm^3 mol^{-1} s^{-1}	0.562	0.316	0.132	0.0562	0.0316

EXERCISE 57   Questions from A-level Papers

1. A dilute solution of hydrogen peroxide can be used to bleach hair. It decomposes slowly in aqueous solution according to the following equation:

$$2H_2O_2(aq) \longrightarrow 2H_2O(l) + O_2(g)$$

A solution with an original concentration of $3.0 \, mol \, dm^{-3}$ was placed in a bottle contaminated with transition metal ions, which act as catalysts for the decomposition. The rate of decomposition was measured by withdrawing $10 \, cm^3$ portions at various times and titrating with acidified $0.1 \, mol \, dm^{-3} \, KMnO_4(aq)$. (5 moles of peroxide react with 2 moles of $KMnO_4$.) The following results were obtained:

Time/min	Volume of 0.1 mol dm^{-3} KMnO$_4$(aq)/cm^3
0	30.0
5	23.4
10	18.3
15	14.2
20	11.1
25	8.7
30	6.8

a) Confirm that the reaction is first order with respect to the peroxide.

b) Write an expression for the rate equation and calculate the rate constant and half-life.

c) Calculate the concentration of the hydrogen peroxide at the time the first portion was withdrawn. Hence estimate how long the solution had been in the contaminated bottle.

d) Suggest a method whereby the shelf-life of hydrogen peroxide solutions could be increased.   (C91)

2. a) The reaction

$$C_{12}H_{22}O_{11} + H_2O \longrightarrow C_6H_{12}O_6 + C_6H_{12}O_6$$

   Sucrose   Glucose   Fructose

is catalysed by hydrogen ions. It is found that the rate of reaction depends only on the concentration of sucrose, for a given hydrogen ion concentration.

At $25°C$, when the initial concentration of sucrose was $1.00 \, mol \, dm^{-3}$, the following data were obtained.

Time/min	0	60	90	130	180
Sucrose reacted/mol dm^{-3}	0	0.195	0.277	0.373	0.478

   i) By plotting an appropriate graph, show that this is a first-order reaction with respect to sucrose concentration.

   ii) Use your graph to find the half-life and first-order rate constant for the reaction.

   iii) How would halving the initial concentration of sucrose affect 1) the initial rate, 2) the half-life of this reaction?

   iv) Why does this reaction follow a first-order rate law although water enters into the stoichiometric equation?

b) Use the photochemical reaction of chlorine with methane to illustrate and explain the following processes in a chain reaction: i) initiation; ii) propagation; iii) termination.   (O90)

3. Iodine reacts rapidly with propanone (CH_3COCH_3) in acidic or alkaline solution but only very slowly when neutral.

The reaction in acidic solution is

$$CH_3COCH_3(aq) + I_2(aq) \longrightarrow CH_3COCH_2I(aq) + H^+(aq) + I^-(aq)$$

The reaction rate was studied in the following experiment.

0.1 mol of propanone and 0.01 mol of hydrochloric acid in a total volume of 90 cm^3 of water were placed in a flask at constant temperature. 0.0004 mol of iodine (I_2) in 10 cm^3 of water was added and timing was begun.

10 cm^3 samples were periodically withdrawn at 10 minute intervals and were neutralised by adding to excess aqueous sodium hydrogencarbonate. These were then titrated with sodium thiosulphate solution containing 0.01 mol dm^{-3} $Na_2S_2O_3$. The results are shown in the graph below.

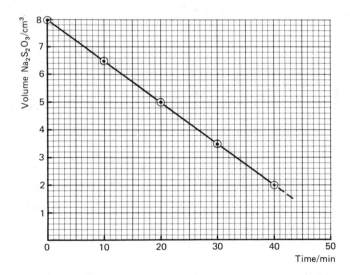

a) What is the order of the reaction with respect to iodine? Explain your answer.

b) Why are the initial concentrations of propanone and acid chosen to be so much greater than that of the iodine?

c) i) Why is the mixture neutralised (before titration)?
    ii) Why would sodium hydroxide be unsuitable for this purpose?

d) What would be the effect on the gradient of the graph of doubling the initial concentration of iodine?

e) If the initial concentration of hydrochloric acid is doubled, the rate of the reaction is also doubled. What is the order of reaction with respect to hydrochloric acid?

f) Indicate, *without* giving experimental detail, how you would show that the species responsible for doubling the rate in e) is the hydrogen ion and not the chloride ion.   (O91,AS)

4. The following data refer to the hydrolysis of 2-bromo-2-methylpropane, $CH_3-C(CH_3)Br-CH_3$, by aqueous sodium hydroxide at $25\,^{\circ}C$.
   The equation is

$$CH_3-C(CH_3)Br-CH_3 \ + \ OH^- \longrightarrow CH_3-C(CH_3)(OH)-CH_3 \ + \ Br^-$$

Concentration of 2-bromo-2-methylpropane /mol dm^{-3}	Concentration of OH$^-$ ions/mol dm^{-3}	Initial rate of hydrolysis /mol dm^{-3} s^{-1}
0.100	0.500	0.0020
0.100	0.250	0.0020
0.075	0.250	0.0015
0.050	0.250	0.0010
0.025	0.250	0.0005

a) What is the order of the reaction with respect to
   i) 2-bromo-2-methylpropane
   ii) hydroxide ion?
   iii) Explain how you arrived at your answers to i) and ii).

b) Write down the rate expression for the reaction.

c) Give the units of the rate constant.

d) Write down a mechanism for the reaction.

e) Suggest a method by which the rate of this hydrolysis might be followed in the laboratory.   (AEB90)

5. a) Explain how ion-exchange may be used for the purification of water on a domestic scale.

   b) i) Explain the phenomenon of osmosis and define osmotic pressure.
      ii) How may sea water be desalinated using osmosis?

   c) Calculate the age of an ancient papyrus scroll found in Egypt which gave a count rate of $110\ min^{-1}\ g^{-1}$ compared with $150\ min^{-1}\ g^{-1}$ from new papyrus. The half-life of the nuclide $^{14}_{6}C$ is 5600 years.
      Explain the assumptions made in this dating procedure.   (O91)

6. a) Distinguish between *order* and *molecularity* of a reaction.

b) For the gas phase isomerisation reaction

$$A \longrightarrow B$$

the following data were obtained at 700 K.

Time/hour	0	2	5	10	20	30
Fraction of A unreacted	1.0	0.91	0.79	0.63	0.40	0.25

i) Plot an appropriate graph to show that this is a first-order reaction.

ii) Use your graph to find the *half-life* and *first-order rate constant* for the reaction.

iii) How would trebling the initial concentration of A affect 1) the half-life, 2) the initial rate of this reaction?

c) Discuss briefly the factors which are responsible for the increase of a reaction rate with temperature and indicate their relative importance.   (O91)

*7. The hydrolysis of methyl methanoate proceeds at a measurable rate in the presence of aqueous acid:

$$HCO_2CH_3 + H_2O \longrightarrow HCO_2H + CH_3OH$$

a) What method would you use to investigate how the rate of this reaction varies with concentration of the reactants?

Your answer should include an outline of the experimental procedure and the measurements you would make.

b) In an investigation of this reaction the following results were obtained for the rate of hydrolysis of methyl methanoate at 298 K:

$[HCO_2CH_3]$ /mol dm^{-3}	$[H^+]$ /mol dm^{-3}	Initial rate $\times 10^3$/mol dm^{-3} s^{-1}
0.50	1.00	0.56
1.00	1.00	1.11
2.00	1.00	2.24
2.00	0.50	1.13
2.00	2.00	4.49

Deduce the rate equation for this reaction, and calculate the rate constant at 298 K. Explain how you arrive at your answers.

Using the value of your rate constant at 298 K and the data below, plot a suitable graph and hence find a value for the activation energy of the reaction.

Temperature /K	Rate constant /dm^3 mol^{-1} s^{-1}
313	3.95×10^{-3}
323	8.63×10^{-3} (L90,N)

8. The kinetics of the hydrolysis of the ester methyl ethanoate

$$CH_3CO_2CH_3(l) + H_2O(l) \rightleftharpoons CH_3COOH(aq) + CH_3OH(aq)$$

may be investigated by measuring the concentration of ethanoic acid produced. One such investigation, where 17.8 g of the ester were mixed with 1 cm³ of concentrated hydrochloric acid and sufficient water to raise the volume to 1 dm³ and then kept in a temperature-controlled water bath at 35 °C, gave the following results:

Time/s × 10⁴	Concentration of ethanoic acid/mol dm⁻³
0	0
0.36	0.084
0.72	0.136
1.08	0.172
1.44	0.195

a) Why was a small amount of hydrochloric acid added?

b) i) Calculate the number of moles of ester in 17.8 g.
   ii) Plot a graph to show how the concentration of the ester varies with time.
   iii) Determine the half-life for the reaction.
   iv) Determine the order of the reaction with respect to the ester and give your reasoning.
   v) Calculate the initial rate of reaction.
   vi) Using your value for the initial rate, determine the value of the rate constant, k, for this reaction at 35 °C.

c) In a second investigation the concentration of the hydrochloric acid was doubled and the initial rate was found to be 2.33×10^{-5} mol dm⁻³ s⁻¹. Write the rate equation for this reaction, assuming that the rate of reaction is independent of the amount of water present.

d) i) Why was the reaction mixture kept in a temperature-controlled water bath?
   ii) What would happen to the rate of reaction if the experiment were repeated at 25 °C, all other factors being the same?

e) Describe briefly how you would measure the concentration of the ethanoic acid as part of such an investigation.   (O&C90,AS)

9. The isotopes of some elements emit radiation, characterised by a particular half-life. For example, ^{24}Na is a beta emitter having a half-life of 15.0 hours.

a) Name two other types of radiation which may be emitted by radioisotopes.

b) Which of these has the greater penetrating power?

c) What is meant by the term *half-life*?

d) A patient receives a dose of sodium chloride containing ^{24}Na, giving a reading of 1200 counts s^{-1} in a blood sample. How many hours must pass for the reading of this sample to fall to 75 counts s^{-1}?

(L92,p)

*10. a) Explain the meaning of *each* of the following terms:
   i) order of a reaction
   ii) rate constant
   iii) mechanism
   iv) half-life
   v) activation energy.

b) A first-order gas phase decomposition reaction was carried out in a sealed vessel at constant temperature and the rate of the reaction was monitored by measuring the total pressure in the vessel.

The following results were obtained at 326 K.

Time/minutes	Total pressure/kPa
0	10
10	16
20	20
40	25
60	27.5
80	28.75
300	30
500	30

   i) Suggest a schematic equation for the decomposition.
   ii) Calculate the rate constant and the half-life of the reaction at 326 K.   (AEB91,S)

11. a) For a chemical reaction, state what is meant by the terms *rate of reaction, rate constant* and *transition state*. Distinguish carefully between the first two terms.

b) The following data were obtained with respect to the alkaline hydrolysis of two different halogenoalkanes at 20°C.

Experiment	Initial concentration/mol dm^{-3}		Initial rate /mol dm^{-3} s^{-1}
	$CH_3(CH_2)_3Cl$	OH^-	
A	0.2	0.2	6.0×10^{-6}
B	0.4	0.2	12×10^{-6}
C	0.2	0.8	24×10^{-6}

	$(CH_3)_3CCl$	OH^-	
D	0.1	0.1	3.0×10^{-5}
E	0.2	0.1	6.0×10^{-5}
F	0.1	0.3	3.0×10^{-5}

   i) Deduce the order of reaction with respect to concentration of
      1) halogenoalkane
      2) hydroxide ion
      for the hydrolysis of *both* halogenoalkanes, and explain your
      deductions.
   ii) Write the rate equations for *both* reactions and give the unit of
      the rate constant in *each* case.
   iii) State the mechanism of hydrolysis of 1-chlorobutane (which is
      the same as that of 1-bromobutane) and explain how the results
      above support this mechanism.
   iv) State, giving a reason, whether the mechanism of hydrolysis
      of $(CH_3)_3CCl$ is likely to be the same as that in b) iii) above
      and, if not, suggest a possible mechanism.
c) i) The relative rate constants for the alkaline hydrolysis of
      l-chlorobutane, (chloromethyl)benzene and chlorobenzene are
      in the ratio of

$$1 : 200 : 0.001$$

   respectively. Discuss the reasons for the differences.
   ii) Suggest *one* method of measuring the rate of the reactions in
      c) i) experimentally.
d) Here are two statements relevant to the problem of obtaining
   maximum yields in industrial processes as quickly as possible.

   'The equilibrium yields of exothermic reactions decrease as the
   temperature increases.'

   'The rates of all chemical reactions increase with increasing temperature.'

   Explain why these statements are correct, discuss the apparent
   conflict between them, and state how the industrialist deals with
   this.   (WJEC92)

12. The reaction between manganate(VII) ions and ethanedioate ions in
   acid solution is described by the equation:

$$2MnO_4^- + 16H^+ + 5C_2O_4^{2-} \longrightarrow 2Mn^{2+} + 8H_2O + 10CO_2$$

   If potassium manganate(VII) is reacted with ethanedioic acid in the
   presence of sulphuric acid the product is a mixture of manganese(II)
   sulphate and potassium sulphate solutions. The extent of the
   reaction may be followed by measuring the concentration of the

manganate(VII) ions using a colorimeter. In Experiment 1, $100 \, cm^3$ of potassium manganate(VII), of concentration $0.02 \, mol \, dm^{-3}$, were mixed with $100 \, cm^3$ of ethanedioic acid, $0.10 \, mol \, dm^{-3}$, and $50 \, cm^3$ of sulphuric acid, $1 \, mol \, dm^{-3}$, and a sample was placed in the colorimeter. In Experiment 2, the same mixture was made but this time a little solid manganese(II) sulphate was also added. Both experiments were conducted under the same conditions of temperature and pressure. The following results were noted.

Experiment 1		Experiment 2	
Time/s	Concentration of manganate(VII) $/10^{-3} \, mol \, dm^{-3}$	Time/s	Concentration of manganate(VII) $/10^{-3} \, mol \, dm^{-3}$
0	8.0	0	8.0
400	7.9	400	5.8
800	7.7	800	4.0
1200	6.8	1200	2.8
1600	4.0	1600	2.0
2000	2.0	2000	1.4
2400	1.0	2400	1.0

a) Consider first Experiment 2.
  i) On graph paper plot a graph of concentration of manganate against time and label it appropriately.
  ii) On lined paper *sketch* a graph of rate against time and indicate how you arrived at that conclusion.
  iii) From either of the two graphs, or by any other method, determine the order of the reactions with respect to the manganate(VII) ion, clearly showing your reasoning.

b) Consider Experiment 1.
  i) On the graph produced in a) i) *sketch* a graph of concentration of manganate(VII) against time.
  ii) On lined paper *sketch* a graph of rate against time.
  iii) How do the two graphs of rate against time differ?
  iv) What can you deduce about the role of manganese(II) sulphate in this reaction? (O&C91,AS)

*13. a) Discuss *three* different ways in which the rates of simple chemical reactions can be altered. Relate your discussion to a generalised rate equation and sketch graphs, as appropriate, to illustrate your answer.

b) Consider a reaction between two species P and Q which is first order with respect to each of these two components. When the concentrations of P and Q are of the same order of magnitude, overall second-order kinetics will be observed experimentally. But if either P or Q is present in considerable excess, the reaction will show an apparent overall order of one, called the *pseudo order*.

Explain why this is so and relate the *pseudo* rate constant k_{ps} to the *true* rate constant k for the case where component P is present initially in considerable excess.

c) Propanone reacts with iodine as follows:

$$CH_3COCH_3 \; + \; I_2 \; \longrightarrow \; CH_3COCH_2I \; + \; HI$$

The data in the table below give the concentration of iodine at different times t for a reaction at $25\,^\circ C$ between propanone $(7.0\,M)$ and iodine $(0.11\,M)$ to which hydrochloric acid $(0.05\,M)$ has been added.

$[I_2]/mol\,l^{-1}$	0.089	0.068	0.047	0.026	0.005
t/s	150	300	450	600	750

Plot a graph of concentration against time and from it determine the apparent order of reaction with respect to iodine and also the value of the pseudo rate constant, stating its units.

In a further series of experiments, it was found that the reaction was first order with respect to propanone and also first order with respect to hydrogen ions. Write the full rate equation for this reaction and calculate the value of the true rate constant, stating its units.

Explain why the data from which you plotted your graph cannot, by themselves, give you information about the influence of propanone concentration *or* of pH on the rate of reaction, and hence why extra experiments are needed to characterise the full rate equation.

(JMB91,S)

*14. A study of the decomposition of ammonia into its elements on the surface of a tungsten wire at two temperatures and less than atmospheric pressure gave the following results.

Time/seconds	Total pressure/N m⁻² at 1129 K	Total pressure/N m⁻² at 1148 K
0	2.66×10^4	2.66×10^4
100	2.81×10^4	2.87×10^4
200	2.97×10^4	3.08×10^4
300	3.13×10^4	3.29×10^4
400	3.29×10^4	3.50×10^4
500	3.45×10^4	3.71×10^4

a) Interpret these results graphically to determine:
   i) the order of the reaction
   ii) the rate constants for the reaction at 1129 K and 1148 K.

b) Describe how the catalyst in this reaction functions at the molecular level.

c) i) Suggest an explanation for the observed order of the reaction in these experiments.

 ii) At very low pressures the observed order of the reaction changes. Suggest what the order of the reaction might become at very low pressures and offer an explanation for the change in order.

d) The rate constant, k, varies with temperature according to the Arrhenius equation, which can be written

$$\ln k = \ln A - \frac{E}{RT}$$

where A is a constant, E is the activation energy, T is the temperature in K, and R is the gas constant ($8.31 \, J \, K^{-1} mol^{-1}$).

 i) Calculate the activation energy for the decomposition of ammonia on tungsten.

 ii) The standard enthalpy of formation of ammonia is $-46 \, kJ \, mol^{-1}$. Calculate the activation energy for the formation of ammonia from its elements.

e) In both experiments the total pressure eventually reaches the same constant value. By reference to the stoichiometry of the reaction, suggest what this constant value might be. Explain your answer and justify any assumptions which you make.   (L92,S)

15 Equilibria

CHEMICAL EQUILIBRIUM

An example of a reversible reaction between gases is the reaction between hydrogen and iodine to form hydrogen iodide:

$$H_2(g) \ + \ I_2(g) \ \rightleftharpoons \ 2HI(g)$$

If the reaction takes place in a closed vessel, the combination of hydrogen and iodine gradually slows down as the concentrations of these gases decrease. At first, there is very little decomposition of hydrogen iodide into hydrogen and iodine, but, as the concentration of hydrogen iodide increases, the rate of decomposition of hydrogen iodide into hydrogen and iodine increases until the rates of the forward and reverse reactions are equal, and the concentration of each species is constant.

An example of a reversible reaction which takes place in solution is the reaction between ethanoic acid and ethanol to form ethyl ethanoate:

$$CH_3CO_2H(l) \ + \ C_2H_5OH(l) \ \rightleftharpoons \ CH_3CO_2C_2H_5(l) \ + \ H_2O(l)$$

As the concentrations of ester and water increase, the reverse reaction — hydrolysis of the ester to form the acid and alcohol — speeds up. At equilibrium, the rate of the forward reaction is equal to the rate of the reverse reaction. Esterification is catalysed by inorganic acids. The presence of a catalyst speeds up the rate at which equilibrium is established.

THE EQUILIBRIUM LAW

If a reversible reaction is allowed to reach equilibrium, it is found that the product of the concentrations of the products divided by the product of the concentrations of the reactants has a constant value at a particular temperature. In the esterification reaction,

$$CH_3CO_2H(l) \ + \ C_2H_5OH(l) \ \rightleftharpoons \ CH_3CO_2C_2H_5(l) \ + \ H_2O(l)$$

it is found that

$$\frac{[CH_3CO_2C_2H_5]\,[H_2O]}{[CH_3CO_2H]\,[C_2H_5OH]} = K_c$$

where K_c is the equilibrium constant for the reaction in terms of concentration. In the reaction between hydrogen and iodine,

$$H_2(g) \ + \ I_2(g) \ \rightleftharpoons \ 2HI(g)$$

and

$$\frac{[HI]^2}{[H_2]\,[I_2]} = K_c$$

Since this is a reaction between gases, the concentration of each gas can be expressed as a partial pressure. Then,

$$\frac{p_{HI}^2}{p_{H_2} \times p_{I_2}} = K_p$$

K_p is the equilibrium constant in terms of partial pressures. In the reaction between iron and steam,

$$3Fe(s) \; + \; 4H_2O(g) \; \rightleftharpoons \; Fe_3O_4(s) \; + \; 4H_2(g)$$

The equilibrium constant is given by

$$\frac{p_{H_2}^4}{p_{H_2O}^4} = K_p$$

The solids do not appear in the expression. Their vapour pressures are constant (at a constant temperature) as long as there is some of each solid present. These constant vapour pressures are incorporated into the value of the constant K_p.

Another type of reaction which reaches an equilibrium position is thermal dissociation. For example, when phosphorus(V) chloride is heated, it dissociates partially to form phosphorus(III) chloride and chlorine:

$$PCl_5(g) \; \rightleftharpoons \; PCl_3(g) \; + \; Cl_2(g)$$

As explained in Chapter 9 (pp. 104–14) the dissociation increases the number of moles of substance present and causes an increase in volume, or, if the volume is kept constant, an increase in pressure. The result is that the experimental determinations of molar mass give an unexpectedly low value. The degree of dissociation, α, can be obtained from the ratio,

$$\frac{\text{Molar mass calculated in the absence of dissociation}}{\text{Experimentally determined molar mass}} = 1 + \alpha$$

Inserting the value for α into the expression for the equilibrium constant, and putting c = initial concentration of PCl_5, we get

$$K_c = \frac{[Cl_2][PCl_3]}{[PCl_5]}$$

$$K_c = \frac{\alpha c \times \alpha c}{(1-\alpha)c} = \frac{\alpha^2 c}{1-\alpha}$$

If the total pressure $= P$, the partial pressures of PCl_3 and Cl_2 are $P\alpha/(1+\alpha)$, the partial pressure of PCl_5 is $P(1-\alpha)/(1+\alpha)$, and

$$K_p = \frac{P^2\alpha^2/(1+\alpha)^2}{P(1-\alpha)/(1+\alpha)} = \frac{\alpha^2 P}{1-\alpha^2}$$

EXAMPLE 1 1.00 mole of ethanoic acid was allowed to react with: **a)** 0.50 mole, **b)** 1.00 mole, **c)** 2.00 mole, and **d)** 4.00 mole of ethanol. At equilibrium, the amount of acid remaining was a) 0.58 mole, b) 0.33 mole, c) 0.15 mole and d) 0.07 mole. Calculate the equilibrium constant for the esterification reaction.

METHOD If the original amounts of acid and ethanol are a mol and b mol, then, at equilibrium, the amount of ester formed is x mol, and the amounts of acid and ethanol remaining are $(a-x)$ and $(b-x)$ mol.

$$CH_3CO_2H(l) + C_2H_5OH(l) \rightleftharpoons CH_3CO_2C_2H_5(l) + H_2O(l)$$
$$(a-x) \qquad\quad (b-x) \qquad\qquad\qquad x \qquad\qquad\quad x$$

Since the equilibrium constant is given by

$$\frac{[CH_3CO_2C_2H_5]\,[H_2O]}{[CH_3CO_2H]\,[C_2H_5OH]} = K_c$$

then $$\frac{(x/V)(x/V)}{[(a-x)/V]\,[(b-x)/V]} = K_c \quad \text{or} \quad \frac{x^2}{(a-x)(b-x)} = K_c$$

In reaction **a)**

$$(a-x) = 0.58; \quad x = 0.42; \quad (b-x) = 0.08$$

∴ $$\frac{(0.42)^2}{0.58 \times 0.08} = K_c = 3.8$$

Substituting the other values of a, b and x in the equation gives the following values of K:

a	b	$a-x$	x	$b-x$	K_c
1.00	0.50	0.58	0.42	0.08	3.8
1.00	1.00	0.33	0.67	0.33	4.1
1.00	2.00	0.15	0.85	1.15	4.4
1.00	4.00	0.07	0.93	3.07	4.0

ANSWER The average value of the equilibrium constant is 4.1.

EXAMPLE 2 Calculate the amount of ethyl ethanoate formed when 1 mole of ethanoic acid and 3 moles of ethanol and 3 moles of water are allowed to come to equilibrium. The equilibrium constant for the reaction is 4.0.

METHOD Let the amount of ethyl ethanoate $= x$ mol
Then   Equilibrium amount of acid $= (1-x)$ mol
        Equilibrium amount of ethanol $= (3-x)$ mol
        Equilibrium amount of water $= (3+x)$ mol

Then, since $\dfrac{[CH_3CO_2C_2H_5]\,[H_2O]}{[CH_3CO_2H]\,[C_2H_5OH]} = K_c = 4.0$

$$\frac{x(3+x)}{(1-x)(3-x)} = 4.0$$

$$3x^2 - 19x + 12 = 0$$

Solving this quadratic equation (see p. 13) gives

$$x = 5.6 \quad \text{or} \quad 0.71$$

The value $x = 5.6$ can be excluded because it is higher than the number of moles of ethanoic acid present initially. The solution $x = 0.71$ must be the practical one.

ANSWER   The amount of ethyl ethanoate formed is 0.71 mol.

EXAMPLE 3   A mixture of iron and steam is allowed to come to equilibrium at 600 °C. The equilibrium pressures of hydrogen and steam are 3.2 kPa and 2.4 kPa. Calculate the equilibrium constant K_p for the reaction.

METHOD   The reaction is

$$3Fe(s) + 4H_2O(g) \rightleftharpoons 4H_2(g) + Fe_3O_4(s)$$

The equilibrium constant is given by

$$K_p = \frac{p_{H_2}{}^4}{p_{H_2O}{}^4}$$

Substituting in this equation gives

$$K_p = \left(\frac{3.2}{2.4}\right)^4$$

ANSWER $$K_p = 3.1$$

EXAMPLE 4   A molar mass determination on dinitrogen tetraoxide, N_2O_4, gave a value of $60\,g\,mol^{-1}$ at 50 °C and $1.01 \times 10^5\,Pa$. Find the equilibrium constant for the dissociation

$$N_2O_4(g) \rightleftharpoons 2NO_2(g)$$

METHOD   If the degree of dissociation is α, then a total of $1 + \alpha$ moles of particles are formed from 1 mole of N_2O_4. P is the total pressure.

$$\frac{\text{Molar mass}}{\text{Experimentally determined molar mass}} = \frac{92}{60} = 1 + \alpha$$

$$\alpha = 0.53$$

Since $$K_p = \frac{p_{NO_2}{}^2}{p_{N_2O_4}}$$

and $\qquad p_{NO_2} = \dfrac{2\alpha}{1+\alpha} P$ and $p_{N_2O_4} = \dfrac{1-\alpha}{1+\alpha} P$

$$K_p = \frac{4\alpha^2}{1-\alpha^2} \times P = \frac{4(0.53)^2}{1-(0.53)^2} \times 1.01 \times 10^5\,Pa$$

ANSWER $\qquad K_p = 1.58 \times 10^5\,Pa$

EXERCISE 58   Problems on Equilibria

1. Write an expression for the equilibrium constant for the reaction,
   A + B $\rightleftharpoons$ C + D:
   a) when A, B, C and D are gases
   b) when A, B, C and D are solutions
   c) when A and C are solids, B and D are gases
   d) when C is a solid, and A, B and D are solutions.

2. In the equilibrium
   $$N_2O_4(g) \rightleftharpoons 2NO_2(g)$$
   3.20 g of dinitrogen tetraoxide occupy a volume 1.00 dm³ at 1.00×10^5 Pa and 25 °C. Calculate: a) the degree of dissociation, and b) the equilibrium constant.

3. Sulphur dichloride dioxide has an apparent molar mass of 75 g mol⁻¹ at 400 °C and a pressure of 10^5 N m⁻². Calculate the equilibrium constant K_p for the reaction
   $$SO_2Cl_2(g) \rightleftharpoons SO_2(g) + Cl_2(g)$$
   at this temperature.

4. The equilibrium constant for the reaction
   $$CO(g) + H_2O(g) \rightleftharpoons CO_2(g) + H_2(g)$$
   is 4.00. If 1 mole of CO and 1 mole of H_2O are allowed to come to equilibrium, what fraction of the carbon monoxide will remain?

5. 1 dm³ of dinitrogen tetraoxide, N_2O_4, weighs 2.50 g at 60 °C and 1.01×10^5 N m⁻² pressure. Find a) the degree of dissociation into NO_2 and b) the value of K_p.

6. Equimolar amounts of hydrogen and iodine are allowed to reach equilibrium:
   $$H_2(g) + I_2(g) \rightleftharpoons 2HI(g)$$
   If 80% of the hydrogen can be converted to hydrogen iodide, what is the value of K_p at this temperature?

7. A mixture of 3 moles of hydrogen and 1 mole of nitrogen is allowed to reach equilibrium at a pressure of $5 \times 10^6 \, N \, m^{-2}$. The composition of the gaseous mixture is then 8% NH_3, 23% N_2, 69% H_2 by volume. Calculate K_p.

8. a) Hydrogen and iodine react to form hydrogen iodide. When the initial concentrations of both reactants are $0.11 \, mol \, dm^{-3}$, the initial rate of reaction is $2.42 \times 10^{-5} \, mol \, dm^{-3} \, s^{-1}$. What is the rate constant for this second-order reaction?

   b) When 1 mol of hydrogen and 1 mol of iodine are allowed to reach equilibrium at 600 K, the equilibrium mixture contains 1.6 mol hydrogen iodide. Calculate the equilibrium constant for the reaction

$$H_2(g) \; + \; I_2(g) \; \rightleftharpoons \; 2HI(g)$$

9. At a certain temperature the reaction

$$C_2H_5CO_2H(l) \; + \; C_2H_5OH(l) \; \rightleftharpoons \; C_2H_5CO_2C_2H_5(l) \; + \; H_2O(l)$$

has an equilibrium constant of 6.

   a) Calculate the amount of ester (in moles) that will be formed at equilibrium from 1 mol acid and 1 mol alcohol.

   b) What mass of propanoic acid must react with 46 g ethanol in order to give 51 g ethylpropanoate at equilibrium?

10. The equilibrium, $H_2(g) + Cl_2(g) \; \rightleftharpoons \; 2HCl(g)$ has $K_p = 2.5 \times 10^4$ at 1500 K. What percentage of hydrogen is converted to hydrogen chloride at this temperature in a 1:1 mixture of hydrogen and chlorine?

11. The equilibrium constant for the reaction $H_2(g) + I_2(g) \; \rightleftharpoons \; 2HI(g)$ is 60 at 450 °C. The number of moles of hydrogen iodide in equilibrium with 2 mol of hydrogen and 0.3 mol of iodine at 450 °C is:

   a 1/100   b 1/10   c 6   d 36   e 3

12. The esterification reaction

$$RCO_2H \; + \; R'OH \; \rightleftharpoons \; RCO_2R' \; + \; H_2O$$

has an equilibrium constant of 10.0 at 25 °C. If 1 mole of ester is dissolved in 5 moles of water, what are the equilibrium amounts of:
a) acid, b) alcohol, c) ester and d) water?

13. A mixture of 1 mol nitrogen and 3 mol hydrogen is allowed to reach equilibrium at $1.0 \times 10^7 \, Pa$ and 500 °C. The equilibrium mixture contains 20% of ammonia by volume. Calculate the value of K_p for the reaction

$$N_2(g) \; + \; 3H_2(g) \; \rightleftharpoons \; 2NH_3(g)$$

at 500 °C.

14. Sulphur dioxide and oxygen in the ratio 2 mol : 1 mol are allowed to reach equilibrium in the presence of a catalyst, at a pressure of 5 atm. At equilibrium, $\frac{1}{3}$ of the SO_2 was converted to SO_3. Calculate the equilibrium constant for the reaction

$$2SO_2(g) + O_2(g) \rightleftharpoons 2SO_3(g)$$

15. The equilibrium constant K_p for the reaction

$$CO_2(g) + H_2(g) \rightleftharpoons CO(g) + H_2O(g)$$

is 0.72 at $1000\,°C$. Calculate the composition of the mixture which results when:

a) 0.5 mole CO_2 and 0.5 mole H_2 are mixed at a pressure of 1 atm and $1000\,°C$.

b) 5 moles CO_2 and 1 mole H_2 are mixed at a pressure of 1 atm and $1000\,°C$.

16. The oxidation of sulphur dioxide is a reversible process:

$$2SO_2(g) + O_2(g) \rightleftharpoons 2SO_3(g)$$

Calculate the value of the equilibrium constant, K_p, in terms of partial pressures from the following data, which were obtained at $1000\,K$:

Partial pressures/N m^{-2}		
p_{SO_2}	p_{O_2}	p_{SO_3}
10 000	68 800	80 100

EXERCISE 59   Questions from A-level Papers

1. For an industrial process represented by the equilibrium

$$2A(g) + B(g) \rightleftharpoons 2C(g)$$

the following data were obtained for
1. the variation of the relative rate of the forward reaction, k_{rel}, with temperature
2. the variation of the fractional conversion, f, to C, at equilibrium.

T/K	$\log_{10} k_{rel}$	f, fractional conversion to C at equilibrium
600	10.16	0.997
650	10.90	0.988
700	11.59	0.967
750	12.14	0.930
800	12.66	0.875
850	13.11	0.798
900	13.50	0.708

a) On graph paper plot
  i) the variation of $\log_{10} k_{rel}$ with T
  ii) the variation of f with T
  iii) the variation of the product, $(\log_{10} k_{rel} \times f)$, with T.

b) State and explain what conclusions may be drawn from the plots in a) above concerning the optimal conditions for the production of C.

c)   i) State and briefly explain the effect on the position of equilibrium of
   1) increase in total pressure
   2) increase in temperature.
  ii) Under industrial conditions a catalyst is used to facilitate the production of C. State what is the effect of the catalyst on
   1) the value of K_p
   2) the value of E_A, the activation energy of the forward reaction.

d) Consider the data given for the fractional conversion to C at equilibrium as a function of temperature. State and explain what can be deduced about $\Delta H^{\ominus}$ for the equilibrium.   (WJEC91)

2. The following equilibrium is established when hydrogen and nitrogen are passed over heated iron

$$N_2 \; + \; 3H_2 \; \rightleftharpoons \; 2NH_3$$

a) Express the equilibrium constant K_p, in terms of the equilibrium partial pressures, p_{N_2}, p_{H_2}, p_{NH_3}, of the three species.

b) If the nitrogen and hydrogen were initially in the molar ratio $1:3$ and the fraction of ammonia at equilibrium is x, obtain expressions for the equilibrium partial pressures, p_{N_2}, p_{H_2}, and p_{NH_3}, in terms of x and the equilibrium total pressure P.

c) Name and state the law used in b).

d) At $400\,°C$, $x = 0.0385$ and $P = 10\,atm$. Calculate K_p.

e) If the total pressure P were increased to $50\,atm$ and the temperature kept at $400\,°C$, indicate, without calculation, the effect on
  i) x, ii) K_p. Give explanations for your answers.

f) When the temperature is increased to $500\,°C$, K_p decreases. What can be deduced about the sign of the enthalpy of formation in NH_3 in this reaction? Give an explanation for your answer.

g) What is the role of iron in this reaction?   (O90)

3. The following chemical equilibria occur in limestone areas subject to rainfall:

$$CO_2(g) \; \rightleftharpoons \; CO_2(aq)$$

$$CO_2(aq) + H_2O(l) + CaCO_3(s) \; \rightleftharpoons \; Ca^{2+}(aq) + 2HCO_3^-(aq)$$

a) By applying Le Chatelier's principle to these equilibria, explain qualitatively how rainwater passing through limestone rock and then dripping from the roof of a cave can produce pillars, stalagmites and stalactites of ever-increasing thickness.

b) Water saturated with *pure* carbon dioxide at atmospheric pressure contains 0.15% by mass of dissolved CO_2.

Calculate the concentration, in $mol\,dm^{-3}$, of dissolved CO_2 in water, $[CO_2(aq)]$, which is in equilibrium with *air* containing 1% of carbon dioxide.

c) Write an expression for the equilibrium constant for the second reaction given above. By using the value of this equilibrium constant $(4.7 \times 10^{-5}\,mol^2\,dm^{-6})$ and the $[CO_2(aq)]$ you calculated in b), estimate the maximum value of $[Ca(HCO_3)_2(aq)]$ that could occur in water passing through limestone rock.   (C91)

4. a) For the industrially important reaction

$$2SO_2(g) \ + \ O_2(g) \ \rightleftharpoons \ 2SO_3(g)$$

$\Delta H(298\,K) = -94.5\,kJ\,mol^{-1}$.

Describe, giving reasons, the effect on the position of equilibrium of:
  i) increase of temperature
  ii) decrease of pressure
  iii) a platinum catalyst
  iv) excess oxygen.

At 1300 K and a total pressure of 1 atm, the partial pressures at equilibrium are 0.27 atm for SO_2 and 0.41 atm for O_2.
Calculate the equilibrium constant K_p. Be careful to give the units of K_p.

b) Discuss the application of the Equilibrium Law to the equilibrium

$$CaCO_3(s) \ \rightleftharpoons \ CaO(s) \ + \ CO_2(g)$$

at 700 K, and explain what would happen if carbon dioxide were added to the system.   (O91)

5. The equilibrium between hydrogen, iodine and hydrogen iodide can be investigated by sealing hydrogen iodide in glass tubes and heating them at known temperatures until equilibrium is reached. The equation for the reaction is

$$2HI(g) \ \rightleftharpoons \ H_2(g) \ + \ I_2(g)$$

and the equilibrium constant $K_c = 0.019\,K$ at 698 K.

The tubes are *rapidly* cooled and then opened under potassium iodide solution when the iodine and hydrogen iodide dissolve.

a) i) Why are the tubes rapidly cooled?
   ii) Describe how the appearance of the contents of a tube would change as it was cooled.

iii) Outline a practical procedure for measuring the amount of iodine dissolved from a tube.

b)　i) Write an expression for the equilibrium constant, K_c.

　ii) A sample tube is found to contain iodine at a concentration of $4.8 \times 10^{-4}\,mol\,dm^{-3}$. Using the equation, deduce the equilibrium concentration of hydrogen.

　iii) Calculate the equilibrium concentration of hydrogen iodide.

(L92,N,AS)

6. When a mixture of ethanol, ethanoic acid and concentrated sulphuric acid is warmed, the reaction represented by the following equation occurs:

$$C_2H_5OH \;+\; CH_3COOH \longrightarrow CH_3COOC_2H_5 \;+\; H_2O$$

a) Give the name and structural formula of the compound with the formula $CH_3COOC_2H_5$.

b) Why is concentrated sulphuric acid included in the mixture?

c) When the reaction mixture is poured into cold water a smell due to the compound $CH_3COOC_2H_5$ is detected. Describe this smell.

d) The results of an experiment show that an equilibrium mixture at 300 K contains the following:

0.33 mol　C_2H_5OH
0.33 mol　CH_3COOH
0.66 mol　$CH_3COOC_2H_5$
0.66 mol　H_2O

Calculate the value of the equilibrium constant K_c at 300 K for the reaction

$$C_2H_5OH \;+\; CH_3COOH \rightleftharpoons CH_3COOC_2H_5 \;+\; H_2O$$

e) What, if any, is the effect of adding more ethanol to the equilibrium mixture on
　i) the value of K_c,
　ii) the composition of the equilibrium mixture?

f) Use the relationship $\Delta G^{\ominus} = -RT \ln K$ to calculate the standard molar free energy change for the reaction at 300 K and comment on the energetic feasibility of the reaction at this temperature. ($R = 8.31\,J\,K^{-1}\,mol^{-1}$.)

(AEB91,AS)

7. a)　i) Explain the concept of *chemical equilibrium.*

　ii) Define the term *equilibrium constant (K)* and explain why it is useful.

　iii) Write down the *simplest* possible expressions for equilibrium constants (K_p or K_c as appropriate) for the following equilibria and give the units of K in each case.
　　1) $H_2(g) + \frac{1}{2}O_2(g) \rightleftharpoons H_2O(g)$
　　2) $CaCO_3(s) \rightleftharpoons CaO(s) + CO_2(g)$
　　3) $2Fe^{3+}(aq) + Sn^{2+}(aq) \rightleftharpoons 2Fe^{2+}(aq) + Sn^{4+}(aq)$

b) In a closed-vessel experiment on the Haber process, nitrogen at 50 atm pressure and hydrogen at 150 atm pressure were reacted together at constant temperature. After a certain time interval it was found that the ammonia formed had a pressure of 40 atm. Given that the equilibrium constant, K_p, at the reaction temperature is 7.316×10^{-5} atm^{-2}, calculate whether or not the system had reached equilibrium.

c) State and discuss the factors which govern the rates at which chemical reactions occur.   (WJEC91)

8. At room temperature, gaseous dinitrogen tetraoxide and nitrogen dioxide are in dynamic equilibrium according to the following equation:

$$N_2O_4(g) \rightleftharpoons 2NO_2(g); \quad \Delta H = +58 \text{ kJ mol}^{-1}.$$

a) Explain what is meant by the term *dynamic equilibrium,* and write the expression for the equilibrium constant, K_p, for this reaction.

b) At a temperature of 25°C (298 K), 1.00 g of a mixture of these two gases takes up a volume of 3.17×10^{-4} m^3 at a pressure of 101 kPa (1.01×10^5 N m^{-2}). Calculate the average relative molecular mass of the mixture.

c) State *Le Chatelier's principle*, and use it to deduce qualitatively the effect on the average relative molecular mass of this gaseous mixture of increasing
   i) the pressure
   ii) the temperature.

d) Nitrogen dioxide (from car exhaust fumes) can react with sulphur dioxide (from the burning of fossil fuels) in the presence of water vapour in the atmosphere to produce sulphuric acid (acid rain) and nitrogen monoxide, NO. The nitrogen monoxide is rapidly reoxidised to nitrogen dioxide by oxygen.

   Construct balanced equations for these two reactions and hence suggest the role played by nitrogen dioxide in the overall process.
                                              (C92)

9. Hydrogen is manufactured nowadays from oil, but earlier this century a major method for the production of hydrogen was the Bosch process. This was a two-stage process starting from steam and coke (about 80% carbon):

Stage 1 at $\sim$1500 °C: the production of 'water-gas', a mixture of carbon monoxide and hydrogen:

$$C(s) + H_2O(g) \longrightarrow CO(g) + H_2(g)$$

Stage 2 at $\sim$500 °C: the mixture of carbon monoxide and hydrogen from Stage 1 is mixed with more steam before undergoing the 'water-gas shift' reaction, involving an iron catalyst:

$$CO(g) + H_2O(g) \rightleftharpoons CO_2(g) + H_2(g)$$

The value of K_p for the water-gas shift reaction at 500 °C is 10. In the process the amount of carbon monoxide in the final gas mixture, after removal of excess steam, had to be kept below 2% by volume.

a) Calculate the ratio by volume required for the steam/water-gas mixture (at 500 °C) to achieve this.

b) Calculate the *total* mass of steam required for the whole process per tonne of coke used for the production of water-gas.

c) Calculate the percentage of the total steam used which is converted to hydrogen.   (L90,N,S)

16 Organic Chemistry

All the techniques you need to enable you to tackle problems in organic chemistry have been covered in Chapter 2 in the sections on empirical formulae, calculations based on chemical equations and reacting volumes of gases, and in Chapter 3 on volumetric analysis.

Numerical problems in organic chemistry give you some quantitative data and ask you to use it in conjunction with your knowledge of the reactions of organic compounds. There is no set pattern for tackling such problems. They are solved by a combination of calculation, familiarity with the reactions of the compounds involved and logic. The following examples and problems will show you what to expect.

EXAMPLE 1 When $0.2500\,g$ of a hydrocarbon X burns in a stream of oxygen, it forms $0.7860\,g$ of carbon dioxide and $0.3210\,g$ of water. When $0.2500\,g$ of X is vaporised, the volume which it occupies (corrected to s.t.p.) is $80.0\,cm^3$. Deduce the molecular formula of X.

METHOD X burns to form carbon dioxide and water.

$$\text{Mass of C in } 0.7860\,g \text{ of } CO_2 = \frac{12.0}{44.0} \times 0.7860 = 0.2143\,g$$

$$\text{Mass of H in } 0.3210\,g \text{ of } H_2O = \frac{2.02}{18.0} \times 0.3210 = 0.0360\,g$$

Therefore $0.2500\,g$ of X contains $0.2143\,g$ of C and $0.0360\,g$ of H

These masses give the molar ratio for $C:H$ of $\dfrac{0.2143}{12.0}$ to $\dfrac{0.0360}{1.01}$

$$= 0.0178 \text{ to } 0.0360 = 1 \text{ to } 2$$

Thus, the empirical formula is CH_2.

Since $80.0\,cm^3$ is the volume occupied by $0.2500\,g$ of X,

$$22.4\,dm^3 \text{ is occupied by } \frac{22.4}{80.0 \times 10^{-3}} \times 0.2500\,g \text{ of } X = 70.0\,g \text{ of } X$$

The formula mass of CH_2 is 14. To give a molar mass of $70.0\,g\,mol^{-1}$, the empirical formula must be multiplied by 5. Therefore:

ANSWER The molecular formula is C_5H_{10}.

EXAMPLE 2 An organic liquid, P, contains 52.2% carbon, 13.0% hydrogen and 34.8% oxygen by mass. Mild oxidation converts P to Q, and, on further oxidation, R is formed. P and Q react together in the presence of anhydrous calcium chloride to form S, which has a molecular formula of $C_6H_{14}O_2$. P and R react to give T, which has a molecular formula of $C_4H_8O_2$. Identify compounds P to T, and explain the reactions involved.

METHOD   First, calculate the empirical formula of P. This comes to C_2H_6O. This must be the molecular formula also as P and R combine to form T, which has 4C in the molecule. Since P contains one oxygen atom, it is an alcohol, an aldehyde, a ketone or possibly an ether. Other classes of compounds are ruled out by the absence of nitrogen and halogens. Oxidation proceeds in two stages, evidence that P is probably an alcohol, being oxidised first to an aldehyde and then to an acid.

According to the formulae, P would be C_2H_5OH, and Q would be CH_3CHO, and R would be CH_3CO_2H. The reaction between P and Q fits in with this theory as

ANSWER $2C_2H_5OH(l)\ +\ CH_3CHO(l)\ \longrightarrow\ CH_3CH(OC_2H_5)_2(l)\ +\ H_2O(l)$

The reaction between P and R to form T is thus

ANSWER $CH_3CO_2H(l)\ +\ C_2H_5OH(l)\ \longrightarrow\ CH_3CO_2C_2H_5(l)\ +\ H_2O(l)$

The molecular formula $C_6H_{14}O_2$ for S fits $CH_3CH(OC_2H_5)_2$, and the molecular formula $C_4H_8O_2$ for T fits $CH_3CO_2C_2H_5$.

ANSWER $P = C_2H_5OH$, ethanol; $Q = CH_3CHO$, ethanal; $R = CH_3CO_2H$, ethanoic acid; $S = CH_3CH(OC_2H_5)_2$, 1,1-diethoxyethane; $T = CH_3CO_2C_2H_5$, ethyl ethanoate.

EXAMPLE 3 X is a liquid containing 31.5% by mass of C, 5.3% H and 63.2% O. An aqueous solution of X liberates carbon dioxide from sodium carbonate, with the formation of a solution from which a substance Y of formula $C_2H_3O_3Na$ can be obtained. X reacts with phosphorus(V) chloride to give hydrogen chloride and a compound, Z, of molecular formula $C_2H_2OCl_2$. Identify X, Y and Z. Write equations for the two reactions, $X \longrightarrow Y$ and $X \longrightarrow Z$.

METHOD   The empirical formula of X is easily shown to be $C_2H_4O_3$. As X liberates carbon dioxide from a carbonate, it must be an acid. Taking CO_2H from $C_2H_4O_3$ leaves CH_3O as the formula for the rest of the molecule. This could be CH_2OH, making X $HOCH_2CO_2H$.

In the reaction with PCl_5, $C_2H_4O_3$ is converted into $C_2H_2OCl_2$. This would fit in with two hydroxyl groups being replaced by two chlorine atoms. This would be the case for the reaction

ANSWER $HOCH_2CO_2H(l)\ +\ 2PCl_5(l)\ \longrightarrow\ ClCH_2COCl(l)\ +\ 2POCl_3(l)\ +\ 2HCl(g)$

If X is $HOCH_2CO_2H$, its reaction with sodium carbonate has an equation

ANSWER $2HOCH_2CO_2H(aq) + Na_2CO_3(s)\ \longrightarrow\ 2HOCH_2CO_2Na(aq) + CO_2(g) + H_2O(l)$

and Y is $HOCH_2CO_2Na$.

ANSWER   All the information agrees with $X = HOCH_2CO_2H$, hydroxyethanoic acid, $Y = HOCH_2CO_2Na$, sodium hydroxyethanoate; $Z = ClCH_2COCl$, chloroethanoyl chloride.

EXAMPLE 4   1.220 g of a dicarboxylic aliphatic acid is dissolved in water and made up to $250\ cm^3$. A $25.0\ cm^3$ portion of the solution requires $21.0\ cm^3$ of $0.100\ mol\,dm^{-3}$ sodium hydroxide solution for neutralisation. Deduce the molecular formula of the acid, and write a structural formula for it.

METHOD   The equation for the neutralisation is

$$R(CO_2H)_2(aq) + 2NaOH(aq) \longrightarrow R(CO_2Na)_2(aq) + 2H_2O(l)$$

No. of moles of NaOH $= 21.0 \times 10^{-3} \times 0.100 = 2.10 \times 10^{-3}\ mol$

From the equation, No. of moles of acid $= \frac{1}{2} \times$ No. moles of NaOH

$$= 1.05 \times 10^{-3}\ mol$$

Mass of acid $= \frac{1}{10} \times 1.220\,g = 0.1220\,g$

$\therefore \qquad\qquad 1.05 \times 10^{-3}\ mol = 0.1220\,g$

$$1\ mol = 116\,g$$

The molar mass is $116\,g\,mol^{-1}$. Subtracting 90 for $(CO_2H)_2$ leaves $26\,g\,mol^{-1}$ for the rest of the molecule. This is the mass of $(CH_2)_2$. The molecular formula is $C_4H_4O_4$, and the structural formula are

$$
\begin{array}{ccc}
HCCO_2H & and & HO_2CCH \\
\| & & \| \\
HCCO_2H & & HCCO_2H
\end{array}
$$

for *cis-* and *trans*-butenedioic acid.

EXERCISE 60   Problems on Organic Chemistry

1. A compound X contains carbon and hydrogen only. 0.135 g of X, on combustion in a stream of oxygen, gave 0.410 g of carbon dioxide and 0.209 g of water. Calculate the empirical formula of X.

    X is a gas at room temperature, and 0.29 g of the gas occupy $120\ cm^3$ at room temperature and 1 atm. What is the molecular formula of X?

2. A monobasic organic acid C is dissolved in water and titrated with sodium hydroxide solution. 0.388 g of C require $46.5\ cm^3$ of $0.095\ mol\,dm^{-3}$ sodium hydroxide for neutralisation. Calculate the molar mass of C and deduce its formula.

3. An organic compound has a composition by mass of 83.5% C; 6.4% H; 10.1% O. The molar mass is $158\,g\,mol^{-1}$. Find the molecular formula and suggest a structural formula for the compound.

4. A compound A contains 5.20% by mass of nitrogen. The other elements present are carbon, hydrogen and oxygen. Combustion of 0.0850 g of A in a stream of oxygen gave 0.224 g of carbon dioxide and 0.0372 g of water. Calculate the empirical formula of A.

5. An alkene A contains one double bond per molecule. 0.560 g of bromine is required to react completely with 0.294 g of A. When A is treated with ozone and the product is hydrolysed and then oxidised, B, which is a monobasic carboxylic acid is formed. 0.740 g of this acid require 100 cm^3 of 0.100 mol dm^{-3} sodium hydroxide for neutralisation. Deduce the formulae of A and B and the structural formula of A.

6. An organic compound, A, contains 70.6% carbon, 5.88% hydrogen and 25.5% oxygen, by mass. It has a molar mass of 136 g mol^{-1}. When A is refluxed with sodium hydroxide solution, and the resulting liquid is distilled, a liquid B distils over, and a solution of C remains. On addition of dilute hydrochloric acid to C, a white precipitate D forms. When this precipitate is mixed with soda lime and heated, the vapour burns with a smoky flame. B reacts with ethanoyl chloride to give a product with a fruity smell. B does not give a positive result in the iodoform test. Identify A, B and C.

7. A compound contains C, H, N and O. When 0.225 g of the compound was heated with sodium hydroxide solution, the ammonia evolved was passed into 25.0 cm^3 of sulphuric acid of concentration 0.100 mol dm^{-3}. The sulphuric acid that remained required 19.1 cm^3 of 0.100 mol dm^{-3}. sodium hydroxide for neutralisation.

A 0.195 g sample of the compound gave on complete oxidation 0.352 g of carbon dioxide and 0.168 g of water.

A solution of 9.12 g of the compound in 500 cm^3 of water froze at -0.465 °C. The cryoscopic constant for water is 1.86 K kg mol^{-1}.

a) Find the molecular formula of the compound, and b) suggest its identity.

8. Phenol reacts with bromine to form a crystalline product. 25.0 cm^3 of a solution of phenol of concentration 0.100 mol dm^{-3} were added to 30.0 cm^3 of a solution of 0.100 mol dm^{-3} potassium bromate(V). An excess of potassium bromide and hydrochloric acid were added to liberate bromine. The excess bromine was estimated by adding potassium iodide and titrating the iodine displaced with a solution of sodium thiosulphate. 30.0 cm^3 of a 0.100 mol dm^{-3} solution of sodium thiosulphate were required.

a) Find the ratio of moles of Br$_2$:moles of phenol, and b) deduce the equation for the reaction.

9. Measurements on an alkene showed that 100 cm^3 of the gas weighed 0.233 g at 25 °C and 1 atm. 25.0 cm^3 of the alkene reacted with 25.0 cm^3 of hydrogen.

a) Find the molar mass of the alkene, and b) give its molecular formula.

Give the names and structural formulae of two alkenes which have this molecular formula.

10. An organic acid has the percentage composition by mass: C, 41.4%; H, 3.4%; O, 55.2%. A solution containing 0.250 g of the acid, which is dibasic, required 26.6 cm³ of 0.200 mol dm⁻³ sodium hydroxide solution for neutralisation.

Calculate: a) the empirical formula, and b) the molecular formula of the acid. c) Give its name and write its structural formula or formulae.

11. An organic liquid contains carbon, hydrogen and oxygen. On oxidation, 0.250 g of the liquid gave 0.595 g of carbon dioxide and 0.304 g of water. When vaporised, 0.250 g of the liquid occupied 131 cm³ at 200 °C and 1 atm.

Find: a) the empirical formula, and b) the molecular formula of the liquid. c) Write the structural formulae of compounds with this molecular formula.

12. A is an organic compound with the percentage composition by mass C, 71.1%; N, 10.4%; O, 11.8%; H, 6.7%, and a molar mass of 135 g mol⁻¹.

On hydrolysis by aqueous sodium hydroxide, A gives an oily liquid, B. B has the percentage composition by mass: C, 77.1%; N, 15.1%; H, 7.5%, and a molar mass of 93 g mol⁻¹. B is basic and gives a precipitate with bromine water.

Find the molecular formulae for A and B. From their reactions, deduce the identity of A and B.

EXERCISE 61   Questions from A-level Papers

1. Ripening tomatoes produce a gaseous hydrocarbon, A, which itself assists the ripening process. The gas A reacts in a 1 : 1 mole ratio with hydrogen bromide to give a liquid, B. Treatment of 1.0000 g of B with hot aqueous sodium hydroxide yields a volatile product, C. Acidification of the residual alkaline solution with dilute nitric acid and the addition of excess silver nitrate solution affords 1.7230 g of silver bromide. Mild oxidation of C yields a volatile product, D, which on treatment with ammoniacal silver nitrate produces a silver mirror.

Compound A may be converted industrially into C by the reversible gas phase addition of one molecule of water. For the equilibrium

$$A(g) \; + \; H_2O(g) \; \rightleftharpoons \; C(g) \qquad \Delta H^{\ominus} \; = \; -46.0 \, \text{kJ mol}^{-1}$$

$$(A_r(H) \; = \; 1.01, \quad A_r(C) \; = \; 12.01, \quad A_r(Br) \; = \; 79.91,$$

$$A_r(Ag) \; = \; 107.87)$$

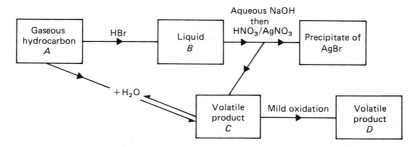

a) i) State the functional group present in each of *A, B, C* and *D*.
   ii) Give your reasoning for these conclusions.
   iii) Calculate the number of moles of silver bromide produced.
   iv) Hence calculate the relative molecular mass of *B* and deduce
       the identity of the gas *A*.

b) i) *Briefly* describe a process for the direct conversion of *A* to *C*,
        giving the appropriate conditions.
   ii) Describe and explain the effects of variation of temperature
       and pressure on the equilibrium yield of *C* obtained.
   iii) Indicate how the use of excess water (as steam) might influence
        the equilibrium yield of *C*.   (WJEC92)

2. Compound *B*, a diacid that occurs in apples and other fruit, has the
   following composition by mass:

   C, 35.8%
   H, 4.5%
   O, 59.7%

   B reacts with ethanol in the presence of concentrated sulphuric acid
   under reflux to give *C*, $C_8H_{14}O_5$. Compound *C* evolves hydrogen gas
   when treated with sodium metal and reacts with acidified potassium
   dichromate(VI) to give compound *D*. Compound *D* produces an
   orange precipitate with 2,4-dinitrophenylhydrazine but has no reaction
   with Fehling's or Tollens' reagent.

   a) Calculate the empirical formulae of *B*.

   b) Suggest structures for compounds *B, C* and *D* and explain the
      reactions described.   (C92)

3. Hippuric acid, an organic substance, was first obtained from horse's
   urine (Greek, *hippos* = horse).

   Historically, organic substances were believed to need a 'vital force' to
   be made. Today, hippuric acid is made in the laboratory by the following overall reaction:

   | COCl | | CONHCH$_2$COOH | |

   Benzoyl chloride   + NH$_2$CH$_2$COOH ⟶   Hippuric acid   + HCl

   Benzoyl chloride   Glycine   Hippuric acid

a) Suggest a present-day definition of an organic substance.

b) What is the name given to the following functional group?

$$\underset{\displaystyle -\overset{\textstyle O}{\overset{\|}{C}}-Cl}{}$$

c) Hippuric acid is made in the laboratory as follows:

Dissolve 25 g (0.33 mol) of glycine in 250 cm³ of 10% sodium hydroxide solution in a conical flask. Add 45 cm³ (46 g, 0.33 mol) of benzoyl chloride. Stopper the flask and shake vigorously until all the chloride has reacted. Transfer the solution to a beaker and add concentrated hydrochloric acid. Collect the precipitate of hippuric acid by suction filtration. Recrystallise the hippuric acid from boiling water. The yield is 45 g and the melting point 187°C.

   i) Write a balanced equation for the reaction of sodium hydroxide with glycine.
   ii) Benzoyl chloride is insoluble in water. How will you know when all of it has reacted?
   iii) Explain why hydrochloric acid is added to the solution.

d) Hippuric acid is purified by recrystallisation.
   i) Explain how you would recrystallise hippuric acid in the laboratory.
   ii) Explain how you would determine the melting point of hippuric acid.
   iii) If the melting point of dry hippuric acid were found to be 180–184°C what would you conclude?

e) Calculate the % yield of hippuric acid based on the amount of glycine used. (Relative atomic masses: $H = 1$, $C = 12$, $N = 14$, $O = 16$.)
(NI92)

4. Liquid K, containing a benzene ring (which is *not* involved in any of the reactions which follow) and two functional groups, and having the composition by mass: $C = 75.0\%$; $H = 6.8\%$; $O = 18.2\%$, is slowly hydrolysed when boiled with aqueous potassium hydroxide solution. Acidification of the reaction mixture leads to the precipitation of a white crystalline solid, L, having the composition: $C = 73.0\%$; $H = 5.4\%$; $O = 21.6\%$. The filtrate, after removal of L, contains ethanol.

L reacts at room temperature with hydrogen bromide forming two compounds, M and N, both of which are chiral and have the molecular formula $C_9H_9O_2Br$.

L loses carbon dioxide on heating to form a hydrocarbon, O, of molecular formula C_8H_8 which rapidly decolorises a solution of bromine in 1,1,1-trichloroethane to give P, which is also chiral.

O yields benzoic acid on oxidation with alkaline potassium manganate(VII).

a) Deduce the structures of K to P, explaining your reasoning.

b) Draw and label the isomers of K and explain how they arise.

c) Give the structures of two isomers of K which are *not* consistent with the given reaction scheme.

d) i) Draw the structural formulae for M and N, identifying the asymmetric carbon atoms.
   ii) Would you expect equal quantities of M and N to be formed when L reacts with hydrogen bromide? Explain your answer.

e) Suggest a synthesis of O starting from any compound containing eight carbon atoms. Name the type of reaction which occurs.

f) Write down the structural formula of the product obtained when L is reduced with
   i) hydrogen in the presence of a platinum catalyst
   ii) lithium aluminium hydride. (L92)

*5. When 0.440 g of a solid, A, is completely burned in oxygen, 0.590 g of carbon dioxide, 0.240 g of water and 0.094 g of nitrogen are produced. Calculate the empirical formula of A.

When A is refluxed with moderately concentrated hydrochloric acid just one substance, B, is formed, with the molecular formula $C_2H_5O_2N$.

B, on treatment with lithium tetrahydridoaluminate, $LiAlH_4$, forms C, C_2H_7ON.

C, on refluxing with sodium bromide and sulphuric acid forms D, C_2H_6BrN.

D, on heating under pressure with alcoholic ammonia forms E, $C_2H_8N_2$.

Deduce the structural formulae and names of the compounds B to E and suggest a displayed formula for A.

Suggest the chemical and physical properties of E. In particular, suggest how E will react with

a) Cu^{2+} ions

b) decanedioyl dichloride.

Give practical observations and chemical formulae where appropriate.
(L91,N,S)

*6. a) The organic compounds A to F, each of which contains either seven or eight carbon atoms per molecule, consist of carbon and hydrogen only with no carbon–carbon triple bonds. The table below shows the results obtained when the given masses of compounds A to F were reacted with an excess of hydrogen, and the actual masses of the hydrogen required for their complete hydrogenation. Also shown is the amount of heat, Q, evolved at constant pressure during each hydrogenation involving the reacting amounts shown.

Compound	Mass of compound taken/g	Mass of hydrogen used/g	Heat Q evolved/kJ
A	90.0	5.09	344
B	75.0	5.77	389
C	100.0	5.66	195
D	90.0	5.87	396
E	90.0	6.92	296
F	100.0	6.52	225

Draw up a table showing the mass of each compound A to F which reacts with one mole of hydrogen, and also the heat evolved per mole of hydrogen consumed. Use this information to deduce the relative molecular masses (which are *integral* to within $\pm 0.1\%$) and hence the possible structural formulae of these compounds. Explain clearly why the heat evolved per mole of hydrogen consumed is not the same in each case. You will need to use the following data.

	Benzene(g)	Cyclohexene(g)	Cyclohexane(g)
$\Delta H_F^{\ominus}$/kJ mol^{-1}	+83	+11	−124

Identify one of the compounds A to F which has four isomeric forms. Draw a structure for each of the four isomers.

b) Compound G has ten carbon atoms per molecule. Full hydrogenation of 2.56 g of G using exactly 0.200 g of hydrogen produces sufficient heat to raise the temperature of 500 g of water by 3.29 K. Deduce the molecular formula of G and draw its structural formula.·

(Specific heat capacity of water = 4.2 J K^{-1}g^{-1}.)   (JMB90,S)

7. a) Consider the reaction scheme shown opposite.

The compounds, A–F, each contain only one functional group; R represents an alkyl group.
   i) Identify the functional groups present in compounds B, C and D respectively.
   ii) Identify the reagents whereby the conversions $A \rightarrow B$, $D \rightarrow E$, $E \rightarrow F$, $F \rightarrow A$, and $A \rightarrow F$ may be brought about, giving *brief* reasoning in *each* case.

b) Use *all* the quantitative information given below to identify fully the compounds A–F above.
   i) When 0.2500 g of A, a monobasic carboxylic acid, was titrated against 0.1000 mol dm^{-3} sodium hydroxide solution, 41.63 cm^3 of the latter was required for complete reaction.

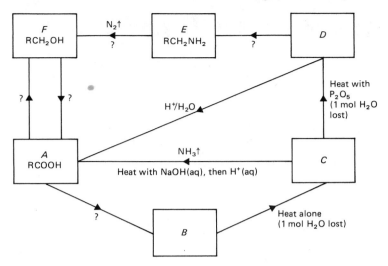

ii) When C is heated, as shown, with sodium hydroxide solution, ammonia is evolved in a 1:1 mole ratio to C. When 0.3000 g of C was thus treated, and the ammonia released absorbed in water, the resulting solution required 25.39 cm³ of 0.2000 mol dm⁻³ hydrochloric acid solution for complete reaction.

iii) When E reacts to form F, as shown above, nitrogen is evolved in a 1:1 mole ratio. When 0.2000 g of E was thus treated, 108.44 cm³ of nitrogen gas, measured at 298 K and 1.01×10^5 Pa, was evolved.

$(A_r(C) = 12.01, \quad A_r(H) = 1.01, \quad A_r(N) = 14.01, \quad A_r(O) = 16.00;$
1 mol of a gas occupies 2.24×10^4 cm³ at 273 K and 1.01×10^5 Pa.)
(WJEC91)

8. This question concerns the organic compound, X, which contains carbon, hydrogen and oxygen.

a) X contains C = 66.67%; H = 11.11%; O = 22.22% by mass. Show that the empirical formula of X is C_4H_8O. (Relative atomic masses: H = 1, C = 12, O = 16.)

b) The relative molecular mass of X is 72. What is the molecular formula of X?

c) Give the structures of the non-cyclic isomers of X which do not react with bromine water.

d) Select an isomer of X which can be readily oxidised and describe its behaviour with a named oxidising agent. Give the reaction conditions and the structures of the organic product(s).

e) Outline the method you would use in order to identify X.   (L91)

Answers to Exercises

The examination boards accept no responsibility whatsoever for the accuracy of the answers given.

CHAPTER 1

Practice 1

1. Density $= \dfrac{\text{Mass}}{\text{Volume}}$ Volume $= \dfrac{\text{Mass}}{\text{Density}}$

2. a) $I = \dfrac{V}{R}$ b) $R = \dfrac{V}{I}$

3. a) $R = \dfrac{V}{I}$ b) $I = \dfrac{V}{R}$

4. a) Mass of solute = Concentration $\times$ Volume of solution

   b) Volume of solution $= \dfrac{\text{Mass of solute}}{\text{Concentration}}$

5. a) $Q = \dfrac{P}{R}$ b) $R = \dfrac{P}{Q}$

Practice 2

1. a) $T = \dfrac{PV}{R}$ b) $V = \dfrac{RT}{P}$

2. a) $\rho = \dfrac{RA}{l}$ b) $A = \dfrac{\rho l}{R}$

3. a) $p = \dfrac{a \times b \times q \times r}{c}$ b) $q = \dfrac{p \times c}{a \times b \times r}$

Practice 3

1. $24\,\text{dm}^3$ 2. $11\,\text{g}$ 3. 300 tonnes

Exercise 1

1. a) 2.3678×10^4 b) 4.376×10^2 c) 1.69×10^{-2} d) 3.45×10^{-4}
   e) 6.72891×10^5
2. a) 5.85×10^4 b) 2.66×10^6 c) 6.35×10^7 d) 1.21×10
   e) 1.34×10^2
3. a) 3.32×10^6 b) 2.72×10^5 c) 1.86×10^{-4} d) 6.44×10^{-5}
   e) 6.11×10^{-3}
4. a) 2.001×10^4 b) 5.648×10^3 c) 1.29×10^5 d) -1.12×10^{-3}
   e) 6.252×10^4
5. a) 4×10^{10} b) 2×10^2 c) 5×10^8 d) 1×10^3
   e) 2×10^{10}
6. a) 3.6753   b) 3.7052   c) -2.8771 d) 1.0033
   e) -5.6356
7. a) 2.862×10^3 b) 1.135   c) 6.969×10^7 d) 3.3791×10^{-7}
   e) 8.7680×10^{-3}
8. a) 4.264×10^{-3} b) 2.867×10^{-7} c) 4.037×10^2 d) 2.055×10^{-10}
   e) 3.781×10^4
9. a) $x = 7$ or -13 b) $x = 4$ or -0.4 c) $y = 5$ d) $z = 14$ or 2
   e) $x = -\frac{1}{2}$ or $-4\frac{1}{2}$

CHAPTER 2

Exercise 2

1. a) $H_2(g) + CuO(s) \longrightarrow Cu(s) + H_2O(g)$
   b) $C(s) + CO_2(g) \longrightarrow 2CO(g)$
   c) $C(s) + O_2(g) \longrightarrow CO_2(g)$
   d) $Mg(s) + H_2SO_4(aq) \longrightarrow H_2(g) + MgSO_4(aq)$
   e) $Cu(s) + Cl_2(g) \longrightarrow CuCl_2(s)$
2. a) $Ca(s) + 2H_2O(l) \longrightarrow H_2(g) + Ca(OH)_2(aq)$
   b) $2Cu(s) + O_2(g) \longrightarrow 2CuO(s)$
   c) $4Na(s) + O_2(g) \longrightarrow 2Na_2O(s)$
   d) $Fe(s) + 2HCl(aq) \longrightarrow FeCl_2(aq) + H_2(g)$
   e) $2Fe(s) + 3Cl_2(g) \longrightarrow 2FeCl_3(s)$
3. a) $Na_2O(s) + H_2O(l) \longrightarrow 2NaOH(aq)$
   b) $2KClO_3(s) \longrightarrow 2KCl(s) + 3O_2(g)$
   c) $2H_2O_2(aq) \longrightarrow 2H_2O(l) + O_2(g)$
   d) $3Fe(s) + 2O_2(g) \longrightarrow Fe_3O_4(s)$
   e) $3Mg(s) + N_2(g) \longrightarrow Mg_3N_2(s)$
   f) $4NH_3(g) + 3O_2(g) \longrightarrow 2N_2(g) + 6H_2O(g)$
   g) $3Fe(s) + 4H_2O(g) \longrightarrow Fe_3O_4(s) + 4H_2(g)$
   h) $2H_2S(g) + 3O_2(g) \longrightarrow 2H_2O(g) + 2SO_2(g)$
   i) $2H_2S(g) + SO_2(g) \longrightarrow 2H_2O(l) + 3S(s)$

CHAPTER 3

Exercise 3

64	40	101
84	278	95
148	99	161
98	63	246
136	685	142
106	74	123.5
159.5	162	249.5
400	286	278

Exercise 4

SECTION 1

1. a) C = 80%   H = 20%
   b) Na = 57.5%   O = 40%
      H = 2.5%
   c) S = 40%   O = 60%
   d) C = 90%   H = 10%
2. a) C = 84%   H = 16%
   b) Mg = 72%   N = 28%
   c) Na = 15.3%   I = 84.7%
   d) Ca = 20%   Br = 80%

SECTION 2

1. a) C = 85.7%   H = 14.3%
   b) N = 35%   H = 5%
      O = 60%
   c) Fe = 62.2%   O = 35.6%
      H = 2.2%
   d) C = 26.7%   H = 2.2%
      O = 71.1%

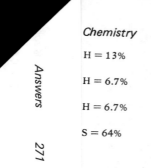

H = 13%

H = 6.7%

H = 6.7%

S = 64%

SECTION 1

1. a) 26 g   b) 8 g   c) 4 g   d) 6 g
   e) 4 g   f) 8 g
2. a) 2.0   b) 2.0   c) 0.25   d) 0.10
   e) 0.25 mol
3. a) 2070 g   b) 10.6 g   c) 25.4 g   d) 20.0 g
   e) 10.0 g   f) 40.0 g   g) 42.0 g   h) 13.0 g
   i) 35.5 g   j) 2.00 g
4. a) 1.00   b) 0.25   c) 0.50   d) 0.20
   e) 0.20   f) 3.0   g) 0.10   h) 2.0
5. a) 6×10^{23} b) 6×10^{23} c) 6×10^{22} d) 3.6×10^{24}
   e) 1.2×10^{24} f) 6×10^{22} g) 1.5×10^{22} h) 1.2×10^{24}
6. a) 65 g   b) 0.065 g
7. a) 9.0 g   b) 0.027 g
8. a) 12 g   b) 0.040 g
9. a) 20.0 g   b) 12.0 g   c) 16.25 g   d) 115 g

SECTION 2

1. £1.25×10^{8}
2. a) 0.2 mol   b) 0.4 mol   c) 1.2 mol   d) 0.2 mol
3. 1.0×10^{23} 4. 55.5 mol   5. 6.3 mol   6. 3×10^{-23} g
7. 2.92 mol

CHAPTER 5

Exercise 6

SECTION 1

1. 1250 tonnes   2. 0.05 g   3. 0.26 g
4. a) $2Al(OH)_3, 3H_2SO_4, 3H_2O$ b) i) 0.46 kg   ii) 0.86 kg
5. Loss of 83p   6. a) 72 g   7. 19 kg/year
8. Yes, 1 mol C (12 g) combines with 4 mol F (4×19 g)
9. 7.1 tonne
10. a) 0.01 mol   b) 0.02 mol   c) 2 mol
   d) $Zn(s) + 2Ag^{+}(aq) \longrightarrow Zn^{2+}(aq) + 2Ag(s)$

SECTION 2

1. 127
2. a) 4.05 tonnes   b) 8.33 tonnes
3. a) i) 1060 tonnes   ii) 1030 tonnes
   b) natural limestone; manufactured ammonia   c) Ammonium sulphate is a fertiliser
4. 9.78 kg   5. 522.7 tonnes   6. 0.1435 g   7. 2.808 g
8. a) 7.00 tonnes   b) 7.24 tonnes   9. 36.46 g   10. 4.481 g
11. 40% $CaCO_3$ 60% $MgCO_3$ 12. 3.5000 g $NaHCO_3$ 6.5000 g Na_2CO_3

Exercise 7

1. $S_2O_8^{2-}(aq) + 2I^-(aq) \longrightarrow I_2(aq) + 2SO_4^{2-}(aq)$
2. $H_2NSO_3^-(aq) + OH^-(aq) \longrightarrow NH_3(g) + SO_4^{2-}(aq)$
3. $Na_2S_2O_3(aq) + AgCl(s) \longrightarrow NaCl(aq) + NaAgS_2O_3(aq)$
4. $C_6H_8 + 2Br_2 \longrightarrow C_6H_8Br_4$
5. $C_6H_5NH_2 + 3Br_2 \longrightarrow C_6H_2Br_3NH_2 + 3HBr$

Exercise 8

1. 92.9%	2. 90.5%	3. 89.0%	4. 91.0%
5. 99.2%			

Exercise 9

1. 436 tonnes	2. 46 kg	3. 2.7 kg	4. 304 kg
5. 93.5%			

CHAPTER 6

Exercise 10

SECTION 1

1. Mg_3N_2	2. Fe_3O_4	3. Al_2O_2	4. $BaCl_2 \cdot 2H_2O$
5. PbO_2			
6. a) P_2O_3	b) NH_3	c) Pb_3O_4	d) SiO_2
e) MnO_2	f) N_2O_5	g) $CrCl_3$	

7. $A = C_2F_4$ $B = C_4H_8O_2$
    $C = C_2H_6$ $D = C_6H_6$
    $E = C_3H_6$ $F = C_2H_6O_2$
    $G = C_2H_4Cl_2$ $H = C_6H_3N_3O_6$

8. D 9. MO_2

SECTION 2

1. a) MgO	b) $CaCl_2$	c) $FeCl_3$	d) CuS
e) LiH			
2. a) FeO	b) Fe_2O_3	c) Fe_3O_4	d) K_2CrO_4
e) $K_2Cr_2O_7$	f) CH	g) C_3H_8	
3. a) $a = 5$	b) $b = 6$	c) $c = 2$	d) $d = 3$
e) $e = 6$	f) $f = 12$		
4. a) $C_5H_{10}O$	b) $C_5H_{10}O$	5. a) C_2H_4O	b) C_2H_4O
6. a) $C_5H_{10}O_2$	b) $C_5H_{10}O_2$	7. $C_9H_{10}O_2$	

CHAPTER 7

Exercise 11

SECTION 1

1. a) $2C_2H_6(g) + 7O_2(g) \longrightarrow 4CO_2(g) + 6H_2O(g)$
   b) 30 cm³   c) 40%
2. C   3. C
4. b) i) 579 cm³   ii) 308 cm³   c) 44, CO_2

SECTION 2

1. 20 cm³ ethane + 10 cm³ ethene
2. a) 2 dm³   b) 750 cm³   c) 625 cm³   d) 937.5 cm³
   e) 2 dm³
3. 500 cm³ SO_2 4. 50%   5. d

Exercise 12

1. C_3H_8 **2.** C_4H_6 **3.** $a = 30 \text{ cm}^3$, $b = 40 \text{ cm}^3$

4. $CH_4 + 2O_2 \longrightarrow CO_2 + 2H_2O$; CH_4

Exercise 13

SECTION 1

1. 25 g, 6.0 dm^3 **2.** 3250 g, 1120 dm^3 **3.** 560 cm^3, 1120 cm^3

4. a) $2H_2O$ on LHS   b) 1.33 g

5. a) $2H_2O$ on RHS   b) i) 16.0 g   ii) 33.3 g

SECTION 2

1. a) KO_2 b) $4KO_2 + 2CO_2 \longrightarrow 3O_2 + 2K_2CO_3$ c) 237 dm^3

2. 3.5 g   **3.** £90 daily   **4.** 267 dm^3 **5.** 3.50 dm^3

6. 1.107 g   **7.** 2.388 g   **8.** 3.646 g   **9.** 11.5 dm^3

10. 2460 dm^3

Exercise 14

1. b) $CaFe_2O_4$ oxidation no. of $Fe = +3$

2. a) S_2Cl_2

b) B is S_4N_4 C is $H-S \equiv N$

c) $S_4N_4 + 2SnCl_2 + 4HCl \longrightarrow 4HSN + 2SnCl_4$

d) A large volume of gas, SO_2 and NO_x, is formed when the solid is ignited

3. A is CrO_3 B is Cr_2O_3 C is $(NH_4)_2Cr_2O_7$ D is CrO_2Cl_2

$CrO_3 + H_2O \longrightarrow H_2CrO_4$

$H_2CrO_4 + 2NaOH \longrightarrow Na_2CrO_4 + 2H_2O$

$4CrO_3 \longrightarrow 2Cr_2O_3 + 3O_2$

$2CrO_3 + 2NH_3 + H_2O \longrightarrow (NH_4)_2Cr_2O_7$

$(NH_4)_2Cr_2O_7 \longrightarrow N_2 + 4H_2O + Cr_2O_3$

$CrO_3 + 2HCl \longrightarrow CrO_2Cl_2 + H_2O$

4. c) i) 694 kg   ii) $69\,400 \text{ dm}^3$ **5.** b) ii) 74.8%   iii) $C_2H_4I_2$

6. c) 1.63 tonne   d) iv) 9380 litres   **7.** d) ii) 21.8%

8. A and B are pentaamminenitrocobalt(III) sulphate

$$[Co^{III}(NH_3)_5NO_2]^{2+}SO_4^{2+}$$

(co-ordination through N of the $-NO_2^-$ group)

and pentaamminenitritocobalt(III) sulphate

$$[Co^{III}NH_3)_5ONO]^{2+}SO_4^{2-}$$

(co-ordination through O of the $-O-N=O$ group)

Another isomer is

$$[Co^{III}(NH_3)_5SO_4]^+NO_2^-$$

9. a) $CsICl_2$ b) $[Cl-I-Cl]^-$ c) linear   d) +1

e) $ICl_2^-(aq) + SO_2(g) + 2H_2O(l) \longrightarrow I^-(aq) + 2Cl^-(aq) + SO_4^{2-}(aq) + 4H^+(aq)$

g) i) $AgCl(s) + 2NH_3(aq) \longrightarrow Ag(NH_3)_2Cl(aq)$

ii) $2AgI(s) + 2H_2SO_4(aq) \longrightarrow I_2(s) + SO_2(g) + Ag_2SO_4(s) + 2H_2O(l)$

10. b) *A* is PF_3Cl_2 *B* is $[PCl_4]^+[PF_6]^-$ *C* is $PFCl_4$ *D* is $Na^+PF_6^-$

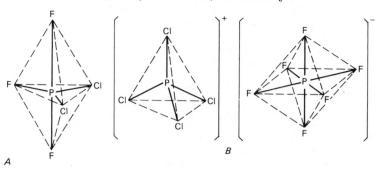

CHAPTER 8

Exercise 15

1. a) $0.0500 \, mol \, dm^{-3}$ b) $1.00 \, mol \, dm^{-3}$ c) $0.250 \, mol \, dm^{-3}$ d) $2.00 \, mol \, dm^{-3}$
  e) $0.100 \, mol \, dm^{-3}$ f) $0.125 \, mol \, dm^{-3}$ g) $0.250 \, mol \, dm^{-3}$ h) $0.200 \, mol \, dm^{-3}$
2. a) $0.250 \, mol$ b) $0.125 \, mol$ c) $0.005 \, 00 \, mol$ d) $2.50 \, mol$
  e) $0.0500 \, mol$ f) $0.0250 \, mol$ g) $0.370 \, mol$ h) $1.125 \, mol$
3. a) $5.90 \, g$ b) $5.30 \, g$ c) $9.45 \, g$ d) $42.0 \, g$
  e) $19.1 \, g$
4. a) $0.500 \, dm^3$ b) $0.075 \, 0 \, dm^3$ or $75.0 \, cm^3$ c) $11.4 \, cm^3$
  d) $192 \, cm^3$ e) $192 \, cm^3$

Exercise 16

SECTION 1
1. a) True   b) False   c) True   d) False
  e) True   f) False
2. a) True   b) False   c) True   d) False
  e) True   f) False
3. a) False   b) False   c) True   d) True
  e) True   f) False
4. a) True   b) False   c) False   d) False
  e) True   f) True
5. 75%   **6.** a) $1.1 \, mol \, dm^{-3}$ b) 6.6%   **7.** $0.145 \, mol \, dm^{-3}$
8. 84%   **9.** b) $3360 \, dm^3$ **10.** a) $1.5 \times 10^{-6} \, g$ b) 1/30
11. a) 0.050%   b) $3.3 \times 10^{-3} \, mol \, dm^{-3}$ **12.** $0.109 \, mol \, dm^{-3}$
13. a) $1.20 \times 10^{-2} \, mol$ b) $1.20 \times 10^{-2} \, mol$ c) $1.20 \times 10^{-2} \, mol$ d) $0.276 \, g$
  e) 9.2%

SECTION 2
1. 95.0%   **2.** 87.6%   **3.** 90.6%   **4.** 18.0%
5. 46.1%   **6.** 3   **7.** 10

Exercise 17

1. a) $+2$ b) $+2$ c) 0   d) $+1$
  e) $+3$ f) $+1$ g) $+3$ h) $+5$
  i) $+5$ j) $+4$ k) 0   l) $+7$
  m) $+2$ n) -1 o) $+3$ p) $+6$
  q) $+5$ r) -1 s) $+6$ t) $+4$
  u) -1 v) $+6$ w) $+4$ x) $+1$
  y) $+4$ z) $+5$

2. a) Sn is oxidised from $+2$ to $+4$; Pb is reduced from $+4$ to $+2$.

  b) Mn is oxidised from $+2$ to $+7$; Bi is reduced from $+5$ to $+3$.

  c) As is oxidised from $+3$ to $+7$; Mn is reduced from $+7$ to $+2$.

3. a) F is reduced from 0 to -1.

  b) Cl disproportionates from 0 to $+5$ and -1.

  c) N disproportionates from -3 and $+5$ to -1.

  d) Cr is reduced from $+6$ to $+3$.

  e) C is oxidised from $+3$ to $+4$.

4. a) -2 b) $+2$

5. a) $IO_4^- + 7I^- + 8H^+ \longrightarrow 4I_2 + 4H_2O$

  b) $BrO_3^- + 6I^- + 6H^+ \longrightarrow Br^- + 3I_2 + 3H_2O$

  c) $2V^{3+} + H_2O_2 \longrightarrow 2VO^{2+} + 2H^+$

  d) $SO_2 + 2H_2O + Br_2 \longrightarrow 4H^+ + SO_4^{2-} + 2Br^-$

  e) $4NH_3 + 3O_2 \longrightarrow 2N_2 + 6H_2O$

  f) $2NH_3 + 2O_2 \longrightarrow N_2O + 3H_2O$

  g) $4NH_3 + 5O_2 \longrightarrow 4NO + 6H_2O$

  h) $Fe^{2+}C_2O_4^{2-} + 3Ce^{3+} \longrightarrow 2CO_2 + 3Ce^{2+} + Fe^{3+}$

6. $Cr_2O_7^{2-} + 6I^- + 14H^+ \longrightarrow 3I_2 + 2Cr^{3+} + 7H_2O$

Exercise 18

1. a) $NO_2^- + H_2O \longrightarrow NO_3^- + 2H^+ + 2e^-$

  b) $AsO_3^{3-} + H_2O \longrightarrow AsO_4^{3-} + 2H^+ + 2e^-$

  c) $Hg_2^{2+} \longrightarrow 2Hg^{2+} + 2e^-$

  d) $H_2O_2 \longrightarrow 2H^+ + 2e^- + O_2$

  e) $V^{3+} + H_2O \longrightarrow VO^{2+} + 2H^+ + e^-$

2. a) $NO_3^- + 2H^+ + e^- \longrightarrow NO_2 + H_2O$

  b) $NO_3^- + 4H^+ + 3e^- \longrightarrow NO + 2H_2O$

  c) $NO_3^- + 10H^+ + 8e^- \longrightarrow NH_4^+ + 3H_2O$

  d) $2BrO_3^- + 12H^+ + 10e^- \longrightarrow Br_2 + 6H_2O$

  e) $PbO_2 + 4H^+ + 2e^- \longrightarrow Pb^{2+} + 2H_2O$

3. a) $2MnO_4^-(aq) + 5H_2O_2(aq) + 6H^+(aq) \longrightarrow 5O_2(g) + 2Mn^{2+}(aq) + 8H_2O(l)$

  b) $MnO_2(s) + 4H^+(aq) + 2Cl^-(aq) \longrightarrow Mn^{2+}(aq) + Cl_2(g) + 2H_2O(l)$

  c) $2MnO_4^-(aq) + 5C_2O_4^{2-}(aq) + 16H^+(aq) \longrightarrow 2Mn^{2+}(aq) + 10CO_2(g) + 8H_2O(l)$

  d) $Cr_2O_7^{2-}(aq) + 3C_2O_4^{2-}(aq) + 14H^+(aq) \longrightarrow 2Cr^{3+}(aq) + 6CO_2(g) + 7H_2O(l)$

  e) $Cr_2O_7^{2-}(aq) + 6I^-(aq) + 14H^+(aq) \longrightarrow 2Cr^{3+}(aq) + 3I_2(aq) + 7H_2O(l)$

  f) $H_2O_2(aq) + NO_2^-(aq) \longrightarrow NO_3^-(aq) + H_2O(l)$

4. a) 1.5×10^{-2} mol   b) 7.5×10^{-3} mol   c) 7.5×10^{-3} mol   d) 7.5×10^{-3} mol

  e) 1.5×10^{-2} mol

5. a) 6.0×10^{-4} mol   b) 3.0×10^{-4} mol   c) 6.0×10^{-4} mol   d) 3.0×10^{-4} mol

  e) 3.0×10^{-4} mol

6. a) 4.0×10^{-3} mol   b) 2.0×10^{-3} mol   c) 2.0×10^{-3} mol   d) 4.0×10^{-3} mol

  e) 6.7×10^{-4} mol

7. a) 62.5 cm^3 b) 250 cm^3 c) 5.00 cm^3 d) 12.5 cm^3

  e) 8.3 cm^3

8. a) 45.0 cm^3 b) 12.0 cm^3 c) 7.2 cm^3 d) 4.50 cm^3

  e) 9.0 cm^3

9. 0.090 mol dm^{-3} **10.** 0.0195 mol dm^{-3} **11.** 0.0894 mol dm^{-3} **12.** 99.5%

13. 90.6% **14.** 1.64×10^{-2} mol dm^{-3} **15.** 0.103 mol dm^{-3}

16. a) $[Fe^{2+}] = 0.0600$ mol dm^{-3} b) $[Fe^{3+}] = 0.0160$ mol dm^{-3}

17. a) 20.0 cm^3 b) 22.4 cm^3 **18.** 7.63×10^{-2} mol dm^{-3}

19. $+4$ **20.** 2.7×10^{-2} mol dm^{-3}

21. a) $1NH_2OH : 2Fe^{3+}$ b) -1 c) -1 d) $+1$

  e) N_2O f) $2NH_2OH + 4Fe^{3+} \longrightarrow 4Fe^{2+} + N_2O + H_2O + 4H^+$

22. 82.3% **23.** b) 0.74 mol dm^{-3}

Exercise 19

1. 9.37×10^{-3} mol dm^{-3} **2.** 78 ppm

3. 60.0% CaO 40.0% MgO   **4.** 18

Exercise 20

1. $1.48 \times 10^{-2}\,\text{mol dm}^{-3}$
2. 49.7%
3. 34.6% NaCl   65.4% NaBr
4. $2.77 \times 10^{-2}\,\text{mol dm}^{-3}$
5. a) $1.58 \times 10^{-2}\,\text{mol dm}^{-3}$
   b) $4.97 \times 10^{-2}\,\text{mol dm}^{-3}$
6. 55.3%

Exercise 21

1. c) $5.0 \times 10^{-8}\,\text{mol l}^{-1}$
2. g) $1.0 \times 10^{-2}\,\text{mol}$ h) i) 14%   ii) 34%
3. b) i) $MnO_4^-(aq) + 8H^+(aq) + 5e^- \longrightarrow Mn^{2+}(aq) + 4H_2O(l)$
      ii) $Fe^{2+}(aq) \longrightarrow Fe^{3+}(aq) + e^-$
      iii) $C_2O_4^{2-}(aq) \longrightarrow 2CO_2(g) + 2e^-$
      $3MnO_4^- + 5Fe^{2+} + 5C_2O_4^{2-} + 24H^+ \longrightarrow 3Mn^{2+} + 5Fe^{3+} + 10CO_2 + 12H_2O$
      Volume $= 41.7\,\text{cm}^3$
4. b) i) $+5$
      ii) 1) $2IO_3^- + 12H^+ + 10e^- \longrightarrow I_2 + 6H_2O$
         2) $2I^- \longrightarrow I_2 + 2e^-$
         $IO_3^- + 5I^- + 6H^+ \longrightarrow 3I_2 + 3H_2O$
      iii) $0.383\,\text{g}$
5. c) $Fe^{2+}(aq) \longrightarrow Fe^{3+}(aq) + e^-$
      $MnO_4^-(aq) + 8H^+(aq) + 5e^- \longrightarrow Mn^{2+}(aq) + 4H_2O(l)$
      $MnO_4^-(aq) + 8H^+(aq) + 5Fe^{2+}(aq) \longrightarrow Mn^{2+}(aq) + 5Fe^{3+}(aq) + 4H_2O(l)$
   d) 4.67%
6. b) $ICl + KI \longrightarrow I_2 + KCl$ c) $4.0 \times 10^{-3}\,\text{mol}$
   d) $2.0 \times 10^{-3}\,\text{mol } I_2,\ 2.0 \times 10^{-3}\,\text{mol ICl}$ e) $0.5 \times 10^{-3}\,\text{mol ICl}$
   f) 100
7. b) 63.5%
8. a) P is $NaClO_2$ $3NaClO_2 \longrightarrow NaCl + 2NaClO_3$
   b) First: $2NO_2^-(aq) + 4H^+(aq) + 2I^-(aq) \longrightarrow I_2(aq) + 2NO(g) + 2H_2O(l)$
      Second: $2NO_3^-(aq) + 8H^+(aq) + 6I^-(aq) \longrightarrow 3I_2(aq) + 2NO(g) + 2H_2O(l)$
      Air oxidises NO_2^- to NO_3^-, from $+3$ oxidation state to $+5$ oxidation state. Instead
      of changing from $+3$ in NO_2^- to $+2$ in NO, N changes from $+5$ in NO_3^- to $+2$ in NO,
      and can therefore oxidise three times the amount of iodide ion.
9. d) M^{2+}
10. a) A is ClO_2
    b) In 2. chlorate(III), ClO_2^-, is formed. In 5. ClO_2 disproportionates to form HCl and
       $HClO_3$, chloric(V) acid
    c) i) 2. $ClO_2 + e^- \longrightarrow ClO_2^-$
          $(+4)$ $(+3)$

          $O_2^{2-} \longrightarrow O_2 + 2e^-$
          (-1) (0)

          Adding the two half-equations,
          $2ClO_2 + O_2^{2-} \longrightarrow 2ClO_2^- + O_2$

          3. $ClO_2^- + 4H^+ + 4e^- \longrightarrow Cl^- + 2H_2O$
          $(+3)$ (-1)

          $2I^- \longrightarrow I_2 + 2e^-$
          (-1) (0)

          Adding the two half-equations,
          $ClO_2^-(aq) + 4H^+(aq) + 4I^-(aq) \longrightarrow Cl^-(aq) + 2I_2(aq) + 2H_2O(l)$

          4. $2I^- \longrightarrow I_2 + 2e^-$
          (-1) (0)

          $NO_3^- + 2H^+ + e^- \longrightarrow NO_2 + H_2O$
          $(+5)$ $(+4)$
          Adding the two half-equations,
          $2I^-(aq) + 2NO_3^-(aq) + 2H^+(aq) \longrightarrow I_2(g) + 2NO_2(g) + 2H_2O(l)$

5. a) $ClO_2 + 4H^+ + 5e^- \longrightarrow Cl^- + 2H_2O$
   (+4)   (−1)

   $ClO_2 + OH^- \longrightarrow ClO_3^- + H^+ + e^-$
   (+4)   (+5)

   Adding the two equations,
   $6ClO_2 + 5OH^- \longrightarrow Cl^- + 5ClO_3^- + H^+ + 2H_2O$

   b) $ClO_3^- + 6H^+ + 6e^- \longrightarrow Cl^- + 3H_2O$
   (+5)   (−1)

   $2I^- \longrightarrow I_2 + 2e^-$
   (−1)   (0)

   Adding the two equations,
   $ClO_3^-(aq) + 6H^+(aq) + 6I^-(aq) \longrightarrow Cl^- + 3I_2(aq) + 3H_2O(l)$

11. b) ii) 95%

12. a) i) 2.5 mol   ii) 6.0×10^{-4} mol   iii) 1.5×10^{-3} mol   iv) 190 ppm

13. c) $MgBa(CO_3)_2$

14. a) i) 0.0129 mol   ii) 0.0905 mol   iii) 7
    b) i) 65.9 mg $ZnSO_4 \cdot 7H_2O$ ii) 4.58×10^{-2} mol dm^{-3}
    c) i) 1:1   ii) to ensure that edta is ionised   iii) Sodium zincate would be formed

15. a) i) $2MnO_4^- + 16H^+ + 5C_2O_4^{2-} \longrightarrow 2Mn^{2+} + 8H_2O + 10CO_2$ ii) 2.59 g
    b) i) $Mn^{2+}(aq)$ is precipitated as $MnCO_3(s)$. When the carbonate is heated in air, it
       dissociates to form CO_2 and MnO, which is immediately oxidised to MnO_2
       ii) 0.537 g
    c) $K_4Mn(C_2O_4)_3 \cdot 2H_2O$
    d) i) +2   ii)

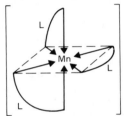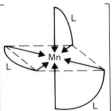

L = Ligand and stands for

16. b) ii) 1) Loss in mass corresponds to $PbCl_4(s) \longrightarrow PbCl_2(s) + Cl_2(g)$
       2) When $PbCl_4(s) + H_2O(l) \longrightarrow A(s) + B(aq)$,
          1.43×10^{-3} mol $PbCl_4$ forms 5.73×10^{-3} mol Cl^-, that is 1 mol $PbCl_4$ forms
          4 mol Cl^-. Therefore $PbCl_4(s) + H_2O(l) \longrightarrow 4HCl(aq) + PbO_2(s)$

17. d) i) 0.83 mol   ii) 16.4-volume

CHAPTER 9

Exercise 22

1. 85.6 u   2. 69.8 u   3. 24.3
4. 3 $^{35}Cl:1\ ^{37}Cl$; 35.5 u   5. 6.93
6. 1, ^1H 2, ^2H 3, $^1H^2H$ 4, $^2H_2$ 17 $^{16}O^1H$ 18, $^{16}O^2H$ and $^1H_2\ ^{16}O$ 19, $^1H^2H^{16}O$
   20, $^2H_2\ ^{16}O$
7. 39.1 u
8. ^{63}Cu ^{65}Cu ^{63}CuO ^{65}CuO $^{63}CuNO_3$ $^{65}CuNO_3$ $^{63}Cu(NO_3)_2$ $^{65}Cu(NO_3)_2$

Exercise 23

1. c) i) A is CH_3CO_2H B is CH_3COCH_3 C is $CH_3CH_2CH_2CH_3$
    ii) A peaks: $60 =$ molecular ion, $45 = CO_2H$, $15 = CH_3$
       B peaks: $58 =$ molecular ion, $43 = CH_3CO$, $28 = CO$, $15 = CH_3$
       C peaks: $58 =$ molecular ion, $43 = CH_3CH_2CH_2$, $29 = CH_3CH_2$, $15 = CH_3$

2. b) ii) X is $\langle\bigcirc\rangle$—CH_2Cl Y is Cl—$\langle\bigcirc\rangle$—CH_3 or the 1, 2- or 1, 3 isomer

   d) Peaks at 126 and 128 for molecular ions, peak at 15 for CH_3 not found in X spectrum, peaks at 35 and 37 for Cl, peak at 76 for C_6H_4, no peaks at 49 and 51 for CH_2Cl as found in the mass spectrum of X
3. d) i) ^{35}Cl, ^{37}Cl, $^{35}Cl_2$, $^{35}Cl^{37}Cl$, $^{37}Cl_2$
    ii) Chlorine consists of more Cl-35 atoms than Cl-37 atoms
   e) The peak at $m/e = 11$ s is $4 \times$ the height of the peak at $m/e = 10$. Peaks are observed for B^{2+} at $m/e = 5$ and 5.5 and for B^{3+} at $m/e = 3.33$ and 3.67
4. b) i) 80.0   ii) Br_2^+ peaks at 158, 160, 162   relative heights $1:2:1$
5. c) Empirical formula C_4H_8O; peaks at $15 = CH_3$, $29 = CH_3CH_2$ or CHO, $43 = CH_3CH_2CH_2$, $(CH_3)_2CH$ or CH_3CO, $57 = CH_3CH_2CO$, $72 =$ molecular ion. Therefore the molecular formula is C_4H_8O. (The compound could be $CH_3CH_2CH_2CHO$, $(CH_3)_2CHCHO$ or $CH_3CH_2COCH_3$. The small peak at 57 shows the structural formula to be $CH_3CH_2COCH_3$.)
6. b) i) $47 = C^{35}Cl$, $49 = C^{37}Cl$, $82 = C^{35}Cl_2$, $84 = C^{35}Cl^{37}Cl$, $86 = C^{37}Cl_2$. The relative heights of the peaks at 82, 84, 86 are $9:6:1$.
    ii) Since ^{35}Cl is 3 times as abundant as ^{37}Cl (the 47 peak is 3 times the height of the 49 peak), $C^{35}Cl_2$ is 9 times as abundant as $C^{37}Cl_2$ and $C^{35}Cl^{37}Cl$ is 3 times as abundant as $C^{37}Cl_2$
    iii) 83 peak is $^{13}C^{35}Cl_2$ 48 peak is $^{13}C^{35}Cl$ iv) $[C^{35}Cl_2]^{2+}$ ion
   c) i) $m/e = 16$ peak is CH_4^+, $m/e = 15$ peak is CH_3^+ ii) $347\ kJ\ mol^{-1}$
   d) $C_2H_5NO_2$ (add $A_r(C) + A_r(N) + A_r(O) + A_r(H)$; the total is 43.005 805; to make up the difference add 32, i.e. COH_4; the total then comes to 75.032 015)
    Structural formula $H_2NCH_2CO_2H$

Exercise 24

1. a) $a = 1$, $b = 0$ b) $a = 17$, $b = 8$ c) $a = 4$, $b = 2$ d) $a = 210$, $b = 83$
   e) $a = 4$, $b = 2$ f) $a = 4$, $b = 2$ g) $a = 14$, $b = 6$ h) $a = 16$, $b = 7$
   i) $a = 207$, $b = 83$ j) $a = 4$, $b = 11$ k) $q = 0$, $p = 1$, $r = 35$
   l) $c = 3$, $d = 1$

Exercise 25

1. a) alpha particle $4m_p$, $+2e$ beta particle $m_p/2000$, $-e$
   b) i) Mass number decreases by 4; atomic number decreases by 2.
    ii) No change in mass number; atomic number increases by 1.
   c) i) Mass number of $X = 207$ ii) Atomic number of $X = 82$ iii) X is lead
2. a) ii) 24.328
   b) i) Q is ^{234}Th R is ^{41}K ii) ionisation; $^{40}K \longrightarrow {}^{40}K^+ + e^-$

3. b)

	X	Y	Z
Mass number	228	228	228
Atomic number	88	89	90

   c) CO_2 peaks at $44 = {}^{12}C^{16}O_2$, $48 = {}^{12}C^{18}O_2$, $46 = {}^{12}C^{16}O^{18}O$ have relative intensities $16:1:4$

CHAPTER 10

Exercise 26

1. a) $186 \, cm^3$　　　b) $387 \, cm^3$　　　c) $6.23 \, dm^3$　　　d) $132 \, cm^3$
　e) $2.43 \, dm^3$
2. $1.75 \times 10^5 \, N \, m^{-2}$　　3. $0.943 \, dm^3$　　4. $484 \, K$　　5. $586 \, cm^3$
6. a) $185 \, cm^3$　　　b) $36.7 \, cm^3$　　　c) $6.46 \, dm^3$　　　d) $3.83 \, dm^3$
　e) $436 \, cm^3$

Exercise 27

1. a) 46　　　　　　b) NO_2　　　　　　2. $59 \, cm$　　　　3. 160
4. 4　　　　　　　　5. $A = H_2$　$B = O_3$　　6. $48 \, s$
7. $16.3 \, cm^3 \, min^{-1}$　　8. $24.9 \, cm^3$　　　9. $CO, 25\%$　$CO_2, 75\%$
10. $NO_2, 43.7\%$　　$N_2O_4, 56.3\%$

Exercise 28

1. $44.1 \, g \, mol^{-1}$　　2. $39.8 \, g \, mol^{-1}$　　3. $6.01 \, dm^3$　　4. $83.8 \, g \, mol^{-1}$
5. $0.583 \, mol$　　　6. $44.0 \, g \, mol^{-1}$　　7. $6.18 \, dm^3$　　8. $7.54 \times 10^{-2} \, mol$

Exercise 29

1. a) $176.5 \, cm^3$　　　b) $207 \, cm^3$　　　c) $26.3 \, dm^3$
2. $2.73 \times 10^5 \, N \, m^{-2}$　　3. $2.53 \times 10^4 \, N \, m^{-2}$　　4. $4.50 \times 10^5 \, N \, m^{-2}$
5. a) $p(N_2) = 3.00 \times 10^4 \, N \, m^{-2}$　$p(O_2) = 2.63 \times 10^4 \, N \, m^{-2}$,　$p(CO_2) = 1.88 \times 10^4 \, N \, m^{-2}$
　b) $p(N_2) = 3.00 \times 10^4 \, N \, m^{-2}$　$p(O_2) = 2.63 \times 10^4 \, N \, m^{-2}$
6. a) $p(NH_3) = 6.00 \times 10^4 \, N \, m^{-2}$　$p(H_2) = 3.75 \times 10^4 \, N \, m^{-2}$　$p(N_2) = 5.25 \times 10^4 \, N \, m^{-2}$
　b) $p(H_2) = 3.75 \times 10^4 \, N \, m^{-2}$　$p(N_2) = 5.25 \times 10^4 \, N \, m^{-2}$

Exercise 30

1. $84.4 \, g \, mol^{-1}$　　2. $1840 \, m \, s^{-1}$　　3. $3410 \, J \, mol^{-1}$　　4. $413 \, m \, s^{-1}$
5. $44.1 \, g \, mol^{-1}$　　6. 2.02　　　7. $242 \, m \, s^{-1}$
8. a) 4.47　　　　b) $5460 \, K$

Exercise 31

1. $4.52 \times 10^{-5} \, g$　　2. b) i) 16　　ii) 64
3. a) $24.9 \, dm^3$　　　b) 46
　c) i) C_2H_6O　　ii) C_2H_5OH and CH_3OCH_3
　　iii) $2C_2H_6O \; + \; 7O_2 \longrightarrow \; 4CO_2 \; + \; 6H_2O$
　d) i) C_2H_5OH　　ii) $2C_2H_5OH(l) \; + \; 2Na(s) \longrightarrow 2C_2H_5ONa(s) \; + \; H_2(g)$
4. a) $2490 \, J \, mol^{-1}$　　b) iii) 36.5
5. a) CH_2　　　　　b) 56, C_4H_8
　c) i)

CHAPTER 11

Exercise 32

1. $343 \, g \, mol^{-1}$　　2. PF_5　　　3. $90 \, g \, mol^{-1}$　　4. $64.6 \, g \, mol^{-1}$
5. $134 \, g \, mol^{-1}$
6. a) $46 \, g \, mol^{-1}$　　　b) $58 \, g \, mol^{-1}$　　　c) $74 \, g \, mol^{-1}$

Exercise 33

1. 0.400   2. 0.500   3. 73.9%   4. 0.300
5. 0.68

Exercise 34

1. 3.39 g   2. 1.98 g   3. $1.62 \times 10^{17}\,cm^{-3}$ 4. 3.17 kPa

Exercise 35

1. a) $5.19 \times 10^5\,N\,m^{-2}$ b) $1.98 \times 10^6\,N\,m^{-2}$ c) $2.44 \times 10^6\,N\,m^{-2}$
2. a) $68.7\,g\,mol^{-1}$ b) $166\,g\,mol^{-1}$ c) $259\,g\,mol^{-1}$
3. 50   4. 2400   5. $1.24 \times 10^3\,N\,m^{-2}$
6. a) $2.01 \times 10^5\,Pa$ b) $5.00\,g\,dm^{-3}$ 7. $0.274\,mol\,dm^{-3}$, $16.0\,g\,dm^{-3}$
8. 1.39×10^4

Exercise 36

1. $20\,690\,N\,m^{-2}$ 2. a) $40\,000\,N\,m^{-2}$ b) 0.300
3. a) $1.60 \times 10^3\,Pa$ b) $6.38 \times 10^2\,Pa$ 4. a) 38.0 kPa   b) 33.6 kPa

Exercise 37

1. 26.9%   2. 27.8%   3. 32.3 g

Exercise 38

1. 13.6 g   2. 4.76 g   3. 20.0   4. 0.0514 mol
5. a) 4.74 g   b) 4.95 g   6. 25.0   7. 1.18×10^{-2}
8. a) 3.0 g   b) 3.5 g   9. a) 4.75 g   b) 4.95 g

Exercise 39

1. c) i) 0.389   ii) p(hexane) $= 7850\,Pa$, p(heptane) $= 3680\,Pa$
   iii) Total $= 11\,530\,Pa$ iv) 0.681
   v) Hexane has a higher vapour pressure than heptane. Fractional distillation
   d) The vapour pressure of the mixture is reduced by forces of attraction between
   molecules of the two compounds in the liquid phase
2. b) Amount of U in solvent/Amount of U in water $= 20$
3. b) i) $74\,g\,mol^{-1}$ ii) Empirical formula $C_4H_{10}O$ molecular formula $C_4H_{10}O$
   iii) a primary alcohol and a secondary alcohol
   iv) A: 31 peak $= CH_2OH$, 43 peak $= C_3H_7$
       B: 29 peak $= C_2H_5$, 45 peak $= CH_3CHOH$, 59 peak $= C_2H_5CHOH$
       Therefore $A = CH_3CH_2CH_2CH_2OH$ and $B = CH_3CH_2CH(OH)CH_3$
4. a) 74
5. b) $M_r = 29.1$. M_r is higher than the expected value of 27.0 because some dimerisation
   of HCN occurs.
6. c) 65 °C   e) 0.23 g   7. b) i) 16.7 g
8. b) 30 mm   c) 87 mm
   d) i) positive
   ii) Intermolecular forces of attraction are lower in the mixture than in the pure liquids
   iii) minimum   iv) e.g. methylbenzene
9. p(methanol) $= 54\,mm$, p(ethanol) $= 15\,mm$
10. c) Add 100 cm³ dichloromethane; then mass of caffeine extracted $= 5.00\,g$. Add 50 cm³
    dichloromethane; then mass of caffeine extracted $= 3.33$ g. Add a further 50 cm³
    solvent and a further 2.22 g of caffeine is extracted, a total of 5.55 g
11. c) 326, Fe_2Cl_6
    d) i) $Fe^{3+} + e^- \longrightarrow Fe^{2+}$ ii) $S^{2-} \longrightarrow S + 2e^-$
12. b) ii) 1) 3.33 mol   2) 3.75 mol

CHAPTER 12

Exercise 40

SECTION 1

1. a)   2. a)   3. b)   4. d)
5. b)

SECTION 2

1. 0.265 g
2. 0.403 g   a) doubled   b) doubled   c) unchanged
3. 1.24 g Ca   2.21 g Cl_2 4. 0.0560 g
5. a) 0.0672 A   b) 23.1 cm^3 c) 0.195 g   6. 268 minutes
7. 0.454 A   8. 1.77 mol dm^{-3} 9. 2482 hours
10. 2.14 dm^3 O_2 4.28 dm^3 H_2 11. 1.84 $\times 10^4$ C   12. 1050 s

Exercise 41

1. 1.75 $\times 10^{-5}$ mol dm^{-3}
2. a) 0.0271   b) 6.78 $\times 10^{-4}$ mol dm^{-3} c) 1.89 $\times 10^{-5}$ mol dm^{-3}
3. a) 0.230   b) 1.37 $\times 10^{-3}$ mol dm^{-3}
4. 1.75 $\times 10^{-4}$ mol dm^{-3}
5. a) 0.0256   b) 2.02 $\times 10^{-5}$ mol dm^{-3}

Exercise 42

1. | pH | pOH | | pH | pOH | | pH | pOH | | pH | pOH |
    |----|-----|--|----|-----|--|----|-----|--|----|-----|
    | a) 8 | 6 | | b) 4 | 10 | | c) 7 | 7 | | d) 2.2 | 11.8 |
    | e) 4.5 | 9.5 | | f) 1.5 | 12.5 | | g) 0.60 | 13.4 | | h) 8.3 | 5.7 |
    | i) 6.2 | 7.8 | | j) 1.0 | 13.0 | | | | | | |

2. a) 12   b) 11   c) 6.0   d) 12.7
   e) 11   f) 12.9   g) 12   h) 9.7
   i) 6.8   j) 4.6
3. In mol dm^{-3}, the values are:
   a) 1.00   b) 5.01 $\times 10^{-5}$ c) 4.47 $\times 10^{-3}$ d) 0.0132
   e) 7.08 $\times 10^{-5}$ f) 1.45 $\times 10^{-8}$ g) 6.17 $\times 10^{-10}$ h) 2.00 $\times 10^{-14}$
   i) 3.16 $\times 10^{-1}$ j) 2.34 $\times 10^{-3}$
4. a) 0.784   b) 1.05   c) 13.3   d) 12.7
   e) 13.4
5. 2.52   6. a) 1.00 $\times 10^{-6}$ mol dm^{-3} b) 6.00
7. 9.92
8. a) 3.7 $\times 10^{-8}$ b) 1.74 $\times 10^{-5}$ c) 3.96 $\times 10^{-10}$ d) 1.3 $\times 10^{-5}$
   (all in mol dm^{-3})
9. a) 1.81 $\times 10^{-5}$ b) 3.97 $\times 10^{-10}$ c) 1.43 $\times 10^{-5}$ d) 2.00 $\times 10^{-9}$
   (all in mol dm^{-3})
10. a) 2.28 $\times 10^{-11}$ b) 5.62 $\times 10^{-10}$ c) 1.86 $\times 10^{-11}$ d) 4.24 $\times 10^{-10}$
    (all in mol dm^{-3})
11. a) 4.46 $\times 10^{-4}$ b) 2.41 $\times 10^{-2}$ c) 1.89 $\times 10^{-2}$ d) 7.93 $\times 10^{-2}$
    (all in mol dm^{-3})
12. a) 2.00   b) 12.0   c) 2.30   13. 0.0110%
14. 11.1   15. a) 10^{-6} b) 6
16. pH = 3.0   a) 9.0 cm^3 b) 0.90 cm^3

Exercise 43

1. a   2. 1.00 mole   3. a) 3.34   b) 3.94
4. a) 4.73   b) 0.117 mol

Exercise 44

1. a) $1.69 \times 10^{-28} \, mol^2 \, dm^{-6}$
   b) $3.99 \times 10^{-19} \, mol^2 \, dm^{-6}$
   c) $5.93 \times 10^{-51} \, mol^3 \, dm^{-9}$
   d) $5.00 \times 10^{-16} \, mol^3 \, dm^{-9}$
2. a) $2.52 \times 10^{-27} \, mol^2 \, dm^{-6}$
   b) $8.29 \times 10^{-17} \, mol^2 \, dm^{-6}$
   c) $1.07 \times 10^{-10} \, mol^2 \, dm^{-6}$
   d) $2.68 \times 10^{-11} \, mol^3 \, dm^{-6}$
   e) $1.25 \times 10^{-16} \, mol^2 \, dm^{-6}$
3. a) $6.3 \times 10^{-34} \, mol \, dm^{-3}$
   b) $6.3 \times 10^{-43} \, mol \, dm^{-3}$
   c) $3.2 \times 10^{-14} \, mol \, dm^{-3}$
   d) $2.2 \times 10^{-30} \, mol \, dm^{-3}$
   e) $1.6 \times 10^{-46} \, mol \, dm^{-3}$
   f) $6.3 \times 10^{-12} \, mol \, dm^{-3}$
4. a) Yes   b) No   c) Yes   d) No
   e) Yes   f) No
5. CdS and NiS
6. $K_{sp}(AgCl) = 1.96 \times 10^{-10} \, mol^2 \, dm^{-6}$
   $K_{sp}(Ag_2CrO_4) = 3.61 \times 10^{-12} \, mol^3 \, dm^{-9}$
   a) $[Ag^+] = 1.96 \times 10^{-9} \, mol \, dm^{-3}$
   b) $[Ag^+] = 2.69 \times 10^{-5} \, mol \, dm^{-3}$
7. $CaCO_3$ 9.99 g   8. $0.3 \, ion \, dm^{-3}$ 9. PbI_2 45.6 g
10. a) $8.12 \times 10^{-3} \, g \, dm^{-3}$ b) $6.38 \times 10^{-8} \, g \, dm^{-3}$ c) $9.63 \times 10^{-4} \, g \, dm^{-3}$
11. a) $6.3 \times 10^{-4} \, mol \, dm^{-3}$ b) $4.0 \times 10^{-6} \, mol \, dm^{-3}$
12. a) 1.2×10^{-3} b) $1.8 \times 10^{-7} \, mol \, dm^{-3}$
13. a) $2.15 \times 10^{-4} \, mol \, dm^{-3}$ b) $4.00 \times 10^{-5} \, mol \, dm^{-3}$
    c) $7.12 \times 10^{-8} \, mol \, dm^{-3}$

Exercise 45

1. Ag, I^- 2. I_2, Fe^{3+}
3. a) $+0.40 \, V$ b) $+0.26 \, V$ c) $-0.27 \, V$
4. a) $+0.94 \, V$ b) $+0.44 \, V$ c) $-0.67 \, V$ d) $+0.78 \, V$
   e) $+0.63 \, V$
5. $Fe(s) + Fe^{3+}(aq) \longrightarrow 2Fe^{2+}(aq)$
6. $Cr_2O_7{}^{2-}(aq) + 14H^+(aq) + 6Fe^{3+}(aq) \longrightarrow 2Cr^{3+}(aq) + 6Fe^{3+}(aq) + 14H_2O$
7. a) $2Fe^{3+}(aq) + 2I^-(aq) \longrightarrow 2Fe^{2+}(aq) + I_2(aq)$
   b) $2Ag^+(aq) + Cu(s) \longrightarrow 2Ag(s) + Cu^{2+}(aq)$
   c) No reaction   d) No reaction
   e) $Br_2(aq) + 2Fe^{2+}(aq) \longrightarrow 2Br^-(aq) + 2Fe^{3+}(aq)$
      i) Cl_2 and Br_2 ii) Cl_2, Br_2 and I_2

Exercise 46

1. c) $-0.010 \, V$ d) from Sn to Pb because Sn is the stronger reducing agent
2. b) ii) $6.6 \times 10^{-16} \, mol^2 \, dm^{-6}$ c) 13.9
3. a) iii) $-0.28 \, V$ iv) from cobalt to copper
   b) i) no change   ii) deposition of copper   c) $+0.34 \, V$
4. a) Curve *I*: *Q* added to *D*
      Curve *II*: *S* added to *A*
      Curve *III*: *S* added to *B*
      Curve *IV*: *T* added to *B*
   b) $K_a = 1.55 \times 10^{-5} \, mol \, dm^{-3}$ (find the gradient $1000\alpha^2/(1/c) = 1000\alpha^2 c$ and use the value in $K_a = \alpha^2 c$)
5. b) $4.80 \times 10^{-11} \, mol^2 \, dm^{-6}$
   c) Amount of acid $= 4.4 \times 10^{-3} \, mol$,
      $[HCO_3{}^-(aq)] = 0.151 \, mol \, dm^{-3}$, $[CO_3{}^{2-}(aq)] = 0.145 \, mol \, dm^{-3}$
6. a) ii) $3.2 \times 10^{-4} \, mol \, dm^{-3}$ b) $0.110 \, mol \, dm^{-3}$
   c) i) weak acid   ii) acid dissociation constant
      iii) $9.1 \times 10^{-7} \, mol \, dm^{-3}$
7. b) i) 1   ii) pH changes from 7.00 to 1.30   iii) 4.74
      iv) pH changes from 4.74 to 4.70

8. a) 1.0×10^{-5} mol dm^{-3} c) 20.3 g

   d) i) 4.98 ii) 2.00

9. d) 10.3 mg

10. b) 3.8

11. a) i) $+6$ ii) 0 b) Na^+ and CrO_4^{2-} e) $n = 3$

   f) $Cr^{3+}(aq) + 3e^- \longrightarrow Cr(s)$

   g) $CrO_4^{2-}(aq) + 8H^+(aq) + 3e^- \longrightarrow Cr^{3+}(aq) + 4H_2O(l)$

12. b) 1.26×10^{-3} mol dm^{-3} c) Different pK_a values

   d) No difference

   e) i) Chloroethanoic acid has a higher K_a than ethanoic acid

      ii) Cl attracts electrons, stabilising the anion

$$Cl \twoheadleftarrow CH_2 \twoheadleftarrow C \left. \begin{matrix} O \\ \vdots \\ O \end{matrix} \right\} -$$

13. a) i) $Zn(s) + CuSO_4(aq) \longrightarrow ZnSO_4(aq) + Cu(s)$ iii) -1.10 V

   iv) Zinc hydroxide and copper(II) hydroxide

   b) i) $Cu(s) \longrightarrow Cu^{2+}(aq) + 2e^-$ ii) $Cu^{2+}(aq) + 2e^- \longrightarrow Cu(s)$

   iii) Copper(II) hydroxide, blue

14. a) ii) ethanoic acid 1.74×10^{-5} mol dm^{-3}, trichloroethanoic acid 0.219 mol dm^{-3}.

      Cl atoms attract electrons and stabilise the anion

$$Cl \twoheadleftarrow \underset{\underset{Cl}{\uparrow}}{\overset{\overset{Cl}{\downarrow}}{C}} - C \left. \begin{matrix} O \\ \vdots \\ O \end{matrix} \right\} -$$

15. a) i) $K_c = \dfrac{[CrO_4^{2-}]^2[H^+]^2}{[Cr_2O_7^{2-}]}$ ii) 2.01×10^{-13} mol^3 l^{-3}

   iii) On addition of $BaCl_2(aq)$, CrO_4^{2-} ions are precipitated as $BaCrO_4(s)$ and $Cr_2O_7^{2-}$
       ions are therefore converted into CrO_4^{2-} ions and H^+ ions to keep K_c constant. The
       addition of $NaCl(aq)$ dilutes the solution and reduces $[H^+]$

   b) The gradient of the graph of r^2 against c is K_c/K_{sp}^2 and $s^2 = K_s$. $s = 3.00 \times 10^{-4}$ mol l^{-1}

16. a) 3 ppm year^{-1} b) The uptake of carbon dioxide by plants is greater in the summer

   c) $[CO_3^{2-}] = 2.5 \times 10^{-6}$ mol dm^{-3}

   d) The value calculated for the product $[Ca^{2+}][CO_3^{2-}]$ exceeds the value of K_{sp}. However,
       some CO_2 comes out of solution in the dissociation

$$H_2CO_3(aq) \rightleftharpoons CO_2(g) + H_2O(l)$$

   and this factor reduces the concentration of CO_3^{2-}

   e) $K = 1.0 \times 10^{-4}$. As more CO_2 dissolves, $[H_2CO_3]$ increases, $[HCO_3^-]$ increases and
       $[H^+]$ increases.

   As $[H^+]$ increases the tendency is for H^+ to react with HCO_3^- to form
   $H_2CO_3 + CO_3^{2-}$ rather than for HCO_3^- to dissociate to form $H^+ + CO_3^{2-}$.
   The equilibrium constants are 1×10^{-4} and 5×10^{-11} mol dm^{-3} respectively. The
   increasing $[H^+]$ is therefore absorbed to form $H_2CO_3 + CO_3^{2-}$, and there is no
   great increase in $[H^+]$ which would lead to dissolving of rocks.

   f) $CO_2 + H_2O \rightleftharpoons H_2CO_3$

$$K = 5 \times 10^{-7}\ \text{mol dm}^{-3}$$

   $K = 5 \times 10^{-11}$ mol dm^{-3} $K = 1.0 \times 10^{-4}$

   $2H^+ + CO_3^{2-} \rightleftharpoons H^+ + HCO_3^- \rightleftharpoons H_2CO_3 + CO_3^{2-}$

   g) Less CO_2 will dissolve; the greenhouse effect will increase. Less $CaCO_3$ will dissolve
       and $[CO_3^{2-}]$ will decrease. As a result, the equilibrium
       $2HCO_3^- \rightleftharpoons H_2CO_3 + CO_3^{2-}$ will move towards the right. As more H_2CO_3
       is formed, it dissociates partially to release CO_2, which will increase the greenhouse
       effect

CHAPTER 13

Exercise 47

1. $-56.6\,kJ\,mol^{-1}$ 2. $-58.4\,kJ\,mol^{-1}$ 3. $-49.3\,kJ\,mol^{-1}$ 4. $-56.4\,kJ\,mol^{-1}$
5. $-59.1\,kJ\,mol^{-1}$

Exercise 48

1. a) $-484\,kJ\,mol^{-1}$ b) $108\,kJ\,mol^{-1}$ c) -75 and $+33\,kJ\,mol^{-1}$
d) $-118\,kJ\,mol^{-1}$ e) $-11\,kJ\,mol^{-1}$ f) -106 and $+504\,kJ\,mol^{-1}$
g) $-251, -246, -471$ and $-152\,kJ\,mol^{-1}$
2. $-78.2\,kJ\,mol^{-1}$ 3. a) -297 b) -394 c) $-262\,kJ\,mol^{-1}$
4. $-64\,kJ\,mol^{-1}$ 5. $-847\,kJ\,mol^{-1}$ exothermic 6. $-0.30\,kJ\,mol^{-1}$
7. $184\,kJ\,mol^{-1}$ 8. $-106\,kJ\,mol^{-1}$ Mg reduces Al_2O_3
9. a) $-890\,kJ\,mol^{-1}$ b) $+317\,kJ\,mol^{-1}$ c) $-55\,kJ\,mol^{-1}$
d) i) $-1.43 \times 10^5\,kJ$ ii) $-2.97 \times 10^4\,kJ$ iii) $-4.84 \times 10^4\,kJ$
10. $416\,kJ\,mol^{-1}$
11. a) $C_2H_6, -76$; $C_2H_4, +48\,kJ\,mol^{-1}$ b) $-95\,kJ\,mol^{-1}$ c) $-200\,kJ\,mol^{-1}$
d) $-343\,kJ\,mol^{-1}$
e) $+264\,kJ\,mol^{-1}$ The difference is the 'bond delocalisation energy' of benzene
f) $+136\,kJ\,mol^{-1}$ $28\,kJ\,mol^{-1}$ higher than the value from combustion Butadiene
is more stable than it is calculated to be because it is stabilised by bond delocalisation
g) i) $+74\,kJ\,mol^{-1}$ ii) $-20\,kJ\,mol^{-1}$ Reaction ii) will occur
12. $-362\,kJ\,mol^{-1}$ 13. $-440\,kJ\,mol^{-1}$ 14. $-372\,kJ\,mol^{-1}$ 15. b) $-380\,kJ\,mol^{-1}$
16. $-775\,kJ\,mol^{-1}$
17. a) $-387\,kJ\,mol^{-1}$ b) $-246\,kJ\,mol^{-1}$
$\Delta H_f^{\ominus}(CaCl_2)$ has a larger negative value than $\Delta H_f^{\ominus}(CaCl)$
18. a) Solubility (LiCl) > Solubility (NaCl) since $(\Delta H_{lattice}^{\ominus} + \Delta H_{hydration}^{\ominus})$ has a more
negative value for LiCl than for NaCl.
b) Solubility (NaCl) > Solubility (NaF) since $(\Delta H_{lattice}^{\ominus} + \Delta H_{hydration}^{\ominus})$ has a positive
value for NaF and a negative value for NaCl.
19. $-367\,kJ\,mol^{-1}$
20. $\Delta H_c^{\ominus}$ is $< 3\Delta H_a^{\ominus}$ because bond delocalisation makes benzene more stable than
calculated from bond energy terms, and less enthalpy than expected is released when
benzene $\longrightarrow$ cyclohexane
The hydrogenation of the first double bond destroys the bond delocalisation in
benzene, and ΔH_b is therefore positive, showing that the enthalpy content of the system
has increased

Exercise 49

1. In $J\,K^{-1}\,mol^{-1}$: a) $+20.0$ b) -199 c) -163.5
d) -121 e) $+176$ f) -90.1
g) $+285$ h) $+681$
2. a) $+$ b) $+$ c) $-$ d) $-$
e) $+$ f) $-$
3. a) $+0.202\,kJ\,mol^{-1}$ at $25\,^{\circ}C$ $+0.127\,kJ\,mol^{-1}$ at $100\,^{\circ}C$ above $227\,^{\circ}C$
b) $+2.56\,kJ\,mol^{-1}$ at $25\,^{\circ}C$ above $491\,^{\circ}C$
4. a) $-2.91\,kJ\,mol^{-1}$ b) $+2.91\,kJ\,mol^{-1}$ trans

Exercise 50

1. b) ii) $I_2(g) \longrightarrow 2I(g)$ $\Delta H^{\ominus} = +151.1\,kJ\,mol^{-1}$
iii) $I_2(s) \longrightarrow I_2(g)$ $\Delta H^{\ominus} = +62.5\,kJ\,mol^{-1}$
2. $-3440\,kJ\,mol^{-1}$. The estimated value of $\Delta H^{\ominus}_c$ is greater than the measured value
because benzene is stabilised by the energy of delocalisation

3. b) i) $+376\,\text{kJ mol}^{-1}$

 ii) $\Delta H^{\ominus}(\text{CaCl}_2) = -940\,\text{kJ mol}^{-1}$, $\Delta H^{\ominus}(\text{CaCl}_3) = +1193\,\text{kJ mol}^{-1}$
 Since $\Delta H^{\ominus}(\text{CaCl}_2)$ is negative and $\Delta H^{\ominus}(\text{CaCl}_3)$ is positive, CaCl_2 is formed

 iii) Lattice-breaking in NaCl(s) is strongly endothermic as the ions must be separated. The enthalpies of hydration of Na^+ and Cl^- are negative, and when NaCl(s) dissolves the negative enthalpies of hydration partially compensate for the positive enthalpy of lattice-breaking

4. 2500 K

5. a) i) $+127.5\,\text{kJ mol}^{-1}$
 Increase in temperature will increase the yield because the reaction is endothermic. Increase in pressure will decrease the yield because the reaction involves an increase in the number of moles of gas

 ii) Reaction (2) requires the input of more heat per mole of $\text{CF}_2{=}\text{CF}_2$. The dangerous pollutant HF is produced

 b) i) $\text{CF}_4(\text{g}) \longrightarrow \text{C}(\text{g}) + 4\text{F}(\text{g})$ $\Delta H^{\ominus} = +1707\,\text{kJ mol}^{-1}$
 $\text{CCl}_4(\text{g}) \longrightarrow \text{C}(\text{g}) + 4\text{Cl}(\text{g})$ $\Delta H^{\ominus} = +1318\,\text{kJ mol}^{-1}$
 $\text{C}-\text{F}$ bond energy $= 426.8\,\text{kJ mol}^{-1}$, $\text{C}-\text{Cl}$ bond energy $= 329.5\,\text{kJ mol}^{-1}$.
 Ease of breaking bonds: $\text{C}-\text{Cl} > \text{C}-\text{H} > \text{C}-\text{F}$

 ii) $\text{C}-\text{Cl}$ bonds are more likely to break, producing $\text{Cl}\cdot$ radicals

6. a) i) $v + w + x + y + z$ ii) $w/2$ iii) $y/2$ iv) $z/2$
 b) i) y ii) x
 c) i) $+312\,\text{kJ mol}^{-1}$ ii) $+156\,\text{kJ mol}^{-1}$
 d) $\text{KH}(\text{s}) + \text{H}_2\text{O}(\text{l}) \longrightarrow \text{H}_2(\text{g}) + \text{KOH}(\text{aq})$ ii) 39

7. b) $\Delta H^{\ominus} = -168\,\text{kJ mol}^{-1}$ of reaction
 c) i) $+28\,\text{kJ mol}^{-1}$
 ii) Since $\Delta H^{\ominus}$ is positive, the reaction is endothermic and less likely to be used than the exothermic dimerisation of ethene
 iii) Less energy is required to break the $\text{C}-\text{C}$ bond in cyclobutane than in butane. This shows that the bonds are under some strain in cyclobutane: the $\text{C}-\text{C}-\text{C}$ bond angle is less than the tetrahedral value

8. b) i) $+720\,\text{kJ mol}^{-1}$ ii) $-83\,\text{kJ mol}^{-1}$ iii) Yes, $\Delta G^{\ominus}$ is negative at this temperature
 iv) CO is poisonous and also a 'greenhouse gas'

9. c) i) 1670 J   ii) $+66.7\,\text{kJ mol}^{-1}$
 d) $+55.4\,\text{kJ mol}^{-1}$
 e) i) The sign of the free energy change determines the feasibility of the reaction; the activation energy determines the rate of the reaction.
 ii) This is true. To break the $\text{C}-\text{Cl}$ bond in halogenoalkanes requires drastic conditions, e.g. refluxing with a concentrated solution of an alkali
 iii) No, a catalyst does not affect the value of $\Delta H^{\ominus}$; it lowers the activation energy

10. a) $2\text{C}_4\text{H}_{10}(\text{g}) + 13\text{O}_2(\text{g}) \longrightarrow 8\text{CO}_2(\text{g}) + 10\text{H}_2\text{O}(\text{l})$
 b) 357 g   c) from the mass
 d) i) $2\text{C}_4\text{H}_{10}(\text{g}) + 5\text{O}_2(\text{g}) \longrightarrow 8\text{C}(\text{s}) + 10\text{H}_2\text{O}(\text{l})$
 ii) Carbon, a product of combustion d) i), can be oxidised further to CO_2 with the evolution of more heat.
 iii) $\Delta H^{\ominus}$ combustion (C) $= -200\,\text{kJ mol}^{-1}$

11. c) i) Step 1: $\Delta H^{\ominus} = +210\,\text{kJ mol}^{-1}$
 Step 2: $\Delta H^{\ominus} = -102\,\text{kJ mol}^{-1}$
 ii) Increase in temperature will raise the yield in Step 1 but will decrease the yield in Step 2. Increase in pressure will decrease the yield in Step 1 but will increase the yield in Step 2
 iii) Heat released in Step 2 can be used to increase the temperature of the reactants in Step 1. Products can be taken off from Step 1 to prevent an increase in pressure and to move the position of equilibrium to the right. These products can be fed into Step 2

12. a) i) $\text{H}(\text{g}) \longrightarrow \text{H}^+(\text{g}) + \text{e}^-$ ii) $-1453\,\text{kJ mol}^{-1}$
 b) i) $-1073\,\text{kJ mol}^{-1}$
 ii) The proton is very much smaller than other ions
 c) i) Ethene reacts with hydrogen

ii) Find the sum of the mean bond enthalpies of the bonds broken and the sum of the mean bond enthalpies of the bonds created. Then $\Delta H^\ominus = +$ (sum of mean bond enthalpies of bonds broken) $-$ (sum of mean bond enthalpies of bonds created)

13. a) iii) $-2154\,\text{kJ}\,\text{mol}^{-1}$

b) i) $SrCl_3$ will not be formed as the energy required is high. Whereas the formation of $SrCl$ is exothermic, the formation of $SrCl_2$ is more exothermic and therefore this is the product that is formed

ii) 1) Sr^{2+} is a much smaller ion than Sr^+; therefore Cl^- approach more closely to Sr^{2+} and the lattice energy of $SrCl_2$ is much greater than that of $SrCl$

2) In forming Sr^{2+} from Sr, the two s-electrons are removed. To form Sr^{3+} a third electron must be removed from Sr; this is a d-electron and much more difficult to remove

c) i) $\Delta S^\ominus = \Delta H^\ominus/T = +63.7\,\text{J}\,\text{K}^{-1}\,\text{mol}^{-1}$ ii) $\Delta S^\ominus = +82.1\,\text{J}\,\text{K}^{-1}\,\text{mol}^{-1}$

iii) $\Delta G^\ominus = \Delta H^\ominus - T\Delta S^\ominus = -5.46\,\text{kJ}\,\text{mol}^{-1}$; since $\Delta G^\ominus$ is negative, the reaction is feasible

14. a) $-316\,\text{kJ}\,\text{mol}^{-1}$

15. b) $ZnO(s) + C(s) \longrightarrow Zn(s) + CO(g)$ $\Delta G^\ominus = +35\,\text{kJ}\,\text{mol}^{-1}$ at 1100 K
Aluminium, yes. Hydrogen, no

c) i) Since 1 volume of gas is converted into a solid, S decreases: $\Delta S^\ominus$ is negative. The value of $-T\Delta S^\ominus$ becomes more positive as T increases; therefore $\Delta G^\ominus$ becomes less negative as T increases

ii) Since 1 volume of gas forms 2 volumes of gas, $\Delta S^\ominus$ is positive; therefore the value of $-T\Delta S^\ominus$ becomes more negative as T increases and $\Delta G^\ominus$ becomes more negative as T increases

16. d) i) $182\,\text{kJ}\,\text{mol}^{-1}$ ii) $-35\,\text{kJ}\,\text{mol}^{-1}$

iii) Reverse the second equation; then add the first equation and the reverse of the second equation to obtain:
$$I_2(g) + Cl_2(g) \longrightarrow 2ICl(g)$$
For this reaction, $\Delta H = \Delta H_1 - \Delta H_2$

CHAPTER 14

Exercise 51

1. 2   **2.** 1 w.r.t. A   2 w.r.t. B   **3.** d

4. $\dfrac{d[X]}{dt} = k[X]^0$ 0   **5.** $10.0\,\text{mol}^{-1}\,\text{dm}^3\,\text{s}^{-1}$

6. $\dfrac{d[P]}{dt} = k[A][B]$ $2.0 \times 10^{-3}\,\text{dm}^3\,\text{mol}^{-1}\,\text{s}^{-1}$

7. a) 1   b) 2 $1.67 \times 10^5\,\text{mol}^{-2}\,\text{dm}^6\,\text{h}^{-1}$ or $46.4\,\text{mol}^{-2}\,\text{dm}^6\,\text{s}^{-1}$

8. $\dfrac{d[N_2O_5]}{dt} = k[N_2O_5]$ a) $0.150\,\text{mol}\,\text{dm}^{-3}\,\text{s}^{-1}$ b) $1.80\,\text{mol}\,\text{dm}^{-3}\,\text{s}^{-1}$

Exercise 52

1. $1.54 \times 10^{-4}\,\text{s}^{-1}$ **2.** $1.10 \times 10^{-5}\,\text{s}^{-1}$

3. Gradients/mol dm^{-3} min^{-1}:   a) -0.842 b) -0.386

c) -0.200 d) 0 $0.365\,\text{min}^{-1}$ or $6.10 \times 10^{-3}\,\text{s}^{-1}$ 1

4. a) Initial rates/mol dm^{-3} min^{-1}:   1) 6.30×10^{-3} 2) 1.27×10^{-2} 3) 1.90×10^{-2}

b) 1   c) $3.75\,\text{min}$ $7.50\,\text{min}$ 1

d) $\dfrac{d[C]}{dt} = k[A][B]$ e) $1.06 \times 10^{-2}\,\text{dm}^3\,\text{mol}^{-1}\,\text{s}^{-1}$

Exercise 53

1. $50.0 \, dm^3 \, mol^{-1} \, s^{-1}$
2. 1 w.r.t. A   2 w.r.t. B   Rate = $k[A][B]^2$ $4.0 \, dm^6 \, mol^{-2} \, s^{-1}$

Exercise 54

1. 170 min   2. 64.5 h
3. An isotope with $t_{1/2} = 0.85$ h decays to form an isotope with $t_{1/2} = 36$ h
4. 14 days   5. a) 0.03125 (= $1/2^5$)   b) 9.3×10^{-10} (= $1/2^{30}$)
6. 832 years   7. 1.00×10^{-5} g
8. a) 75.5%   b) 5.9%   9. 3276 years

Exercise 55

1. $3.85 \times 10^{-4} \, s^{-1}$ 70 min   2. $5.86 \times 10^{-4} \, s^{-1}$
3. a) $1.6 \times 10^{-6} \, s^{-1}$ b) 20 days   4. $2.95 \times 10^{-8} \, mol \, dm^{-3} \, s^{-1}$
5. $1.36 \times 10^{-7} \, mol \, dm^{-3} \, s^{-1}$

6. Experiment 1: $\dfrac{d[A]}{dt} = k[A]^2[B]$ $k = 0.417$

   Experiment 2: $\dfrac{d[A]}{dt} = k[A]^2$ $k = 0.05$ B is adsorbed on the surface

7. a) 1   b) $7.48 \times 10^{-4} \, s^{-1}$

Exercise 56

1. $29.0 \, kJ \, mol^{-1}$ 2. $25.5 \, kJ \, mol^{-1}$ 3. $13.6 \, kJ \, mol^{-1}$

Exercise 57

1. a) Time interval between 30 cm³ and 15 cm³ = 14 min = time interval between 15 cm³ and 7.5 cm³
   b) Rate = $k[H_2O_2]$, $k = 0.0495 \, min^{-1}$, $t_{1/2} = 14$ min
   c) At $t = 0$, $[H_2O_2] = 0.75 \, mol \, dm^{-3}$; therefore the solution has been in the bottle for 28 min
2. a) ii) $t_{1/2} = 195$ min, $k = 3.55 \times 10^{-3} \, min^{-1}$
      iii) 1) The initial rate would be halved   2) The half-life would be the same
      iv) The concentration of water is much greater than that of sucrose; therefore only a small fraction of the water is used and the concentration remains effectively constant
3. a) Zero. The rate remains constant as the reaction proceeds.
   b) to find the order w.r.t. iodine
   c) i) to stop the reaction   ii) It reacts with iodine
   d) No effect (zero order)   e) first order
   f) Use different acids
4. a) i) 1   ii) Zero
      iii) The rate doubles when [RBr] doubles and does not change when [OH⁻] changes.
   b) Rate = $k[CH_3C(CH_3)BrCH_3]$ c) s^{-1}
5. c) 2506 years
6. b) i) Time between 1.0 and 0.5 = 15 hours = time from 0.5 to 0.25; therefore the reaction is first order
      ii) $t_{1/2} = 15$ hours, $k = 0.046 \, hour^{-1}$
      iii) 1) No effect on half-life   2) The initial rate would treble
7. b) The reaction is first order in H⁺ and first order in ester
      Rate = $k[H^+][HCO_2CH_3]$
      $k = 1.12 \times 10^{-3} \, mol^{-1} \, dm^3 \, s^{-1}$
      A plot of $\lg k$ against $10^3/T$ gives a straight line with a gradient of 3050 K; therefore $E = 58.4 \, kJ \, mol^{-1}$

8. a) as a catalyst

  b) i) 0.240 mol   iii) 5800 s   iv) first order w.r.t. ester

    v) $2.33 \times 10^{-5}\,mol\,dm^{-3}\,s^{-1}$ vi) $k = 9.7 \times 10^{-5}\,s^{-1}$

  c) Order w.r.t. HCl is zero; Rate $= k\,[CH_3CO_2CH_3]$

9. d) 60 hours

10. b) i) $ABC(g) \longrightarrow A(g) + B(g) + C(g)$

    ii) From the graph, $k = 0.037\,min^{-1}$, $t_{1/2} = 20\,min$

11. b) i) Order w.r.t. $CH_3(CH_2)_3Cl = 1$, order w.r.t. $OH^- = 1$

    Order w.r.t. $(CH_3)_3CCl = 1$, order w.r.t. $OH^- = 0$

    ii) Rate$_1 = k_1[CH_3(CH_2)_3Cl]\,[OH^-]$; $k_1\,mol^{-1}\,dm^3\,s^{-1}$

    Rate$_2 = k_2[(CH_3)_3CCl]$; $k_2\,s^{-1}$

    iii) a bimolecular mechanism

    iv) A unimolecular mechanism is likely

12. a) ii) The rate decreases with time

    iii) Order w.r.t. $MnO_4^- = 1$

  b) The rate increases with time up to the half-life, then decreases. After Mn^{2+} is produced in the early stages of the reaction, it starts to catalyse the reaction, and the rate increases. Then the decrease in $[MnO_4^-]$ makes the reaction begin to slow down. The two experiments reach the same $[MnO_4^-]$ at 2400 s because at this time the Mn^{2+} produced in Experiment 1 has had the same effect as that added in Experiment 2.

13. b) Rate $= k\,[P]\,[Q]$. If P is present in large excess, $[P]$ remains effectively constant. Then, Rate $= k_{ps}[Q]$ and $k_{ps} = k\,[P]$

  c) Order w.r.t. $I_2 = $ zero, $k_{ps} = 2.04 \times 10^{-5}\,s^{-1}$. [Propanone] remains effectively constant, falling from $7.0\,mol\,dm^{-3}$ to $6.98\,mol\,dm^{-3}$. Only one [HCl] has been used

14. a) i) First order

    $k(1129\,K) = 5.98 \times 10^{-4}\,s^{-1}$, $k(1148\,K) = 7.89 \times 10^{-4}\,s^{-1}$

  d) i) $161\,kJ\,mol^{-1}$ ii) $115\,kJ\,mol^{-1}$

  e) Since $2NH_3(g) \longrightarrow N_2(g) + 3H_2(g)$, the amount of gas doubles and the pressure doubles, rising to $5.32 \times 10^4\,N\,m^{-2}$

CHAPTER 15

Exercise 58

1. a) $K_p = \dfrac{p_C \times p_D}{p_A \times p_B}$ b) $K_c = \dfrac{[C]\,[D]}{[A]\,[B]}$ c) $K_{Het} = \dfrac{p_D}{p_B}$ d) $K_{Het} = \dfrac{[D]}{[A]\,[B]}$

2. a) 0.16   b) $1.05 \times 10^4\,Pa$ **3.** $1.78 \times 10^5\,N\,m^{-2}$ **4.** $\frac{1}{3}$

5. a) 0.34   b) $8.1 \times 10^4\,N\,m^{-2}$

6. 64   **7.** $3.39 \times 10^{-15}\,N^{-2}\,m^4$

8. a) $2.00 \times 10^{-3}\,dm^3\,mol^{-1}\,s^{-1}$ b) 64.0

9. a) 0.71 mol   b) 43 g   **10.** 98.8%   **11.** c

12. a) 0.483 mol   b) 0.483 mol   c) 0.517 mol   d) 4.517 mol

13. $9.26 \times 10^{-15}\,Pa^{-2}$ **14.** $0.2\,atm^{-1}$

15. a) CO_2, H_2: 0.27 mol   CO, H_2O: 0.23 mol

  b) CO_2: 4.2 mol H_2: 0.21 mol   CO, H_2O: 0.79 mol

16. $9.33 \times 10^{-4}\,N^{-1}\,m^2$

Exercise 59

1. c) i) Equilibrium position moves towards right-hand side in 1) and towards left-hand side in 2)

    ii) 1) No effect on K_p 2) Decrease in E_A

  d) The system moves towards the left-hand side at higher temperature; therefore the forward reaction is exothermic: $\Delta H^{\ominus}$ for the forward reaction is negative, and $\Delta H^{\ominus}$ for the reverse reaction is positive

2. a) $K_p = \dfrac{p^2 NH_3}{p_{N_2} \times p^3 H_2}$

b) $p_{N_2} = \dfrac{(1-x)P}{4}$ $p_{H_2} = \dfrac{(1-x)P}{1.33}$ $p_{NH_3} = xP$

d) 9.75×10^{-6} atm^{-2}

e) If P increases, i) x increases and ii) K_p remains constant

f) The system absorbs heat by dissociation of NH_3: $\Delta H^{\ominus}$ for the formation of NH_3 is negative

g) catalyst

3. b) 3.41×10^{-4} mol dm^{-3}

c) $K = \dfrac{[Ca^{2+}][HCO_3^-]^2}{[CO_2(aq)]}$, $[Ca(HCO_3)_2(aq)] = 1.59 \times 10^{-3}$ mol dm^{-3}

4. a) i) shifts to left-hand side   **ii)** shifts to left-hand side
iii) no effect   **iv)** shifts to right-hand side
$K_p = 3.43$ atm^{-1}

b) If CO_2 is added, it will combine with CaO to form $CaCO_3$. To promote the dissociation of $CaCO_3$, CO_2 is removed from the equilibrium mixture

5. b) i) $K_c = \dfrac{[H_2][I_2]}{[HI]}$

ii) 4.8×10^{-4} mol dm^{-3} **iii)** 1.21×10^{-5} mol dm^{-3}

6. d) 4.0

e) i) No effect   **ii)** [ester] and [water] increase and [acid] decreases

f) $\Delta G^{\ominus} = -3.46$ kJ mol^{-1}. The negative value shows that the reaction is feasible

7. a) iii) 1) $K_p = \dfrac{p_{H_2O}(g)}{p_{H_2}(g)\sqrt{p_{O_2}(g)}}$ (atm^{-1})

2) $K_p = p_{CO_2}(g)$ (atm)

3) $K_c = \dfrac{[Fe^{2+}(aq)]^2[Sn^{4+}(aq)]}{[Fe^{3+}(aq)]^2[Sn^{2+}(aq)]}$ (dimensionless)

b) The value of $p(NH_3)^2/[p(N_2)p(H_2)^3] = 7.316 \times 10^{-5}$ atm^{-2}. This is the value of K_p, therefore the system is at equilibrium

8. a) $K_p = p^2_{NO_2}/p_{N_2O_4}$ **b)** 77.4

c) i) M increases   **ii)** M decreases

d) $NO_2 + SO_2 + H_2O \longrightarrow NO + H_2SO_4$
$2NO + O_2 \longrightarrow 2NO_2$

9. a) 120   **b)** 140 tonnes   **c)** 2%

CHAPTER 16

Exercise 60

1. C_2H_5 C_4H_{10} **2.** 88 g mol^{-1} $C_3H_7CO_2H$ **3.** $C_{11}H_{10}O$

4. $C_{16}H_{13}O_3N$ **5.** A is $C_6H_{12} = CH_3CH_2CH{=}CHCH_2CH_3$ B is $C_2H_5CO_2H$

6. A is $C_6H_5CO_2CH_3$ B is CH_3OH C is $C_6H_5CO_2Na$ D is $C_6H_5CO_2H$

7. a) C_3H_7ON **b)** propanamide, $C_2H_5CONH_2$

8. a) 3 mol Br$_2$: 1 mol C_6H_5OH

b) $C_6H_5OH + 3Br_2 \longrightarrow C_6H_2Br_3OH + 3HBr$

9. a) 55.9 g mol^{-1} **b)** C_4H_8
but-1-ene, $CH_3CH_2CH{=}CH_2$ and but-2-ene, $CH_3CH{=}CHCH_3$

10. a) CHO   **b)** $C_4H_4O_4$

c) butenedioic acid, *cis* HCCO$_2$H and *trans* HO$_2$CCH
$$\overset{\|}{HCCO_2H} \qquad \overset{\|}{HCCO_2H}$$

11. a) $C_4H_{10}O$ b) $C_4H_{10}O$ c) $C_2H_5OC_2H_5$ $CH_3OCH_2CH_2CH_3$
$CH_3OCH(CH_3)_2$ $CH_3CH_2CH_2CH_2OH$ $CH_3CH_2C(OH)CH_3$
$CH_3CH(CH_3)CH_2OH$ $CH_3CH_2CH(CH_3)OH$

12. A is $C_8H_9ON = C_6H_5NHCOCH_3$ B is $C_6H_7N = C_6H_5NH_2$

Exercise 61

1. a) i) A C=C B $-\overset{|}{\underset{|}{C}}-Br$ C $-\overset{|}{\underset{|}{C}}-OH$ D $-\overset{\overset{H}{|}}{C}=O$

 iii) 9.16×10^{-3} mol   iv) 109, A is C_2H_4

 b) ii) The process is exothermic; therefore a high temperature is avoided and a catalyst is used. There is a decrease in volume; therefore high pressure will favour the formation of C

 iii) Since $K_p = p_C/(p_A \times p_{H_2O})$, then an increase in the partial pressure of water vapour, p_{H_2O}, will increase the partial pressure of C, p_C, that is increase the yield of C.

2. a) $C_4H_6O_5$

 b) B: $\underset{\overset{|}{CO_2H}}{\overset{|}{\underset{\overset{|}{CH_2}}{\underset{\overset{|}{CHOH}}{CO_2H}}}}$ C: $\underset{\overset{|}{CO_2C_2H_5}}{\overset{|}{\underset{\overset{|}{CH_2}}{\underset{\overset{|}{CHOH}}{CO_2C_2H_5}}}}$ D: $\underset{\overset{|}{CO_2C_2H_5}}{\overset{|}{\underset{\overset{|}{CH_2}}{\underset{\overset{|}{C=O}}{CO_2C_2H_5}}}}$

3. e) 75%

4. a) K is $C_6H_5CH=CHCO_2C_2H_5$ L is $C_6H_5CH=CHCO_2H$ M is $C_6H_5CHBrCH_2CO_2H$
N is $C_6H_5CH_2CHBrCO_2H$ O is $C_6H_5CH=CH_2$ P is $C_6H_5CHBrCH_2Br$

 b) *cis*- H\C=C/H   *trans*- C_6H_5\C=C/H
   (structures with C_6H_5 and $CO_2C_2H_5$ groups)

 Rotation cannot occur about the double bond.

 c) $CH_3C_6H_3(CO_2CH_3)CH=CH_2$, $CH_3C_6H_3(CH_2CO_2H)CH=CH_2$ and $CH_3CH=CHC_6H_4CO_2H$ are possible isomers

 d) i) M and N are $C_6H_5C^*HBrCH_2CO_2H$ and $C_6H_5CH_2C^*HBrCO_2H$

 ii) More of the first compound because the $-CO_2H$ group withdraws electrons, giving the C atom attached to the ring a $\delta+$ charge.

 e) e.g. $C_6H_5CHClCH_3$ + KOH(ethanolic)

 f) i) $C_6H_5CH_2CH_2CO_2H$ ii) $C_6H_5CH=CHCH_2OH$

5. A is $C_4H_8O_3N_2$ B is $H_2NCH_2CO_2H$ C is $H_2NCH_2CH_2OH$ D is $H_2NCH_2CH_2Br$
E is $H_2NCH_2CH_2NH_2$ A is $H_2NCH_2CONHCH_2CO_2H$

 a) E will form a complex with Cu^{2+}, acting as a bidentate ligand

 b) E will form a condensation polymer with decanedioyl chloride.

6. a) Find the mass which reacts with 1 mol H_2, then multiply by 3 or 4 to obtain a value for the relative molecular mass which corresponds to 7 or 8 carbon atoms. Heat evolved per mol of $H_2 = 135$ or 69 or 85 kJ mol^{-1}, showing 3 types of double bond. These are C=C alicyclic, C=C aromatic and C=C conjugated with a benzene ring
A is C_8H_{10}, B is C_8H_8, C is C_8H_{10}, D is C_7H_8, E is C_8H_8, F is C_7H_8
Structural formulae:

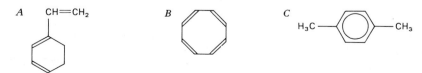

A CH=CH₂   B   C H₃C—⟨○⟩—CH₃

D   E —CH=CH$_2$ F —CH$_3$

Isomers of A :

b) G is C$_{10}$H$_8$:

7. a) i) B is RCO$_2$NH$_4$ C is RCONH$_2$ D is RCN
   b) i) $M_r(A) = 60$: A is CH$_3$CO$_2$H
      ii) $M_r(C) = 59$: C is CH$_3$CONH$_2$, D is CH$_3$CN and B is CH$_3$CO$_2$NH$_4$
      iii) $M_r(E) = 45$: E is CH$_3$CH$_2$NH$_2$ and F is CH$_3$CH$_2$OH
8. b) C$_4$H$_8$O
   c) CH$_3$CH$_2$CH$_2$CHO, (CH$_3$)$_2$CHCHO and CH$_3$CH$_2$COCH$_3$
   d) e.g. CH$_3$CH$_2$CH$_2$CHO $\longrightarrow$ CH$_3$CH$_2$CH$_2$CO$_2$H

Table of Relative Atomic Masses

Element	Symbol	Atomic number	Relative atomic mass	Element	Symbol	Atomic number	Relative atomic mass
Aluminium	Al	13	27.0	Lead	Pb	82	207
Antimony	Sb	51	122	Lithium	Li	3	6.94
Argon	Ar	18	40.0	Magnesium	Mg	12	24.3
Arsenic	As	33	75.0	Manganese	Mn	25	54.9
Barium	Ba	56	137	Mercury	Hg	80	201
Beryllium	Be	4	9.0	Neon	Ne	10	20.2
Bismuth	Bi	83	209	Nickel	Ni	28	58.7
Boron	B	5	10.8	Nitrogen	N	7	14.0
Bromine	Br	35	80.0	Oxygen	O	8	16.0
Cadmium	Cd	48	112.5	Phosphorus	P	15	31.0
Calcium	Ca	20	40.1	Platinum	Pt	78	195
Carbon	C	6	12.0	Potassium	K	19	39.1
Chlorine	Cl	17	35.5	Selenium	Se	34	79.0
Chromium	Cr	24	52.0	Silicon	Si	14	28.1
Cobalt	Co	27	59.0	Silver	Ag	47	108
Copper	Cu	29	63.5	Sodium	Na	11	23.0
Fluorine	F	9	19.0	Strontium	Sr	38	87.6
Germanium	Ge	32	72.5	Sulphur	S	16	32.1
Gold	Au	79	197	Tin	Sn	50	119
Helium	He	2	4.00	Titanium	Ti	22	47.9
Hydrogen	H	1	1.01	Vanadium	V	23	50.9
Iodine	I	53	127	Xenon	Xe	54	131
Iron	Fe	26	55.8	Zinc	Zn	30	65.4
Krypton	Kr	36	83.8				

Periodic Table of the Elements

Key:

Atomic number	Relative atomic mass
11	23.0
Na	
Sodium	

1	2	3	4	5	6	7	8
1 — 1.000 **H** Hydrogen							2 — 4.00 **He** Helium
3 — 6.90 **Li** Lithium	4 — 9.00 **Be** Beryllium	5 — 10.8 **B** Boron	6 — 12.0 **C** Carbon	7 — 14.0 **N** Nitrogen	8 — 16.0 **O** Oxygen	9 — 19.0 **F** Fluorine	10 — 20.2 **Ne** Neon
11 — 23.0 **Na** Sodium	12 — 24.3 **Mg** Magnesium	13 — 27.0 **Al** Aluminium	14 — 28.1 **Si** Silicon	15 — 31.0 **P** Phosphorus	16 — 32.1 **S** Sulphur	17 — 35.5 **Cl** Chlorine	18 — 39.9 **Ar** Argon
19 — 39.1 **K** Potassium	20 — 40.1 **Ca** Calcium	31 — 69.7 **Ga** Gallium	32 — 72.6 **Ge** Germanium	33 — 74.9 **As** Arsenic	34 — 79.0 **Se** Selenium	35 — 79.9 **Br** Bromine	36 — 83.8 **Kr** Krypton
37 — 85.5 **Rb** Rubidium	38 — 87.6 **Sr** Strontium	49 — 115 **In** Indium	50 — 119 **Sn** Tin	51 — 122 **Sb** Antimony	52 — 128 **Te** Tellurium	53 — 127 **I** Iodine	54 — 131 **Xe** Xenon
55 — 133 **Cs** Caesium	56 — 137 **Ba** Barium	81 — 204 **Tl** Thallium	82 — 207 **Pb** Lead	83 — 209 **Bi** Bismuth	84 — 210 **Po** Polonium	85 — 210 **At** Astatine	86 — 222 **Rn** Radon

21 45.0 Sc Scandium	22 47.9 Ti Titanium	23 50.9 V Vanadium	24 52.0 Cr Chromium	25 54.9 Mn Manganese	26 55.8 Fe Iron	27 58.9 Co Cobalt	28 58.7 Ni Nickel	29 63.5 Cu Copper	30 65.4 Zn Zinc
39 88.9 Y Yttrium	40 91.2 Zr Zirconium	41 92.9 Nb Niobium	42 95.9 Mo Molybdenum	43 99 Tc Technetium	44 101 Ru Ruthenium	45 103 Rh Rhodium	46 106 Pd Palladium	47 108 Ag Silver	48 112 Cd Cadmium
57 139 La Lanthanum	72 178 Hf Hafnium	73 181 Ta Tantalum	74 184 W Tungsten	75 186 Re Rhenium	76 190 Os Osmium	77 192 Ir Iridium	78 195 Pt Platinum	79 197 Au Gold	80 201 Hg Mercury

Index